PHILOSOPHIES OF NATURE: THE HUMAN DIMENSION

BOSTON STUDIES IN THE PHILOSOPHY OF SCIENCE

VOLUME 195

Erazim Kohák, 1997

PHILOSOPHIES OF NATURE: THE HUMAN DIMENSION

In Celebration of Erazim Kohák

Edited by

ROBERT S. COHEN
Boston University

and

ALFRED I. TAUBER
Boston University

With a complete bibiliography of Erazim Kohák's works

SPRINGER-SCIENCE+BUSINESS MEDIA, B.V.

Library of Congress Cataloging-in-Publication Data is available.

ISBN 978-90-481-4859-2 ISBN 978-94-017-2614-6 (eBook)
DOI 10.1007/978-94-017-2614-6

Printed on acid-free paper

TABLE OF CONTENTS

II. SELECTED ESSAYS OF ERAZIM KOHÁK

III.

ORIGIN OF THE ESSAYS BY ERAZIM KOHÁK

The following list indicates the first publication forums, if not new, of the essays written by Erazim Kohák included in the present volume. It also lists the essays which appear here for the first time.

* "Varieties of Ecological Experience," *Environmental Ethics 19*.2 (Spring 1997): 153–72.
* "Creation's Orphans: Toward a Philosophy of Artifacts," *The Personalist Forum 1*.1 (Spring 1985): 22–42.
* "Human Rights and Nature's Rightness," *The Lyceum 2*.2 (Fall 1990): 22–36.
* "Transcendental Experience, Everyday Philosophy," in T. Rockmore and W. Zeman, editors, *Transcendental Philosophy and Everyday Experience*, The Humanities Press, New York, 1997; 43–61.

These previously published essays appear here with the permission of the respective copyright owners. These permissions are most gratefully acknowledged.

New are the essays:

* "Phenomenology and Ecology: Dependence and Co-Dependence"
* "Should Auld Acquaintance Be Forgot ...?"
* "Nature as Presence and Experience"
* "The True and the Good"
* "The Ecological Dilemma"

PREFACE

With this book, we celebrate Erazim Kohák, for more than three and a half decades our colleague in the Boston University Department of Philosophy.

In Part I, we offer the principal contributions to the Symposium 'in honor of Erazim Kohák', which was held November 13–17, 1995 under the title 'Philosophies of Nature' by the Boston Colloquium for the Philosophy of Science. The 24 speakers, all teachers and scholars at Boston University, presented essays on Nature and on Human Nature, classical and modern, natural and supernatural, Christian, Taoist, Buddhist, scientific and ethical. Our theme overall was humanist, exploring 'the human dimension' through philosophical, scientific and comparative religious reflections on nature. Indeed, Professor Kohák's personal, evocative work, *The Embers and the Stars,* set the task before us, each in our separate ways, with his challenging sub-title: a philosophical inquiry into the moral sense of nature.

The issue was joined most directly by Lawrence Cahoone, whose profound paper has the title 'Whose nature? Which morality? On Kohák's moral sense of nature'. We include Kohák's succinct response to Cahoone. Our other contributors, going well beyond Kohák's texts, comment, articulate, and then wrestle with many perspectives on what in these times might be termed 'philosophical ecology'.

In Part II, with his advice, we have selected nine of Erazim Kohák's essays. And in Part III, we have a Kohák Bibliography, nearly up-to-date, philosophical, political, journalistic, religious ... a catalogue and a guide to this public philosopher.

The book is introduced by our Czech and Canadian friend, Vladimir Zeman, whose sketch of the mind and the achievement of Professor Kohák is insightful, both for the story of the development of Kohák's thought and for his social and political context.

* * *

A few phases from my own time with Kohák resonate within me; they reach across the sometimes difficult gap that seems to remain, stubbornly, between 'men of good will' (as Jules Romain lamented in his novels of the First World War of our century). Kohák: a Christian, of the Czech Brethren, phenomenologist, personalist, genuine European and true patriot, a democratic socialist, and always seeking the natural mysticism of the fulfillment of life. And I: scientist, a Jewish humanist and atheist, sceptic, logical empiricist, short-term

R.S. Cohen and A.I. Tauber (eds.), Philosophies of Nature: The Human Dimension, xi–xii.

pessimist and occasional long-term optimist. I note, for all these years, how easily Kohák speaks to me, and to all who will join me to listen to him. A few reminders:

(1) ... thinkers whose insight withstood the test of time? their virtue was naiveté, a willingness to *see* before theorizing ...

(2) Is the Person, is matter in motion, the root metaphor of thought and practice? That answered, all else follows ... and Nature lends itself to either interpretation.

(3) ... in the world in which, willy-nilly, we live, breathe, and have our being, doubt presses in and will not be ignored ... our suspicion that things are not what they seem, the actual is not the real, our consciousness illusory, the lives we live and know, themselves are but an impoverished reflection of true reality.

(4) I cannot accept either strategy, that of despair or that of rage; not because I am unaware of the reality of evil ...

(5) Arguments are reason for *not disbelieving* what we believe because we have seen

(6) ... nature as the poetry of God (Thales ... the verb *potein* referred generically to the activity of making, molding, shaping ... thus nature as God's works and as his poetizing)

(7) The question ... is not whether there is a god but rather "What does the Lord require of me?"

(8) ... the driving force of the revolutionary hope ... a transformation in the hearts and minds of people and their relation to each other and their world; ever that one powerful hope, the hope of a personalization of our life world, not of tinkering with the abstract impersonal world of social artefacts.

(9) ... the agonized cry with which humans protest the unacceptability of the inevitable ... so human – so vain in the order of time.

(10) The fulfillment of time cannot be in its future. That future is always an end ... but not to wonder that something perishes. The point of life is not that it should last forever. The wonder is that once it was; there once was a man, there once was a raccoon. That is the miracle.

(11) All Saints' Day ... not a day to grieve for all who have died ... a day to rejoice in all who have lived.

(12) In the pained cherishing of that transient world, the human, a dweller between the embers and the stars, can raise it up to eternity. That is the task of humans. The moral sense of nature is that it can teach us to cherish time, and to look to eternity within it.

Boston University
August 1998

Robert S. Cohen

INTRODUCTION

Erazim Kohák – *vivere per philosophare*

Professor Erazim Kohák's long and still growing *curriculum vitae* is atypical when compared to what is usually expected of a professor of philosophy. Most of his life he has been a hard-working exile, socially and politically engaged, and, in the end, lucky. His life has been between Bohemia, his native country (and, since 1995, once more his homeland) and the USA, his adoptive country for almost half a century. In the USA he received his higher education, established himself professionally, and also worked systematically toward change in the political situation in Czechoslovakia, the change which allowed his to close the circle of his long journey. Since 1995 emeritus professor of philosophy at Boston University, where he taught for 35 years, and since 1991 professor of ethics and ontology at the Charles University in Prague, Kohák remains active.

Kohák has always stressed the overall unity of his life; his personal attitudes, convictions and objectives, viewed in their interplay with the social and political developments, form the core of his professional life as a philosopher. Neither his choices as a translator nor his topics as an author can be viewed as accidental or as determined by the demands of his professional career alone. Here we will review the inner logic of his professional development. But to evaluate the work of a thinker still at work is extremely risky. The thoughtful life of a creative philosopher cannot be encapsulated within a simple formula. As the whole tradition of philosophical writings pioneered by Schilpp's *Library of Living Philosophers* convincingly and fruitfully demonstrated, the subjects of such evaluative attempts tend to disagree with them, if not to be outright displeased. So, what follows cannot be more than a preliminary survey of a process still underway.

The first problem lies with the very character and status of contemporary philosophy. Some twenty years ago, when Nelson Goodman tried to characterize the development of modern philosophy, he saw it as "the movement ... from unique truth and a world fixed and found to a diversity of right and even conflicting versions of worlds in the making" (Foreword to *Ways of Worldmaking*). The arena of philosophy in the past quarter-century appears to have

R.S. Cohen and A.I. Tauber (eds.), Philosophies of Nature: The Human Dimension, xiii–xvii.

confirmed such a judgment. However, those acquainted with Kohák are well aware than neither his overall philosophical orientation nor his philosophical development shown in his publications fit such a general picture. While it would be unfair to view him as subscribing to a fixed world, in particular one separated from us, even standing against us, nevertheless at the very foundation of Kohák's philosophical work is an anti-relativist conviction about the universal character of such a world, and the values we draw from it. Borrowing a term from social and political philosophy, we may see Kohák's philosophy working toward a "participatory universalistic view of the world". For Kohák the philosopher (and theologian), the basic features of such a world are given but not imposed; *pace* Goodman, we do not 'make' our worlds.

From this angle, Kohák must appear first and foremost as a teacher in the broadest sense of the word. Following the teaching and religious orientation of his great countryman, Jan Amos Comenius, the 'Teacher of Nations', Kohák organizes his many activities to go beyond any narrow conception of scholar and educator; he orients his efforts toward spiritual awakening, providing spiritual comfort as well. Such breadth was crucial, indeed typical, for his communications, particularly to his Czech compatriots. Prior to the 'Velvet Revolution' of 1989, he regularly presented his mostly historico-philosophical analyses (in English or Czech) at meetings of learned societies outside his old country; while he regularly broadcast his philosophical-political analysis across the 'Iron Curtain'. Since 1989, his activities have taken place primarily within the varied teaching spaces he has created for himself in Prague: from his well-attended lectures and seminars at the Charles University to public interventions in the course of general discussions of the political and cultural orientation of his native land, in forums, radio and TV, journalism and books. In a country recovering from a planned *Gleichschaltung*, and from decades of unplanned even careless physical devastation, Kohák's long-standing commitment to ecological philosophy became even more central: the philosophical reveries of the 'solitary dweller' in the woods of New Hampshire have lost their tranquility. Now dwelling in one of the prefabricated suburbs of Prague, as a citizen and philosopher, he must look toward globally-oriented social solutions to his country's serious ecological problems. It comes as no surprise that his search seems to be turning more toward 'deep ecology', in which individual values and social goals are expected to intersect, leading to a radical change in our approach to both the civilizational and the natural environment.

Most visible in Kohák's professional development is his many-sided and continuous encounter with phenomenology. His was the first English translation of Ricoeur's *Le volontaire et l'involontaire* (1967). Beside Kohák's devoted involvement in translating phenomenological writings into Czech language, his translations of three books by Jan Patočka into English were welcome last year. In 1978, Kohák broadened his presentation of phenomenological ideas to the English-speaking world by publishing *Idea and Experience*, his ground-breaking commentary on a basic work of Husserl's phenomenology, the *Ideas I* of 1913. In addition, Kohák's critical tribute to his compatriot, and Husserl's student, Jan Patočka, appeared in 1989 under the title *Jan Patočka: His*

Thought and Writings. Not surprisingly, throughout all of his work, Husserl's transcendental phenomenology is Kohák's "rigorous and noetic basis". It is Kohák's conviction that without such a basis, philosophy would not be able to take on the form and the role he envisages for it ... it would turn into an unchecked fabulation (*'bezbrehe pabeni'*). Phenomenology in general, and Husserl's *Wiener Vortrag* in particular, help Kohák even in his most recent attempts to re-formulate the philosophical basis of ecological thought by freeing it from its mainly romantic aspects, and by high-lighting and clarifying the role of human responsibility as "responsibility before the fact".

As suggested earlier, while Kohák's own philosophy is decidedly not relativistic, it is, in its sources, directions, and proposed solutions, of a rather polyphonic character. Even in his phenomenological analysis of human experience, the human he deals with is neither an abstract 'I' nor an empirically delimited individual: it is a person. Following in the steps of his late senior colleague in Boston, Peter Bertocci, himself successor to earlier Boston University philosophical theologians Edgar Brightman and Borden Parker Bowne, Kohák has been an active member of the now rather limited circle of American personalists. Personalism has always been a starting-point for his studies, beautifully articulated in *The Embers and the Stars.* The personalist perspective influenced his interpretation of Thomas G. Masaryk's philosophy and of Masaryk's late 19th century systematic criticism of Karl Marx's political philosophy (translated and edited by Kohák as *Masaryk on Marx*, 1972): and personalism also influenced his critical acceptance of the philosophy of one of Masaryk's most influential disciplines, Emanuel Rádl. It is within this context that we have to look for the roots of Kohák's own humanism. Precisely this humanism, together with his religious commitment, allows him to accept Husserl's critical 'timely considerations' expressed in *The Crisis of European Science*, without falling into romantic pessimism, not uncommon among ecological thinkers and some phenomenologists.

While all his works were critically acclaimed, most crucial for understanding Kohák's philosophy is undoubtedly the most personal among them, *The Embers and the Stars* (1984). When, one and a half centuries ago, Thoreau left for the cabin by Walden Pond, he did it to prove his political point, to show that one could become economically self-sufficient, and to see whether one might also benefit philosophically from such a self-imposed, although temporary, solitude. Kohák's intention is more central to philosophy: to search for the sense of human life and for its unity. Philosophy is shown as still capable of performing basic reflections in the service of life. The resulting appeal to the reader is not to follow Kohák back to the woods, nor to prove that one can still create one's own 'dwelling' with one's own hands while in harmony with the natural worlds surrounding us, admirable and interesting as this achievement might be. The external circumstances of any philosopher's life cannot but remain partly accidental, to serve only as catalyst for the philosopher's work. What was at first Kohák's practical search for his own, his personal, rediscovery of his own life's sense, grew into a philosopher's search for the moral sense of life in general; to be found at the end in the incorporation of human beings into the all-encompassing unity of nature.

"To recover the moral sense of our humanity, we would need to recover first the moral sense of nature" (p. 13).

The general appeal of such an attempt is obvious. Its success, however, depends upon the acceptance of Kohák's premise that

"philosophy ... is, most fundamentally, the art of the intermediate vision" (p. 32)

and on his own capacity to lead us, in his mostly metaphorical rather than argumentative way, along the path provided in this work.

As personal history, *The Embers and the Stars* remains an interesting and inspiring work. However the philosophical reader may ask whether this ambitious attempt to weave together such different threads as those of phenomenology and personalism, with others of theological origin, succeeds. Kohák conceives reality in a comprehensively personalistic way: humans are part of nature, and in its turn nature is the source of values, for it is the provider of sense and meaning. Is this based upon premises not everywhere acceptable? Yet from now on, for Kohák himself, as for any good philosopher, once the eco-philosophical perspective has been recognized to be central within philosophy, and unifying as well, it must remain the focus of his attention. On the surface, the most evident aspect of Kohák's continuing effort is the search for liberation of eco-philosophy from soft romantic inclinations, including his own. To reach this goal, he has to lay out anew the philosophical underpinning of ecology. Not surprising, he does it this time in what is a more narrow and a deeper way, returning to Husserl mainly by re-assessing the relevant ideas from *The Crisis*. In the Czech article of 1997, 'Phenomenology and Ecology – Mutual Support and Dependence' (*Fenomenologie a ekologie – opora a spoluzavislost*, in *Filosoficky Casopis 45*), Kohák formulates his position as follows:

Ecological philosophy appears possible to me only within the frame of reality viewed phenomenologically, as a meaningfully established lifeworld. Moral consequences of such a conception necessarily lead to ecological conclusions. (p. 363)

Since, in his return to Husserl's imperative '*zu den Sachen selbst*' Kohák views

lived experience of phenomenal goal-oriented activity ... as determinant for the human form of existence (p. 364)

phenomenology appears to him primarily as

a decision for living in responsibility for our own freedom (p. 365).

In this sense,

ecology, equally as a phenomenology, is an appeal toward life in responsibility (p. 368).

By this move, if it be as convincing and as acceptable as Kohák believes, ecophilosophy would be freed from its still strong romantic and therefore contingent foundation, while at the same time supported by the authority of Husserlian phenomenology in one of its interpretations. Phenomenology which makes its subject the meaning of what is given in immediate experience rather than in some 'authentic Being', is able to ground ecology without any romantic and thus at the end purely nostalgic and irresponsible inclinations.

Is there a leading idea which would interconnect all of Kohák's theoretical works? As a socially engaged philosopher, he has always looked for a unifying perspective. Even when he deals with the 'technical' aspects of the discipline of philosophy, it is in the same context of unifying meaning. In his *Philosophy of Symbolic Forms* (vol. 3, p. 143), Ernst Cassirer claimed, in 1929, that "the center of gravity has shifted from the philosophy of nature to the philosophy of culture". Kohák's is indeed a philosophy of culture. But clearly, for him, there is no essential conflict between culture and nature. Here, even as a phenomenologist, Kohák remains a Christian personalist (and indirectly the disciple of Masaryk). The unity we can find in his work is, to this observer, not the unity of a philosophical system but a continuously evolving search for the unity of theory with actions based on it.

Concordia University **Vladimir Zeman**
Montréal

I.

THE SYMPOSIUM

KLAUS BRINKMANN

THE NATURAL AND THE SUPERNATURAL IN HUMAN NATURE: HEGEL ON THE SOUL

INTRODUCTION

Looking at contemporary philosophy of mind, one has the impression that philosophers no longer know, nor want to know, nor would find intelligible the desire to know, what the soul is. The soul, it seems, is no longer a legitimate topic for philosophical discourse. It has largely become a form of locution or a linguistic metaphor without philosophical significance, so that it would be inappropriate even to raise the question as to its nature. Rather than being a manifestation of *Geist* as Hegel would have it, the soul has become somewhat of a ghost. This is unfortunate, because, as I hope to show in what follows, the soul is both a philosophically fascinating as well as a perfectly genuine category in our ontology, at least in the way in which it is analyzed and discussed by Hegel. In particular, it seems that Hegel's categorial explication of the soul has the merit of avoiding both the physicalist reductionism of mind to matter as well as the metaphysical hypostatization of a suprasensible immaterial entity, or so I shall argue. Moreover, Hegel's approach also has the advantage of staying clear of the artificial Cartesian separation of *res cogitans* from *res extensa*. In short, I believe that Hegel found a way of saying something both interesting and important about the soul after Kant's decisive critique of rational psychology in the Paralogisms, and without abandoning the topic entirely to empirical psychology.

In this paper, I want to use contemporary philosophy of mind as a backdrop only against which Hegel's interpretation of the soul will be made to stand out in its richness of the phenomena covered and the wealth and subtlety of its conceptual distinctions. I will presuppose only a general familiarity with current theories which, by and large, are all of them attempts to come to terms with the Cartesian legacy of the mind–body relationship. Since Hegel's treatment of the soul bypasses the Cartesian dichotomy *ab initio*, it can be viewed as a serious alternative to current dualist and monist theories which can all be understood as attempts to overcome the Cartesian paradigm – long after that paradigm had already become history due to the dialectical reinterpretation of the mind–body relationship (and that of nature and spirit generally) effected by Fichte, Schelling, and Hegel.[1] For a brief identification of what the backdrop consists in, the following gloss must suffice.[2]

R.S. Cohen and A.I. Tauber (eds.), Philosophies of Nature: The Human Dimension, 3–18.

In contemporary philosophy of mind, the nondualist theories seem to have prevailed, at least for the time being, over the dualists of both the interactionist and non-interactionist variety. Among nondualist theories, identity theories prevail. Identity theories work on the assumption that each specific mental state is either identical with a specific brain state (type identity) or that each specific mental state can be instantiated by a number of different brain states not specifiable in advance (token identity). In either case, the neuronal state is considered to be fundamental and primary while mental states are either emerging from or supervenient upon those physical structures. In other words, both versions of identity theory are ontologically speaking forms of physicalism. Theories of type identity naturally encourage the attempt to eliminate mental states altogether (eliminative materialism). Reference to them becomes a mere linguistic convention (also known as folk psychology) – and hence scientifically redundant. Theories of token identity presuppose the existence of a mental dimension *sui generis* (characterized by intentional states and propositional attitudes, among other things), but in order to avoid dualism need to make them non-physical properties of a physical substratum. Thus theories of token identity retain a non-dualistic distinction between an ontologically primary substratum and its mental properties. They are examples of non-reductionist physicalism. The most persuasive version of such token identity so far seems to be psychological functionalism which has mental properties of physical events stand in causal relationships among themselves.[3] However, functionalism is unable to say anything more specific about the relationship between neuronal and mental states than that there exists some kind of correlation between them. Hence functionalism is in danger of exchanging one dualism for another by replacing substance dualism with property dualism. Neuronal states would then constitute the basic substratum exhibiting both physical and mental properties. Evidently, property dualism will still not solve the Cartesian mystery concerning the interaction between the physical and the mental. Hence the continued attraction of eliminative materialism.[4]

Given these parameters, what would Hegel's position be in this discussion about the identity or non-identity of mind and body? To this question I shall return towards the end of this paper in greater detail. For now, suffice it to characterize the nature of Hegel's position roughly as that of a non-reductionist monism which differs in interesting ways from that other contemporary non-reductionist monism referred to above, viz. functionalism. Despite the fact that both monisms are thus generically of the same type, they are actually worlds apart. But before we can appreciate the important differences between them, we first need to get a closer look at what Hegel's interpretation of the soul actually consists in. One thing should be emphasized at the outset, however: Hegel is a non-reductionist philosopher *avant la lettre*. He believes in ontological diversity and the most nuanced treatment of what we all too blandly refer to as reality. He helps us to appreciate afresh the enjoyment that is to be derived from perceiving and understanding "the many differences between things," to use Aristotle's phrase.[6] At least in this regard, Hegel is close to another philosopher who added enormously to the descriptive wealth to be discovered in phenomena and who made us see and cherish again the many differences between things. My reference here is to Edmund Husserl, a philosopher dear and close in spirit to the colleague we are honoring with this *Festschrift*. His famous maxim may serve as a motto for what follows: "To the things themselves!" And, may I add, to their many *Abschattungen*!

THE HUMAN SOUL: ITS PLACE IN HEGEL'S SYSTEM

To begin with, there is the differentiation in the *Anthropology* of the *Encyclopedia's* Philosophy of Subjective Spirit, which serves as our principal text, between the animal soul and the human soul. The concept of the soul is first introduced in Hegel's Philosophy of Nature as the capacity for self-feeling or as the sensitivity characteristic of organic life. The animal soul thus stands for the emergence of an element of subjectivity or self-referentiality in nature. It also opens up a dimension of interiority which, in its most developed form, i.e. in mammals, develops into an inner life of instinctual drives, emotions, and memories.[6] However, Hegel also points out that although this soulful life of the higher animals in some of its behavioral and phenomenal aspects *resembles* the inner life of the human soul, the two are actually categorially distinct, that is they belong to two different realms of being and, consequently, to two different spheres of experience. The human soul, Hegel argues, differs intrinsically from the animal soul in that it is potentially reflective thought. This potential for intelligent life, Hegel claims, imbues everything that happens in the human soul and thus sets it apart from the animal soul. The similarities between the two are *conceptual* or generic, not *categorial*.[7] Even Aristotle, whose *De anima* Hegel praises as the best work that has been done on the soul so far[8], did not spot this subtle difference: For him, the *human* soul, although already a complex notion, is not distinctively human in its entirety, because it shares two of its elements, the vegetative and the appetitive, with plants and animals. Consequently, the human soul has sensation, feeling and perception in common with animals, while it shares the nutritive principle with plants. In this respect, Aristotle's definition of the human soul is additive, not specific. And insofar as it *is* distinctively human, it is primarily, although not purely, rational, excluding what Hegel discusses under the title of the soul in the *Anthropology*.[9] Conversely, for Hegel the faculties which the human soul has in common with the animal soul are shared homonymously, not synonymously.

For Hegel, then, the human soul constitutes the crucial link between nature on the one hand and spirit on the other.[10] It is at once nature emerging into spirit and spirit emerging out of nature. As Hegel puts it with his incomparable sense for the suggestive metaphor, the soul is the *sleep* of spirit, spirit insofar as it has not yet risen from its immersion in nature.[11] On the other hand, let it also be noted that insofar as the soul thus represents the *naturalized aspect of spirit*, it is correct to say that spirit, too, has a natural aspect. Spirit exists indeed in a naturalized form in which it can be observed and described and even empirically investigated. As will become apparent later on, the soul marks what might be called the instinctual life of spirit, or its pre-conscious natural life.

The human soul is thus located on a *continuum* of forms of subjectivity, at one end of which stands the emerging sensitivity of organic life in nature. At the other end stands, ultimately, absolute spirit. However, there are numerous intermediate steps, and it bears pointing out that, even before we reach the level of self-consciousness in the *Phenomenology* of the *Encyclopedia*, let alone theoretical and practical spirit in the *Psychology*, Hegel will have taken us through a host of forms of human subjectivity and the correlated spheres of human experience. Thus the section on the soul is followed by that on consciousness, and while analyzing the soul Hegel repeatedly draws attention to the fact that what he says about the soul is not meant to be an account of consciousness as such.[12] Consciousness for Hegel is the title for the epistemic subject insofar as it posits an object-world as its opposite to which it brings order with the help of categories such as universal and particular, thing and attribute, cause and effect, and law of nature. It could therefore be said that the phenomena Hegel associated with the human soul are pre-intentional in the sense that they do not yet belong to the sphere of intentionality as defined by Husserl.

Hegel undergirds this highly differentiated tableau of categorial differentiations with an order of intelligibility by locating these phenomena on a continuum of increasing self-referentiality. From the relatively abstract self-referentiality of the soul we move to the richer structure of self-referentiality exhibited by consciousness, self-consciousness, and subjective spirit proper. I propose to present a brief overview of the wealth of material Hegel incorporates into his account of the soul, before I focus on Hegel's treatment of the issue of mind–body identity.

WHAT IS THE HUMAN SOUL?

What, then, *is* the soul? It is crucial to understand that it is not an immaterial entity. To make the soul an entity, Hegel argues, would *reify* the soul and immediately create a fixed opposition between soul and body and would thus generate the problem of the mind–body relationship with the unnecessary and presumably illegitimate pseudo-problem of a causal interaction between the physical and the mental.[13] Obviously, consciousness, self, subjectivity, or spirit are not entities either. Rather, they represent forms of self-referential intelligent, rational and spiritual life of increasing complexity. Hegel offers the following definition of the human soul:

> [The soul] is not only immaterial in relation to itself, it is the universal immateriality of nature, its simple ideal [*ideelles*] life. Soul is the *substance*, the absolute basis of all the particularizing and individualizing of spirit; it is in the soul that spirit finds the material on which its character is wrought, and the soul remains the pervading, identical ideality of it all. But as it is still thus abstractly, the soul is only the *sleep* of spirit – the passive νους of Aristotle, which is potentially all things (Enc. § 389, p. 29 tr. Wallace; II 3 Petry; my own translation on the basis of both the Wallace and the Petry translations).

This is obviously already a very rich characterization. A general word on the meaning of spirit may be in order here, however, before we address Hegel's initial definition of the soul. Spirit in Hegel generally means whatever we understand when we hear the phrase: "Their performance of the music was full of spirit." The spiritedness that enlivens and energizes all intelligent human activities from creating tools to designing bridges, from composing poems to celebrating mass on Sunday to directing the performance of a symphony – all of these activities are, or at least can and should be, full of spirit. They represent expressions of spirit so that spirit is at work in them as the spirit of an individual person, as the spirit of a community, or as the spirit of a nation or an age. Insofar as we have no problem with understanding the meaning of spirit in these cases, we should not have a problem with Hegel's concept of spirit. More technically speaking, spirit is the guiding principle in all intelligent and spiritual activities, the universal that both inspires and unites individuals in their theoretical and practical pursuits. It is a subject that is alive and conscious of itself, and thus for itself, only in and through the self-knowledge of individual agents as the spirit they share collectively and express individually. Hence it is a dispersed subject, not an individual. And yet at the same time it is that which gives unity, direction, and character to the life of individuals and communities. In other words, it is both what (subjectively) inspires individuals and communities in their activities and the objective towards which they aspire. It is both subject, *spiritus rector*, and substance, *essentia*. Hegel obviously wants to maintain that spirit is both the foundation (substance, essence, goal) and the transindividual subjectivity which animates all human creativity. This transindividual subjectivity, however, is one which, although determined in its content and direction by the collective pursuits of individuals, is perceived by *them* to be objective, and thus not merely contingent upon their good will. Despite its fragmentation in individuals, spirit is a universal *for* individual subjects, i.e. an array of collective norms expressed in laws, customs and culture.[14]

Going back to Hegel's rich characterization of the soul, let me pick out three things. The soul is said to be both the "universal immateriality of nature" and the "substance" of spirit. It is also, and thirdly, said to be the identical ideality of all the "material" for spirit. That is to say, firstly, coming from nature and moving into the realm of spirit (or into the realm of emergent intelligent activity), the soul represents an aspect of or in nature which Hegel calls its immateriality. Rather than contrasting material nature with the immateriality of the soul, Hegel here attributes an element of immateriality to nature itself: corporeal, animated nature is immaterial to the extent that it is a living body which transcends the externality of mechanical systems and chemical processes. Even if we acknowledge that nature in large part consists of so-called dead matter, such inanimate matter must now be viewed as having been incorporated into a nature which is inherently alive. The soul generally therefore represents that aspect within nature at which nature shades over into, and becomes the basis for, organic life. Rather than being juxtaposed with an immaterial, supersensible entity, so-called dead matter such as minerals or rock, as well as the not-so-dead matter of air, water, and earth, has become a "moment" in the self-renewing life-cycle of nature, playing an important part in the universal metabolism of life. In contrast with the soul in general, however, the human soul marks the emergence of the life of the mind and thus the beginning of the liberation, *within* nature, of the intelligent and spiritual life from its immersion in the general planetary life, i.e. the emergence of spirit's *autonomous life* within the medium of heteronomy. There is no mystery here other than that the sheer fact of the *existence* of intelligent life can indeed by called a mystery.[15]

Secondly, and conversely, instead of looking at the human soul from the standpoint of nature and seeing it as the phenomenon that marks the transition from nature to spirit, we may also view it from the standpoint of spirit, as the link in spirit's affiliation with nature. In this respect, the soul is substance in the sense of the Aristotelian *hypokeimenon*, that which underlies all further development towards the more specifically spiritual expressions of spirit. Looked at in this way, the soul represents spirit in its naturalized form, this being merely the flipside of the characterization of the soul as the immateriality of nature. Incidentally, Hegel's comparison of naturalized spirit with Aristotle's passive *nous*, although rich in metaphorical value, should not be taken too literally. Hegel's concept of the naturalized side of spirit is much more comprehensive than Aristotle's passive *nous*, for it includes phenomena which extend far beyond the essences that lie dormant in Aristotle's *nous pathetikos*. But despite this fact, the parallel seems appropriate enough, given a wide interpretation of Aristotle's passive *nous* as including the appetitive and emotive aspects of the soul. Actually, a closer parallel exists with Aristotle's analysis of the soul at the beginning of Book II in the *Nicomachean Ethics*, where Aristotle identifies passions, sensitivities and dispositions as constituting the three elements to be found in the soul from the ethicist's point of view.[16] These and similar traits are indeed latent in the human soul *qua* substance of spirit according to Hegel.

Thirdly, to call the soul the pervading *ideality* of spirit's material is meant to indicate that whatever the material to be found in the soul, it will already be such that it may become an expression, sign or manifestation of intelligent life itself. To put it another way, rather than being a mere facticity, or a brute physical reality, that material is such as to be shaped and transformed by spirit. Spirit will *react towards* that material and make it its own. As it does so, it will successively become a consciousness, a self-consciousness, and, in the *Psychology* of the *Encyclopedia*, theoretical and practical reason. This is Hegel's way of saying that spirit is not a ghost, that it does have roots in nature, or that spirit is at least as real as nature in which it assumes a naturalized existence.[17]

Soul as the naturalized aspect of spirit's existence, soul as potentially intelligent life and soul as the incipient stage of spirit's self-development – or spirit's being-(and becoming)-for-itself – represent just so many aspects of the triadic dialectical structure of Hegel's Notion. In this way, the three aspects of the soul as the immediacy of spirit (spirit naturalized), as the internal opposition of spirit within itself in its immediacy (soul as the in-itself aspect of spirit over against its being-for-itself), and as a mediated self-referentiality in the form of immediacy (soul as incipient spirit) in *Enc.* § 389 serve to organize the structure of the ensuing Anthropology. However, setting aside technicalities of this sort, I propose to think of the human soul primarily as the aspect of *the symbiosis of spirit*, or intelligent life, *with nature*. Since we do need some sort of definition of the soul, if we want to be agreed upon what we are talking about, I offer the following paraphrase of the Hegelian language: the soul according to Hegel represents the pre-conscious or, in any case, the pre-thematic life which is potentially capable of intelligent self-articulation (and will begin to articulate itself in terms of the object-world in the *Phenomenology* section of the *Philosophy of Subjective Spirit*). This way of presenting the matter highlights the *ontological* nature of Hegel's analysis, since it identifies a *region of being* in which nature and spirit overlap, so to speak, and which is to be investigated not so much from the *phenomenological* perspective of a constituting consciousness than from the more neutral, more 'objective,' if you will, less subject-oriented standpoint of *categorial reconstruction*. Note, however, that this does not mean that, viewed from a Husserlian perspective, Hegel would relapse into the natural attitude. In fact, Hegel would agree with Husserl on this that the application of what Husserl calls the natural attitude to the soul leads to at least a partial falsification of the actual phenomena insofar as the soul would then have to be treated as a spatio-temporal psychic entity.[18]

One further result can be secured at this point. When Hegel characterizes the soul as nature in its immaterial aspect, he is implying that the soul represents a form of life which lies beyond the diagnostic framework of physics, chemistry, and biology. To say that the soul is immaterial, then, means that it cannot properly be understood in terms of the laws of physics. So whatever mind–brain identity theories do, they do not deal with the phenomenon of the soul or mind *as they are*, not with them *per se* or *kath' hauto*. On the contrary, they deal with the soul or mind insofar as these can be said to have physical properties *accidentally* or *with reference to a different framework of investigations* (*kath' heteron*) which is foreign to their very nature and being. Seen in this light, the virtue of Hegel's – or, for that matter, Husserl's – treatment of these phenomena is to be able to retain a rapport with our *ordinary understanding* of these realities, a rapport the sciences have, by their very nature, long since had to give up.

WHAT IS IN THE SOUL?

Now that we have a better sense of how to locate Hegel's thoughts about the soul ontologically, and have identified the phenomemon with the help of a preliminary definition (all definitions being preliminary, only fully worked out theories can give us an adequate understanding of phenomena), we can approach the question of what constitutes the actual reality of the soul. From the question "What is the soul?" we move to the question "What are the phenomena typically connected with the soul?" However, rather than attempting a running commentary of the content of the *Anthropology*, I shall briefly identify the chief stages of Hegel's account and then focus on one particular aspect that runs through Hegel's presentation, viz. the soul–body relationship.[19]

To those familiar with the 'mechanics' of the Hegelian system, it will not come as a surprise that the structuring of the *Anthropology* section in the *Philosophy of Subjective Spirit* is triadic and proceeds in accordance with the dialectic of the Notion as developed in the *Science of Logic*. Thus we have, roughly speaking, a stage of immediacy, of internal opposition, and of reflection or sublation of the internal opposition, or the "natural soul," the "feeling soul," and the "actual soul" respectively.

We already saw that the immediacy of potentially intelligent life is represented by the soul. The immediacy of the (human) soul, on the other hand, Hegel calls the *natural soul*. This is the soul insofar as its immateriality, i.e. its emancipation from the realm of nature, is as yet minimal. Consequently, the natural soul is subject to the influences of what Hegel refers to as "the general planetary life," that is, the differences in climate, the changes of the seasons and the cycle of night and day (*Enc.* § 392, Wallace-Miller, p. 36)[20] Hegel incidentally distinguishes this planetary life from the terrestrial life which is a topic of the *Philosophy of Nature* rather than that of spirit (*Enc.* § 392, Zusatz, p. 38) – another example of the differentiation Hegel brings to our understanding of "reality." Hegel presumably would have been pleased to learn that there is such a thing as a biorhythm, because it would have fitted well with his analysis of the symbiosis of the soul with nature.

The natural soul is again structured triadically in such a way that we move from the more universal, widespread phenomena to the more particularized, differentiating characteristics, and finally to the individualized soul. The phenomena of the general planetary life which are correlated with certain universal susceptabilities and sensitivities such as the seasonal change with certain moods, are followed by a description of the diversities of climate which seem to influence ethnic differentiation and the different mentalities of ethnic groups in the northern region of the globe, say, as against those of the southern parts.[21] To complete this account of the natural soul in its immediacy, Hegel discusses the natural endowments of individuals such as natural dispositions, talents, and types of character insofar as the latter are innate rather than acquired.

While these characteristics constitute fixed qualities, the natural soul also displays an aspect of alteration, a kind of natural growth or development. In the next section Hegel therefore groups together phenomena such as ageing, gender difference, and the sleep–waking cycle. Through the alternation of the latter, the soul begins to develop an internal differentiation between an unconscious pre-thematic awareness of its own inner life, and a conscious, but equally pre-thematic inner life (*Enc.* § 398). Hegel is quick to point out that the waking state here is not intended to be the same as the self-conscious waking state in which the individual subject has established a clear boundary between itself and the object-world, this being characteristic of spirit at the level of consciousness.[22] The section on the natural soul concludes with an analysis of what Hegel calls the soul's corporeity. By this is meant not the body as a spatial entity in which the soul is lodged, but the natural system of sensations and emotions which the soul experiences on account of its embodiment and that enable the soul to distinguish, in as yet a very abstract manner, between itself and an other which, again, is not yet our familiar physical object, but rather the sensory experience of it.[23] We have arrived, one might say, at the threshold of the intentionality of consciousness. However, as we will see later, with the system of sensibilities and these rudimentary emotions we will not quite yet emerge into the daylight of consciousness. Comments Hegel:

Sensibility (. . .) is the form of the dull stirring, the inarticulate breathing, of the spirit through its unconscious and unintelligent individuality . . . (Enc. § 400, p. 73).

Next comes a stage at which the soul has become more autonomous in its inner life, and also more self-absorbed and self-centered. This self-absorption entails a turning away from the previous symbiotic life with nature to a symbiotic life of the soul with itself, so to speak. The first phenomenon Hegel speculates about at the beginning of this section entitled the *feeling soul* is the life of the child in the womb of the mother. But very fittingly, it seems to me, this is also the place where Hegel discusses the parapsychological phenomena such as extrasensory perception, clairvoyance, premonition, sleepwalking and hypnosis (which in his time was called animal magnetism). I find this placement of parapsychological phenomena felicitious and persuasive. Where else than in a pre-conscious state of the soul's self-absorption with itself would we expect these phenoma to occur? Incidentally, it is interesting that folk psychology refers to these psychic faculties or states as supernatural. Hegel's ontological placement both justifies and qualifies this characterization. These phenomena are supernatural in that they presuppose a concentration of the as yet not fully conscious soul upon itself and thus are no longer phenomena of the natural soul. For all that, however, Hegel, through assigning them a position within the feeling soul, also make it very clear that they are psychic states located below the level of the conscious self, not, as folk psychology would have it, far above consciousness.

Hegel then discusses mental disorders and mental illnesses. He places them at the threshold of the emerging autonomy of the self in consciousness, and one would have to admit, with good reason. He characterizes mental disorders and illnesses generally as indicative of a failure of the feeling soul to become or remain fully autonomous with regard to its inner life. These states represent a possible outcome of the soul's "struggle with the immediacy of its substantial content to raise itself to the self-related simple subjectivity present in the 'I,' whereby it becomes completely self-possessed and conscious of itself" (Enc. § 408, Zusatz, p. 124).

Just before Hegel leaves the soul for the fully conscious self, he finds it necessary to interpose one more mediating step. Through the self-absorption of the soul with itself, the difference between the soul's corporeality and itself as the universal unifier of its natural pre-conscious and its pre-reflective conscious life is still unresolved. The soul is still captivated by its stirrings, feelings and sensations and so is in disagreement with itself insofar as it is also a simple self-relational structure according to which it is essentially an individual self. In other words, it is as yet only the "nominal universality," the would-be master of its inner life (cf. Enc. § 409, p. 139).

So we may well ask ourselves: How does Hegel propose to remove this disagreement of the soul with itself? At this primitive level, Hegel suggests, the intelligent self *converts* its embodiment into a vehicle for easy self-expression. In other words, the emerging self develops *habits*, beginning, one would have to suppose, with very basic measures of self-control such as the coordination of body movements. In other words, the emerging but as yet preconscious self begins to assert itself as the *causa finalis* of the soul's development. Habit marks the disappearance of the finitude of the body as a real limitation to the self-realization of intelligent life. Ironically, that very habit is also what will eventually stifle the life of spirit in the individual and lead to its ultimate sklerosis in death, as Hegel notes in more places than one.[24] Furthermore, a residual corporeality resurfaces at the level of self-consciousness in the form of desire and lust (Enc § 426, p. 167). It is as if the idealization of the subject's corporeality were to return with a vengence at this higher level of spirit, and as if there were a price to be paid for idealization and sublation.[25]

We have moved from the corporeally mediated qualities of the natural soul and the body as system of sensibilities to the body as an accidental moment of the self. The body has become the *sign* of the soul (Enc. § 411, p. 147), i.e. something that has no *intrinsic* meaning of its own.[26] It is no longer the bodily system of sensory orientation, but the 'idealized,' sublated presence of the self in the world, the bodily locus of consciousness. Our capacity for the appreciation of categorial distinctions may have diminished to the extent that we may find it difficult to grasp the difference between a material body, our own body as a source of self-experience, the corporeally based system of sensitivity, which is not *a* body, but embodied, and our system of corporeal habits. But these distinctions are precisely what Hegel may be able to make us appreciate again.

At the end of the *Anthropology*, the body has actually *become immaterial.* I would submit that this is in fact true to our very experience. As long as our bodies function well, are healthy and refreshed, in short, as long as they are in their *normal state*, they are indeed as good as non-existent to us. We do not pay them much attention under these circumstances. In fact, our bodies need to remind *us* of their continued presence through feelings of hunger or tiredness or in illness and disease. Otherwise, we might just forget about them.

THE SOUL-BODY RELATIONSHIP

Earlier on, we characterized Hegel's position on the relationship between nature and spirit in general and body and soul in particular as a nonreductive monism.[27] At first blush, this seems to make Hegel an interesting partner in the contemporary discussion about the viability of non-reductive monism such as functionalism *versus* eliminative materialisms. And it has indeed been argued that contemporary philosophy of mind could and should reclaim Hegel as an ancestor.[28] The question is whether Hegel's approach is not committed to an entirely different paradigm which is opposed not only to all forms of dualism and to all reductionist theories of mind such as eliminative materialism, but also to the very idea of any *relationship of vertical empirical dependency* between two different levels or strata of being. To the extent that contemporary theories of mind attempt to establish such a vertical relationship, either causally or by recourse to the idea of supervenience, or renounce it altogether in favor of reductionism, they have already accepted the mind–body dualism as their point of departure. In other words, they see themselves as presenting a solution to the mind–body dualism. Hegel, on the other hand, does not see himself in this way. For Hegel, the mind–body dualism was a miscarriage of philosophical speculation to begin with. Miscarriages can be deplored, but they cannot be rectified, or turned into healthy deliveries *post factum*. But that means that Hegel starts out from different presuppositions entirely. He is not suggesting a new way to resolve the initial (Cartesian) dualism, rather, he proposes a way to bypass that dualism from the outset.

Looking more closely at Hegel's text, however, one might conclude that Hegel does in fact start out with a dualism between matter and spirit or body and mind which he then seems to convert into a monistic account. Here is what the text says at Enc. § 389, *Zusatz*, p. 33:

> The opposition of soul and body must, of course, be made. Just as the indeterminate universal soul determines and individualizes itself, just as mind [*Geist*] thereby becomes consciousness . . . so does mind [*Geist*] place itself at the standpoint of opposition between itself and its Other, so does its Other appear to it as a reality, as external to mind [*Geist*] and to itself, as something material. From this standpoint, the question as to the possibility of the community of soul and body is quite natural.

However, two little words should not be overlooked here, the "its" in "its Other" and the "appear" in "so does its Other appear to it as a reality." The "appear" indicates that the relationship between body and soul only *appears* to be a relationship between two entities or two "particulars," as Hegel says a few lines later. And the fact that spirit regards matter or the body as *its* other, rather than as *an* other,[29] indicates that the opposition between mind and body has to be viewed against the background of a prior unity of the two, such that it is essentially negated or, as Hegel's magic word goes, "sublated." This reading is confirmed by the subsequent text:

> . . . the question arose as to how the contradiction, to wit, that entities which are absolutely independent and for themselves are yet in unity with one another, could be solved. The question as thus posed was unanswerable. But it is just this form of the question that must be recognized as inandmissible; for in truth the immaterial is not related to the material as a particular is to a particular, but as the true universal which overarches and embraces particularity is related to the particular; the particular material thing in its isolation has no truth, no independence in face of the immaterial. Consequently, the standpoint which separates them is not to be regarded as final, as absolutely true. On the contrary, the separation of the material and the immaterial can be explained only on the basis of the original unity of both (ibid.).

Although it may not be very explicit, Hegel is in fact distinguishing here between two kinds of opposition of mind (or soul) and body, an admissible and an unadmissible opposition. The prime example of the unadmissible opposition would be the Cartesian substance dualism. In its admissible form, the opposition is not "absolutely true," hence only partially true, and does not amount to *genuine* dualism. Substance dualism, on the other hand, would amount to an insistence upon the irreconcilable hiatus between two irreducibly different realities.

The question is: Have contemporary supervenience theories been able to overcome this Cartesian dualism? I believe that a good measure of skepticism is in order here. To be sure, Cartesian substance dualism has been avoided. But it seems to have been replaced with what in discussions of Spinoza's mind–body identity theory is sometimes called property dualism,[30] or at any rate with what one might term state dualism, i.e. a parallelism of physical and mental states. This is particularly true of theories of supervenience. They maintain a) a parallelism of physical and mental states which are b) correlated by an unspecified, and perhaps unspecifiable, empirical relationship of supervenience. They further claim c) a real vertical dependence of mental upon physical states. The relationship of supervenience, however, although obviously an empirical relationship, is not a causal one, or at least the kind of causality at play is one that eludes us. Supposedly, this causality would have to be one that would enable us to map the causal flow existing between a given physical or neuronal state and its mental correlate. If, however, the relationship is empirical, but the causality involved obscure, we do have a parallelism without a) a strict one to one correlation of mental and physical states and lacking b) a causal account of the dependency itself. I would call that a genuine dualism.[31] As long as we are unable to describe the causal flow between physical and mental states, we are, to paraphrase Aristotle, still at the level of mere experience rather than that of science, i.e. we know, say, *that* a certain drug works in certain cases, but we do not know exactly *why*. It does not necessarily improve the situation that in functionalism we simply *ignore* the question of the precise nature of the correlation between mental and physical states. It thus becomes a somewhat dubious proposition to assert that mental attitudes are causally dependent. But causally dependent on what? On other mental attitudes? On physical states? What kind of causality exactly are we invoking here? These are not so much criticisms as skeptical observations regarding unanswered questions.

What, then, is the relationship between soul and body according to Hegel? To begin with, it is clearly not an empirical, but a dialectical, or conceptual, relationship. And although there is a dependence of one level upon the other, its direction is exactly the reverse of the empirical: spirit "posits" a body as its other in order to objectify or materialize itself, so that it may then "return back into itself" out of its other into itself (and thus be reconciled with itself). If this were to be understood as a causal or, as in current supervenience theory, a quasi-causal relationship, it could not but result in bad metaphysics – as it often has in the accounts of Hegel interpreters.[32] Hegel's point, however, seems to be that for purposes of intelligibility we need to reconstruct the many layers and facets of the real in such a way that the antecedents of a given category or phenomenon are successively redefined as constitutive moments of a higher unity on which they depend. We reconstruct an ontological genealogy for an end-product which we assume to be the result of a conceptual, not a real, development. Thus the meaning of the antecedents is reversed, once they have given way to their "over-arching universal" which integrates and sustains them. Instead of being logically prior to their result, they become dependent on it, and instead of producing the result, they turn out to be *prerequisites* for it, i.e. something which is required for something else to be possible, but which does not by itself produce or generate that for which it is required. For Hegel, as for Aristotle, the whole is prior to its parts, and the parts loose their meaning, once separated from the whole. Thus Hegel can say that the unity of the material and the immaterial, or of the body and the soul

> . . . must not be taken as something neutral in which two extremes of equal significance and independence are united, *since the material has absolutely no meaning beyond that of being a negative over against spirit and over against itself*, or must be described – in the words of Plato and other ancient philosophers – as 'the Other of itself,' whereas the nature of spirit is to be recognized as the positive, as the speculative, because the material, *which lacks independence in face of spirit*, is freely pervaded by the latter which overarches this its Other, does not account it as something truly real but reduces it to an ideal moment and to something mediated (ibid., my italics).

It may be possible to call this a form of supervenience of soul upon body, but a supervenience in which the soul figures non-causally as *causa finalis*.

CONCLUSION

There are, then, reasons to remain skeptical of the claim that Hegel is indeed an ancestor of (some positions in) contemporary philosophy of mind.[33] I would like to conclude by making an attempt to determine in a more concise fashion the peculiar way in which Hegel views the mind–body relationship. The body does not figure as a spatial entity in the Anthropology at all. If we want to talk about the body as a spatial object subject to the laws of physics and chemistry, we would have to go back to the earlier stages in the Philosophy of Nature where Hegel discusses the emergence of individual material bodies from the antecedent categories of matter, motion and gravity, and to the processes involved in the chemical interactions and transformations of substances.[34] If the body plays a role in the discussion of the soul in the Anthropology, as we have seen that it does, "body" must have a very different meaning here than in the early stages of Philosophy of Nature. We have encountered the body in the form of the mind's sensory awareness and its system of habitualities. In the latter respect, the body had "sublimated" into a mere sign of the soul, to use an expression from the vocabulary of the physicist. This suggests that it is more appropriate to speak about the soul's *corporeality* rather than the body as a spatial object (as Hegel indeed does). Far from functioning as the physical embodiment of the soul, the body *qua* corporeality *of the soul* here represents the domain of the pre-conscious, pre-thematic, pre-intentional life of the mind. It is as though consciousness or the mind had an embodiment of their own, viz. the pre-conscious soul. Insofar as the soul contains within itself the unifying self, and is thus the emergent *ego* of consciousness over against its own corporeality, it could be said that in relation to consciousness the soul as such comes to function as the corporeality of the mind, or the *psyche* of spirit. The soul, then, is nothing but the mind's 'body.'

Department of Philosophy, Boston University

NOTES

[1] Schelling's and Hegel's contributions in this regard are well known and much commented on and debated. For two more recent studies of the Hegelian approach see Willem de Vries, *Hegel's Theory of Mental Activity*, Ithaca – London (Cornell University Press) 1988 and Michael Wolff, *Das Körper-Seele-Problem. Kommentar zu Hegel, Enzyklopädie (1830), § 389*, Frankfurt am Main (Klostermann) 1992. Fichte's role in this development was less well understood until the publication of Reinhard Lauth's important study *Die transzendentale Naturlehre Fichtes nach den Prinzipien der Wissenschaftslehre*, Hamburg (Meiner) 1984.

[2] For an overview of current theories see N. Block, *Readings in Philosophy of Psychology*, Cambridge, Mass., 1980, vol. 1, and Jaegwon Kim, *Philosophy of Mind*. Westview Press 1996.

[3] A concise explanation of metaphysical functionalism is offered in N. Block, *Readings in Philosophy*, vol. 1. pp. 173 ff.

[4] For a spirited defense of eliminative materialism and an informative critique of functionalism see Paul Churchland, "Eliminative Materialism and the Propositional Attitudes," in T*he Philosophy of Science*, ed. R. Boyd, P. Gasper, J.D. Trout, Cambridge (MIT Press) 1991, pp. 615–630.

[5] See the opening paragraph of Aristotle's *Metaphysics* (Met. I 1, 980a 22ff.).

[6] For a detailed analysis of Hegel's theory of the animal soul and the animal organism see my article "Hegel on the Animal Organism," *Laval théologique et philosophique*, 1996.

[7] See Hegel's *Philosophy of Mind*, Enc. § 392, p. 37 (tr. Wallace). References to Hegel's text will be to the 1830 *Encyclopedia*. I shall quote the German text from *Georg Friedrich Wilhelm Hegel, Werke*, ed. E. Moldenhauer, K.M. Michel, Frankfurt am Main (Suhrkamp) 1970. Engish quotations will be from *Hegel's Philosophy of Mind, Part Three of the Encyclopedia*, tr. W. Wallace with *Zusätze* tr. by A.V. Miller, foreword by J.N. Findlay, Oxford (Clarendon Press) 1971 and *Hegel's Philosophy of Subjective Spirit*, 3 vols., edited and translated with an introduction and explanatory notes by M.J. Petry, Dordrecht-Boston (D. Reidel) 1978. Occasionally, I have taken the liberty of offering my own translation.

[8] See *Enc*. § 378, p. 3 (tr. Wallace).

[9] See Aristotle, *De an*. II 3. Note that Aristotle does not draw a clear boundary between perception in animals and in humans (compare *De an*. II 3 with II 5). The situation is different, however, in the case of the analysis of the soul in *Eth. Nic*. 1 13, where Aristotle attributes a disposition towards rationality to the desirous or emotional part of the human soul. Thus within the ethical context, the human soul seems to be specifically distinct from the animal soul.

[10]. See *Enc* § 391, *Zusatz*, p. 36 (tr. Wallace): "The soul stands midway between Nature which lies behind her, on the one hand, and the world of ethical freedom which extricates itself from natural mind [spirit] on the other hand."

[11] *Enc*. § p. 29, tr. Wallace. Spirit in the form of soul is "still in the grip of Nature and connected with its corporeity," says Hegel (*Enc*. § 387, *Zusatz*, p. 27, Tr. Wallace).

[12] See Hegel's frequent remarks that although an anticipation of later stages cannot always be avoided in subjective spirit, these anticipations are to be clearly understood to be such (e.g., *Enc* § 380, p. 7 tr. Wallace).

[13] Following Gilbert Ryle, some have blamed the problem on a category mistake. For Hegel's more extensive treatment of the mind–body dualism see *Enc*. § 389, *Zusatz* (Wallace-Miller 30–34), where Hegel addresses, and dismisses, the traditional mind–body dualism. He praises Spinoza and in particular Leibniz for having made creditable attempts at resolving the dualism. We will come back to this issue. On this question, see also Michael Wolff's excellent commentary and discussion

of Hegel's theory of the sould–body identity: *Das Leib-Seele-Problem. Kommentar zu Hegel, Enzyklopädie (1839) § 389*, Frankfurt a.M. (Klostermann) 1992, p. 15, pp. 156ff.

14 For Hegel's account of spirit see the opening sections of the Philosophy of Spirit in *Enc.* §§377–385. The above characterization relies more on the following passages from *Hegel's Phenomenology of Spirit*, tr. A.V. Miller with foreword by J.N. Findlay, Oxford (Clarendon Press) 1977, p. 264: "Spirit, being the *substance* and the universal, self-identical, and abiding essence, is the unmoved solid *ground* and *starting-point* for the action of all, and it is their purpose and goal, the in-itself of every self-consciousness expressed in thought. This substance is equally the universal *work* produced by the action of all and each as their unity and identity, for it is the *being-for-self*, the self, action. As *substance*, Spirit is unshaken righteous identity; but as *being-for-self* [i.e., as subject] it is a fragmented being, self-sacrificing and benevolent, in which each accomplishes his own work, rends asunder the universal being, and takes from it his own share."

15 The fact that nature is generally alive, renewing itself, and displaying universally the myriad activities called life constitutes for Hegel an aspect of nature's *ideality*. Ideality is the more general term compared to immateriality and would already be a characteristic of organic life as such. The soul *qua* ideality of nature is nature come alive in the form of plant and animal life (cf. *Enc.* § 337).

16 See Aristotle, NE II 5, 105b 19ff.

17 Note the passage where Hegel says that a man's moral principles, his ideas of right and wrong, his religious beliefs must not be mere cerebral attitudes, but that for them to be real they must pervade the person as a whole and find expression in his emotions and actions, not just in his reasonings (cf. *Enc.* § 400, p. 73, tr. Wallace).

18 Just as Hegel so would Husserl repudiate any philosophical liability to account for the interaction of two allegedly separate entities called mind and body. For him, too, the reification of the soul would be a falsification of the actual phenomenon. See, e.g., E. Husserl, *Phenomenological Psychology*, tr. J. Scanlon, The Hague (Martinus Nijhoff) 1977, p. 81 [*Husserliana* IX 107]: "My psychic life is for me quite directly and in the strictest sense of the word perceived, and directly perceived not alongside my body but as animating it. Here alone I experience in an originally perceptual manner this unity of body and psyche, this intermingling of bodily and animating happening." Or, again: The body is not merely a spatial thing, it "is that too, but still more. If the body becomes a theme of physiology, then it is taken simply as an existentially posited natural object. The physiologist stands upon the ground of the experience of nature. In phenomenology the perceiving, the experiencing of the body is the theme, and the body itself purely as experienced therein, as perceptive meaning" (p. 152). For Husserl's critique of the Cartesian mind–body dualism and its detrimental consequences see *The Crisis of European Sciences and Transcendental Phenomenology*, translated with introduction by D. Carr, Evanston (North-Western University Press 1970), §§ 10ff.

19 For good running commentaries with simultaneous critical analysis and evaluation see e.g. J. Findlay, *Hegel: A Re-examination*, London-New York 1958 and, more recently, Murray Greene, *Hegel on the Soul. A Speculative Anthropology*, The Hague (Martinus Nijhoff) 1972.

20 Unless otherwise indicated, references to Hegel's text will from now on be to the Wallace-Miller translation. Hegel also identifies the general planetary life with the *anima mundi* ("world-soul") of the Ancients. For bibliographical references concerning the background of Hegel's use of world-soul see Petry's commentary (Petry II 439, *ad lineam* 23).

21 See *Enc.* § 393–394 and their extended *Zusätze*. About the mentality of the Germans Hegel offers this comment: "As for the Germans, they usually think of themselves last, either from modesty or because one saves the best till the end." (*Enc.* § 394, *Zusatz*, p. 50).

22 *Enc.* § 398, p. 65: "The characterization [of the waking state] given in [this] section is abstract; it primarily treats waking merely as a natural fact, containing the mental element *implicite* but no yet as invested with a special being of its own. If we are to speak more concretely of this distinction . . ., we must take the self-existence of the individual soul in its higher aspects as the Ego of consciousness and as intelligent mind [spirit]."

23 There is another parallel here with Husserl worth pointing out. Husserl mentions the kinesthetic experience that the body affords in particular, but also calls the body a "subjective internality" which coexists with the body as "a spatial externality." As subjective internality, the body is "an integral body, an organ, a habitual system of subjective functions" (*Phenomenological Psychology*, p. 151). The latter characterization echoes, unwittingly no doubt, Hegel's determination of the final stage of the so-called feeling soul as a system of habitualities (cf. *Enc.*§ 409).

24 See, e.g., *Enc.* § 375 *Zusatz*: "Das Lebendige als Einzelne stirbt an der Gewohnheit des Lebens, indem es sich in seinen Körper, seine Realität hineinlebt (*Georg Friedrich Wilhelm Hegel, Werke*, ed. E. Moldenhauer, K.M. Michel, Frankfurt am Main (Suhrkamp) 1986, vol. IX, p. 536) (Live things *qua* individuals die from the habit of living, through settling into their bodies, their reality).

[25] As Vittorio Hösle has pointed out, the development in the soul's emancipation from its immersion in nature does not only have the meaning of moving away from nature but concurs with a "re-objectification" of the soul in and through its peculiar corporeality (see V. Hösle, *Hegel's System*, Hamburg (Meiner), 1988, vol. 2 p. 353).

[26] For an illuminating treatment of the body as sign in Hegel's Philosophy of Subjective Spirit see J. Dodd, "The Body as 'Sign and Tool' in Hegel's Encyclopedia," *International Studies in Philosophy* 27 (1995) 21–32.

[27] Hegel's monism has been analyzed and compared to Russell's pluralism and Bradley's monism in illuminating ways by Rolf-Dieter Horstmann, *Ontologie und Relationen. Hegel, Bradley, Russell und die Kontroverse über interne und externe Beziehungen*, Königstein/Ts. (Athenäum-Hain) 1984.

[28] See Willem A. deVries, *Hegel's Theory of Mental Activity*, Ithaca and London (Cornell University Press) 1988, p. 32.

[29] For this important distinction which in Hegel generally marks the transition from finitude to infinity or self-relationality (self-containedness), see *Enc.* §§ 94, 95. In the latter section Hegel argues that the tendency to resist this transition typically results in dualism.

[30] See, for instance, Jonathan Bennett, *A Study of Spinoza's* Ethics, (Hackett) 1984, p. 41ff. for property dualism and his excellent analysis of Spinoza's mind–body identity theory at pp. 127–151.

[31] The problem has also been framed in terms of a lack of nomological coextension between mental and physical predicates: see Crawford Elder, *Appropriating Hegel*, (Aberdeen University Press) 1980, pp. 43–49. Elder, too, sees less of a continuity between Hegel and contemporary philosophy of mind and more of a radical alternative in Hegel's treatment of the soul–body unity.

[32] The possibility, and persuasiveness, of a *non-metaphysical*, i.e. categorical and reconstructive, reading of Hegel's philosophy was first demonstrated by Klaus Hartmann: see his "Hegel: A Non-Metaphysical View," in A. MacIntyre (ed.), *Hegel. A Collection of Critical Essays*, New York 1972, pp. 101–124; reprinted in Klaus Hartmann, *Studies in Foundational Philosophy*, Amsterdam-Würzburg 1988. – Hegel did envision a functionalist psychophysiology, as Wolff points out, which would search for "affinities" between emotional states and their bodily expressions (e.g., blushing as a concomitant of an embarassment) (see M. Wolff, *Das Leib-Seele-Problem*, p. 189–191). And although this affinity would not imply any direct causal relationship between a mental and a physical state, it would still constitute some kind of *empirical* relationship. Nonetheless, we should keep the empirical science envisioned here by Hegel distinct from the categorical theory he himself puts forth.

[33] Despite the sympathy I have for deVries's casting of Hegel in the light of Aristotelian teleology, I cannot, therefore, agree with his interpretation of Hegel's position on the soul–body identity question as a weak, i.e. quasi-supervenience monism, unless this supervenience is kept clean of any physicalist interpretation (see W. deVries, *Hegel's Theory of Mental Activity*, pp. 41–46). For the same reason, I am doubtful that, as Wolff claims, Hegel tends towards a functionalist theory of the natural soul, though not at a functionalist theory of the mind–body problem as a whole (see M. Wolff, *Das Leib-Seele-Problem*, p. 166). Note also that, as Crawford Elder has pointed out, Hegel's 'infinite' teleology differs in important respects from Aristotle's 'finite' teleology (C. Elder, *Appropriating Hegel*, p. XIf.).

[34] See the sections entitled Mechanics and Physics, in particular *Enc.* §§ 263ff. and 274ff. Here, too, we come across a number of surprises: For Hegel, the paradigmatic individual spatial material object is not really the abstraction we have come to accept, viz. the three-dimensional spatio-temporal material object. Such objects for Hegel are quite insubstantial, serving merely to instantiate the laws of motion. They have already dissolved at the level of chemical interaction, and thus are ontologically quite insignificant. If we want the genuine article, i.e. a truly individualized spatial object, Hegel implies, we must look at the sun, the planets and the earth. These are individualized bodies worthy of the name.

LAWRENCE CAHOONE

WHOSE NATURE? WHICH MORALITY? ON KOHÁK'S MORAL SENSE OF NATURE

Since Pythagoras made the body (*soma*) the tomb (*sema*) of the soul, it has been common for Western philosophy to understand the human essence as foreign from bodily nature. Though dreary, such dualisms have often seemed the only alternative to a naturalism that, from Democritus to Darwin, has threatened to undermine human uniqueness, and human morality in particular, by absorbing the mind in a nature inhospitable to moral value. The Aristotelian and hence Scholastic philosophy of nature mitigated this problem by suffusing nature with *telē*. But science since Galileo purged nature of purpose. Hobbes drew the ethical implications of modern mechanism with characteristic nonchalence: "whatsoever is the object of any mans Appetite or Desire; that is it, which he for his part calleth *Good*: And the object of his Hate, and Aversion, *Evill* [sic]. . . For these words . . . are ever used with relation to the person that useth them: There being nothing simply and absolutely so . . ." (Hobbes, I, 6, p. 120). Descartes and other dualists struck the bargain of granting to nature what is nature's and to mind what is mind's, accepting the resultant enigmas as the price for avoiding the Hobbesian conclusion. Defenders of mind with more *chutzpah*, however, following the dictum that the best defense is a good offense, declared nature a province of mind, and especially, of *Geist*. German idealism was eventually succeeded by a metaphysically chastened, but no less anti-naturalistic phenomenology in the work of Husserl. Throughout this abbreviated story, one theme remains virtually constant: with few exceptions, all sides in the modern debate have accepted the discontinuity of morality and nature as such. For morality's defenders this discontinuity has been a two-edged sword, since it leaves the moral soul precariously out on a limb, waiting to be hacked off by a new wave of scientific advance. Thus in the twentieth century bedfellows as strange as positivism and existentialism have agreed on the discontinuity of morality and nature – and consequently held morality incapable of rational justification. Note the following remark from one of Europe's foremost thinkers:

> Blind to good and evil, reckless of destruction, omnipotent matter rolls on its relentless way; for Man, condemned to-day to lose his dearest, to-morrow himself to pass through the gate of darkness, it remains only to cherish, ere yet the blow falls, the lofty thoughts that ennoble his little day . . . proudly defiant of the irresistable forces that tolerate, for a moment, his knowledge and his condemnation, to sustain alone, a weary but unyielding Atlas, the world that his own ideals have fashioned despite the trampling march of unconscious power. (Russell, p. 47)

R.S. Cohen and A.I. Tauber (eds.), Philosophies of Nature: The Human Dimension, 19–33.

Cognoscenti will recognize these as the words of the well-known existentialist, Bertrand Russell. The image of Promethean Man in Purposeless Nature, which Russell shares with Camus and Sartre, continues the tradition of discontinuity.

It is this tradition that Erazim Kohák opposes. He does so by reviving the premodern understanding of nature as teleological, albeit through the very modern method of Husserlian phenomenology, interpreted so as to avoid idealism. In answering the question whether nature and morality are continuous, Kohák contends that if our primary method for understanding nature is modern science, then the answer would indeed be no; but if our method is phenomenological, if we take our pre-scientific lived experience of the world as primary, then the answer is yes. For him, the correct method is the latter. There are few philosophical questions as important as determining whether he is right. For what is at stake is not merely ethical responsibilty toward non-human beings, nor the status of morality in nature, but further, whether human being is of one fabric, or rent beyond repair. Can our moral life, hence our spirit, be fully at home in the nature our bodies inhabit? Or are we suspended in permanent tension between two ultimate contexts, the natural and the non-natural, children of a cosmic broken home?

The subtitle of Kohák 1984 book, *The Embers and the Stars: A Philosophical Inquiry into the Moral Sense of Nature*, part philosophical treatise and part diary of life in the New Hampshire forest, contains an instructive ambiguity. That nature exhibits a moral *sensus* can mean either that nature has *moral value* or that nature *values morally*. It is one thing to say that non-human nature – nature independent of human utilization – has moral value, yet another to say that non-human nature is itself moral. Kohák is making both claims.

I will disagree with both. The discontinuity of humanity and nature can be healed in two ways: we can either personalize nature or naturalize persons. Kohák opts for the former. Mine, were I to pursue it, would be the latter, since I believe that persons are not more fundamental than impersonal nature, and that phenomenology's truth is not more fundamental than that of science. Kohák and I each admit a significant discontinuity between human and non-human; for me *moral sense* is part of that discontinuity. Nature is value-laden, but the values with which it is laden are not moral ones. Non-human nature exhibits neither morality nor moral value. We exhibit the latter, and occasionally the former.

Borrowing the Husserlian notion, within what he calls the "brackets of the forest," that is, given the practical *epochē* of our scientific and technological understanding of nature induced by life in wild nature, Kohák argues that the universe is revealed as a value-laden *kosmos*. We thereby experience the truth of the Augustinian principle, *esse qua esse bonum est*, being as being is good, not value-neutral. We recognize as well that non-human beings are *persons*, that is, beings with moral value. Human morality itself emerges from and is validated by our experience of nature. This emergent morality includes responsibility "to and for nature." Last, experience reveals that the temporal cycle of life and death does not exhaust the reality of nature. That being not only is but is *good* is evidence of eternity breaking through the otherwise purely temporal order of *physis*. This fact shows that *God is* (Kohák, 1984, p.82).

It may be well to mention here the related view of Hans Jonas, who argued similarly that *bios* introduces value into being. Living nature is purposive and we cannot deny our intuition of "the superiority of purpose as such over purposelessness" (Jonas, 1984, p.80). The presence of living creatures, for whom their continued life and the environmental conditions this life requires are decidedly not value-neutral, gives being value, "posits it absolutely as the better over against nonbeing." Like Kohák, Jonas affirmed the value, the valuing, and hence the moral sense, of non-human nature.

To answer Kohák we must ask, Does non-human nature exhibit either moral value or a morality with which human morality is continuous and which it must recognize? We can begin by admitting some facts that suggest non-human nature's moral sense.

First, there is our own common experience of non-human nature as *morally valuable*. This experience comes, I believe, in two forms: our sense of identification or kinship with living nature, and our sense of connection or belonging to non-living natural environments.

The sense of kinship, especially with mammals and higher vertebrates, is striking. A visitor for another planet who reviewed the history of religion, myth, and art, or observed contemporary mass culture and our relations with domesticated animals and plants, would be more justified in the conclusion that humans feel a powerful empathic connection to non-human species than that we regard them as value-neutral things for use. Certainly we are not consistent in this; we are capable of empathizing in the most anthropomorphic sense with higher mammals and of righteously condemning their mistreatment, while snacking on their relatives and condoning their use as experimental subjects. Kohák's remarks about the obfuscatory power of modern society are certainly correct: we eat as if our hamburger had no connection with Bossy the cow.

Just as undeniable for many people is the sense of being *at home in* or *belonging to* non-living natural environments. A sense of connection to sea, lake, rock, desert, mountain, and plain, independent of or in addition to their biotic coverings, is common. The feelings generated thereby are different than those of the biotic kinship experience. Such environments are not *kin*, but they can be *home*.

Second, it is conceivable to attribute to non-human nature not just moral value, but *moral significance* and even *morality* itself, if the term is taken broadly enough. The Stoics, for example, found in nature as a whole the embodiment of moral principles from which humans might learn. We might also speak of non-human intra-species behavior as moral. Such attribution is made plausible by the fact of anatomical and behavioral similarity. Whatever the differences in brain activity or meaning, we are driven to eat, socialize, avoid pain, reproduce, and care for our young, as are other species. We share with them a naturalistic continuity of biotic pragmatism. Most relevant here, some of the activities we share are, *for us*, moral. Indeed, some naturalists have held up animal behavior as a model for humans, pointing out, for example, that rutting male elk cease their battles to establish hierarchy short of life-threatening injury, unlike rutting *homo sapiens* with access to nine millimeter handguns. Even if we balk at calling such strictures in non-humans *moral*, we must admit some continuity between social behavior among non-humans that preserves their progeny and communities and analogous behavior in humans that we do call moral.

We must consider whether any of the aforementioned facts can serve to establish continuity between human morality and non-human nature.

Is nature as a whole, or are natural environments, moral? The answer is certainly no. The whole is *amoral*, which is to say, it is morally indifferent. Nature is as horrifying, brutal, cruel, indifferent, rapacious and wasteful as it is beautiful or morally inspiring. When a bear eats a salmon, it first strips the skin – the tastiest part – off the still-living fish. There is a wasp who reproduces by stinging a large beetle into paralyis, then laying her eggs under the beetle's exoskeleton to feed through their larval stage, until the host is consumed. Despite the limited nature of the beetle nervous system, this is an unappealing death. But being eaten alive at a more rapid pace seems only marginally preferable. Simply, all biotic existence, without exception, vegetative and carnivorous, is predicated on the selective destruction of the natural environment and the biotic creatures in it. Even plants live at one another's expense, denying each other nutrients, water, and sunlight, thereby truncating their neighbors' growth, if not ending their life.

As Kohák says, the forest is indeed peaceful, but only *for us*, not *in itself*. It is peaceful for us because we are the pinnacle of the food chain, our abilities to perceive and identify with its most frequent victims are limited, and it lacks the rapid change characteristic of modern human society. Kohák correctly faults our technological culture for obscuring natural processes, but it is just as true that the apparent tranquility of the forest is predicated on our failure to perceive the life and death struggle that goes on either too slowly or among creatures too small for us to notice. He writes that the Judaic Ten Commandments emerge from our experiences of nature, that in the brackets of the forest we see that theft depersonalizes the thief, that wanton killing shatters the peace and makes the perpetrator an outcast. But is the starling depersonalized in stealing the sparrow's nest, or the recently-fed lion an outcast after crushing a turtle in curious exploration? Regarding such matters nature as a whole is indifferent, or if one prefers, tolerant. As Nietzsche wrote of the Stoic injunction to model human life on nature:

> "According to nature" you want to live? O you noble Stoics, what deceptive words these are! Imagine a being like nature, wasteful beyond measure, indifferent beyond measure, without purposes and consideration, without mercy and justice, fertile and desolate and uncertain at the same time; imagine indifference itself as a power – how could you live according to this indifference? (Nietzsche, sec.9)

It is natural for natural beings to struggle to maintain their integrity against assimilation by other natural beings, and to assimilate others who likewise oppose their own assimilation. This struggle is not against nature; it *is* nature. What could any moral principles drawn from this nature look like?

While nature or natural environments are not moral, they can be interpreted as morally significant, as already noted. In its relative timelessness, its massive complexity, its local harmony and self-sustenance, nature can inspire in us endurance, humility, acceptance, patience, and independence from the clamorous but evanescent concerns of human society.

These are arguably moral lessons, and so as their inspiration natural environments can be morally significant. It is in this sense that the *Stoa* found morality in nature. Before them, in the first attribution of moral significance to nature in the history of Western philosophy, Anaximander said that all things arise and disappear with an orderliness designed to make retribution to their opposites for the *adikia* or injustice by which one temporally prevailed at the expense of the other. Older still are the words of Ecclesiastes [3(1–8)], "There is an appointed time for everything . . . A time to be born, and a time to die . . . a time to love and a time to hate; a time of war, and a time of peace." This is a moral lesson in harmony with the metabolic cycles of nature. But note that it is a very particular lesson. Relevant to fellow humans, the message of Ecclesiastes is: sometimes you love them, sometimes you kill them. The moral meaning of natural environments is likewise a particular and limited one, a fact to which we will return.

Can the intra-species behavior of some non-human creatures be considered moral? To answer yes is to ascribe to non-humans intentions and choice, hence join a very complex debate in the philosophy of mind and in science. Let us avoid this debate with a limited admission: some non-human species exhibit something analogous to moral *behavior*, if not moral experience, intention, or decision. As noted earlier, there are for many species limits on ways of using fellow species members as well as positive obligations regarding mates, offspring, and group members. However, the meaning, explanation and extent of these behavioral limits are clearly pragmatic: whichever limits are consonant with the preservation of the species are enforced. While it is true that many species regulate intra-species conflict and consumption of natural resources, they may also leave weak newborns to starve and injured animals to be eaten by predators. That is their "morality," too. In short, we may find these moralities rather immoral. Any analogous human morality would be likewise collectivist and pragmatic, that is, keyed to species survival. So, even ignoring deeper complexities, the behavioral analogy is a limited and particular one.

Let us now turn to the feelings of care and respect we often have for nature and its creatures, rooted, I have suggested, in the feelings of kinship and belonging. Phenomenologically, we may say that these feelings are the recognition of moral value in natural beings, which value demands a moral response. So while these non-human entities may not *be moral* in anything like a human sense, they nevertheless have *moral value*. But real as they are, kinship and belonging cannot serve as a basis for moral dealings even with non-human creatures, let alone fellow humans.

We must admit first of all that our capacity for identification or kinship is limited. Many of us are capable of feeling a sense of kinship with mammals, or at least the higher and larger ones. It is with them that we can most easily feel what Jonas would call their animal "subjectivity." Beyond these creatures are the more difficult cases of birds, reptiles, amphibians, fish, insects, plants, and finally the rudimentary life forms which we are hard pressed to feel as different from non-living matter. I defy anyone to identify with lichen, yeast, or bacteria. As beings with a particular set of traits, each of which represents some community of non-human beings that share the trait, our capacity for empathy is objectively limited. In short, we feel kinship only with *kin*. Is such feeling an adequate basis for morality? We do not think so in human society. I have a young son and, roughly speaking, I would do anything for him. Is this a *moral* relationship? We normally say that it is. But if the only way to save my son were to feed all other children in my neighborhood to wolves, would we still call my fatherly love *moral* after the wolves have had their meal? Or rather, is much of what we call moral responsibility precisely an attempt to establish a sense of duty or empathy *beyond* and even in *opposition to* the always limited circle of kinship?

Furthermore, the fact of predation makes it impossible to derive even a modestly consistent morality from whatever sense of kinship I do have. The kinship I feel for the dove leads me to smile when the hawk's attack is unsuccessful; why do I not fear for the hawk's unfed chicks? Nietzsche would perhaps fault my Judeo-Christian and democratic cultural history, which has led me to identify with the prey, rather than the predator. Whatever the reason, to deal with non-human nature as a whole through felt kinship is impossible, unless one's sense of kinship is not violated by the torture and slaughter of one's kin.

Last, what about my sense of belonging to non-living natural environments? That such feelings are palpable for many, and may sometimes be described in personal terms – as the phrase 'mother earth' implies – does not mean that the objects of these feelings, nor the feelings themselves, are *moral*. Even when we feel this connection, do we regard the valued environments as sharing with us anything like our sense of morality? No; they are *valued*, but not *valuing*. Furthermore, do we value them morally or aesthetically? To value them morally is to make them similar to ourselves, to personalize them, as Kohák would admit. But does the value of such environments hang on their being like us? Are they not valued precisely because in their timelessness and stability they transcend us, hence granting us a sense of connection to the eternal and powerful? Their experienced value, I will suggest, lies not in their personhood, but in their *impersonality*.

Now, it may be said that despite the foregoing, a few non-human species exhibit such complex functioning as rightly to inspire moral obligation. One can argue that dolphins, some primates, and certain whales, both for phenomenological and scientific reasons, are due a respect not owed the rest of nature. Nothing I have said would undermine such an argument, whose validity would have to be judged on a species by species basis. If the ascription of moral significance is made correlative to specific behavioral or morphological traits (like language use or brain use), then a complex discussion would ensue about the physiological and behavioral requirements for morality, which cannot be joined here. Nevertheless, it should be noted that such arguments do not ascribe to nature or life in general a moral significance. They only expand the realm of the moral beyond one species – *homo sapiens* – to include a few others. They leave the vast majority of the world's species, and all non-living nature, outside the moral realm.

The point of these considerations is that, despite the real but limited analogy between human morality and non-human nature, the two are in important respects discontinuous. In particular, one of the most characteristic parts of human morality has no analogue in, does not emerge from, and cannot serve as a model for interacting with virtually all of non-human nature, namely *respect for persons.* I intend this Kantian formulation vaguely, so as not to presume the particulars of deontological ethics. The point is that the concern for the being, integrity, and suffering of the living individual independent of social status or consequences, a concern crucial to many human moralities including those of Christianity and modern liberal humanism, is not to be found in, nor is it continuous with, non-human nature. Simply put, nature is no respecter of persons.

My claim is related to the charge of anthropocentrism levelled by some at environmental ethics. Kohák is familiar with the charge, which raises what he calls the "ecological dilemma" (Kohák, 1994). The dilemma is this: if the human perspective is not privileged and we are treated, following some environmentalists, as merely one species among others, then it would seem that we cannot derive any moral obligations to nature, since there are no moral obligations in non-human nature. Yet if we start with the human valuation of things, thereby making available our sense of moral obligation, we court anthropocentrism, hence stand on the same ground as the anti-environmentalist who happily accepts the primacy of the human utilization of nature. That is, anthropocentrism gives us morality but not necessarily one that values nature, while anti-anthropocentrism puts nature first but leads to no morality at all, hence to no environmental ethics.

Kohák's way out of the dilemma is to accept the view that humans are, in the relevant sense, only one species among others, but to deny that the value of non-human nature is constituted by human experience. Human valuation of non-human nature is but one instance of the valuation of life by life. Following Jonas, being ceases to be value-neutral when life is present, and not just human life. All life values life and the conditions necessary for it.

Kohák rephrases Husserl's answer to the thorny problem intersubjectivity poses for phenomenology by saying that not human life, but *life as such* constitutes the world as valuable (Kohák, 1986, p.7). Hence starting from biocentrism we can derive moral obligations without passing through anthropocentrism. We can show that "we have a prima facie obligation not to destroy the Earth" (Kohák, 1986, p.8).

Now, I share Kohák's view that value is not entirely subjective; things can be valuable in themselves. In the neo-Deweyan account of value which I would give if pressed, experienced value is the registration of a certain element of what the object is in itself in the order of experience; the emergence, if you will, of a potentiality into actuality. This is a metaphysical discussion we cannot pursue here. At any rate, nothing said here presumes that value is the sole province of *anthropos*.

Rather, my objection is to a *particular* anthropomorphism, to the projection of a certain kind of human value onto natural creatures, namely, *moral* value. The spotted owl has no right to live, "right" being a moral category. If the spotted owl does have a right to live, then its destruction by a non-human organism is as unjust as the work of the human poacher or logger. For if moral value belongs *to* or is *in* the creature itself, then there is an offense to moral value regardless of the agent of destruction. If we balk at accusing the non-human predator of the immorality we would ascribe to the human, we reinforce the moral discontinuity of humans from non-human nature. To assert that humans alone are blameworthy for violating ecological balance is to make humans morally unique; as indeed we are. Now, one can try to pare down the moral value ascribed to non-human nature. One might claim that non-humans have moral value but no moral responsibility. Alternately, one may make the species, rather than the individual owl, the bearer of rights. In either case, whatever the outcome of such strategies, the "moral" nature of the value ascribed to the owl is becoming significantly attenuated. Is it still a *moral* value in more than name?

We must simply ask, *pace* Kohák, are non-human entities *persons*? A satisfactory answer would require not only an adequate philosophical definition of *person*, but a review of current scientific research on a variety of species. No doubt some animals exhibit an inner life, albeit one different from our own. Following Kohák and Jonas, we can accept that life evaluates, in however a rudimentary way, and that the most complex and independent forms of life evaluate in ways analogous to human beings. Put in a not unrelated philosophical idiom, not only *Dasein* is *Sorge*, but *Leben* is *Sorge* as well. Care, valuation, and at more complex levels, affectivity characterize biotic existence. Nothing in the present argument presupposes that the domestic cats currently ruining my home are Cartesian machines with fur. Nevertheless, I venture one condition for personhood that non-humans do not meet: to have *moral status* as a person a creature must be at least potentially a *moral subject* in relation to the community of moral subjects. A moral subject is not merely a subject, an experiencer. It is a being who, in the project of controlling its own behavior, can choose to identify with the community of self-controlling subjects. Such beings are then responsible for failing to act in accordance with this identification. But no one proposes that we hold an unscrupulous dolphin morally accountable for infractions of a moral sense, either one shared by human or dolphin, or one confined to dolphins alone. The dolphin is not a *moral peer*. Hence, I doubt the sense of ascribing personhood to non-human creatures.

But, one may still ask, cold argumentation aside, don't those of us not deaf and blind to nature undeniably feel the moral significance of the torment and destruction of some non-human individuals? When Dian Fossey comes upon the headless and handless corpse of a gorilla, butchered by the poacher's greed, can we not share her moral outrage? Do we not feel that those who lack such feelings are dismal as human beings? Didn't Nietzsche himself, whose philosophy of power discounted all pity, famously throw his arms about the neck of a horse being whipped, albeit at the cusp of his looming insanity?

The answer is yes on all counts. But these feelings are an *anthropomorphic projection.* This projection is not, as Freud says of our projection of a Divine Father, an infantile delusion. It is instead a purposeful and essential part of the human moral project. The ascription of moral values to natural beings is – and here the existentialists were half-right, as always – part of that peculiarly human construction, the attempt to *moralize* existence. Our extension of sympathy to a wounded animal reveals not our continuity with nature, but our *discontinuity*, or more precisely, it reveals the naturalness of our disappointment in amoral nature. In caring for the wounded animal, we do what nature will not. All creatures in nature struggle against nature, that is, against the system of organic and inorganic beings that will in the end reabsorb their existence. But we alone can systematically empathize with that struggle beyond the limits of our reproductive circle. We seek to make existence just according to our sense of justice, because nature's "justice" is intolerable for us. We want the wounded animal's pain to matter because for nature it does not matter at all.

Nature is, after all, *not* biocentric. Life is the exception, utterly dependent on the impersonal, indifferent inorganic resources that, statistically, are the universe. Biocentrism is *morally Ptolemaic*; it places the Earth at the center of the universe. Indeed it is far more narrow than that, since it privileges not the Earth as a whole, but only the Earth's surface, and then only its biotic covering. It widens the Protagorean "Man is the measure of all things," to be sure, but stops at: "*Life* is the measure . . . ." In contrast, death, or less dramatically, inorganic matter is the norm. My objection is less to biocentrism's selectivity than to its belief that it has escaped anthropocentrism. The reason for granting priority to *bios* is that it is the layer of nature hermeneutically closest to *anthropos*. Hence biocentrism is only a widened anthropocentrism. It cannot solve the deeper problems which arise with the recognition that the inorganic world has as great a claim on us as does its verdant wrapping.

But if our moral valuation of non-human nature is a projection, do natural things then have no value in themselves? Are we returned by my argument to a value-neutral conception of nature? We are not. The non-pragmatic values of non-human beings are not moral but *aesthetic*. Being as being is good, but it is aesthetically, nor morally, good. This does not mean that nature's value is humanly constructed, unreliable, relativistic, subjective, or a matter of taste. Whatever compels us by its sheer experiential quality, thereby becoming an end in itself, exhibits aesthetic value; this applies not only to the artifacts of SoHo, but to the human form, the pounding surf, and the divine *mysterium tremendum et fascinans*. Aesthetic value compels as a final cause through inspiring enjoyment, awe, fear, love, or contemplation. Moral value, on the other hand, is exhibited only by moral subjects, their acts, decisions and dispositions, and the prescriptions that govern them. Moral value compels, not as a sensuous quality, but through the mediation of social identification, thereby inspiring a practical command. Aesthetic value addresses us as sheer experiences; moral value addresses us as socially identified selves who must choose how to act.

To say that nature's intrinsic value is aesthetic is to say that natural environments and beings exhibit experiential qualities that are compelling as such. Aesthetic values, like pragmatic values, can be the *source* of moral injunctions within human society. An aesthetic valuation of the living individual or of the environmental diversity to which it contributes can ground a moral demand made on humans by other humans, but this does not make the individual moral or morally valuable. That it is morally wrong to waste water in drought does not mean that water has moral value. If it is morally wrong to destroy the last habitat of the spotted owl, it is so because the owl or the biological diversity to which it contributes has aesthetic value. Differences among aesthetic values aside: if it is morally wrong to kill the spotted owl then it is wrong *in the same sense* as it is wrong to deface the *Pietà*.

To make the point further we may consider an example from Jonas. He suggests a marvelous illustration of a being whose sheer existence makes a moral demand on us, a case of the "coincidence of 'is' and 'ought'," namely, a newborn human (Jonas, 1984, p.131). The baby's very being, he claims, makes a moral demand that we cannot doubt; simply "Look and you know," he writes. I grant that most people do find newborns compelling, and that those who do not are affective dullards. But the reason is not that the newborn is moral or makes a moral demand. Our fascination with the newborn is not moral, it is aesthetic, that is, the look and feel of the baby is fascinating, its skin and movements compel us, capture us. We then make moral demands on ourselves with reference to this beautiful being which must be preserved. If the baby were capable of demands, of verbalizing its cry, the result would be something like: Put me back where I was! – a demand we would not honor. But in fact there are no demands in the baby, and no 'I' to make demands, only feelings of comfort and discomfort. It is our moral judgment that makes the child's need moral, that treats the child like *the self it is not yet*. In the same way we humans are the judges of what is morally required in relation to nature, not because we are the appointed rulers of nature, but because we are the only moral beings in town.

My claim is simply that *morality is a human language, one that non-human nature does not speak*. Consequently, with the exceptions noted, there is no moral sense in or of nature, and nature makes no moral demands on us. Nature does not make demands: it says nothing, but only fashions consequences. The moral lessons are ours and ours alone. This means, not that human morality is transcendent in the sense of being other-worldly, but that humans are unique in some respects, and morality is one of those respects. Human morality may be natural, but it is only natural *for humans*.

At this point we are driven to the deepest level of Kohák's argument, to the ultimate home of his morally significant nature. The moral value of non-human nature can be justified only if the universe is, as Thales held, full of gods, that is, if *physis* is infused with or contextualized by *theos*. Kohák writes,

Yet in the stillness of dusk a double order emerge – not only the order of time but the order of eternity. . . It evokes a second recognition – that though we may accept both the life and the death of the chipmunk as at peace in the rhythm of nature, we cannot but rejoice in the one and grieve the other . . . We are beings who are able to see the world in the order of its time . . . Yet we can see it also from the perspective of a perennial order of right and wrong which stands in judgment above the order of it . . . The human is a dweller in time, though not in time alone. He dwells at the intersection of time and eternity (Kohák, 1984, p.82).

Thus, we are not merely time-locked creatures. Kohák admits that in at least this sense human experience and non-human experience are discontinuous, for, "What is crucial is that humans, whether they do so or not, are capable of encountering a moment not simply as a transition between a before and an after but as the miracle of eternity ingressing into time" (Kohák, 1984, p.85). This eternity of moral value, which emerges when nature is experienced *sub specie aeternitatus*, is God.

Now, it may well be true that we humans *need* eternity in order to live in nature. Time is not value-neutral for us, it presents a problem, namely, that goodness and beauty do not last. The fact that time is creator as well as destroyer does not even the score: emergence and alescence are not weighted equally by human experience. That a beloved child was born and lived does not counterbalance the injustice of its early death. We need there to be some dimension of being in which *what is best does not perish*; which is to say, we need the universe to embody a moral order. I concur with Kohák that only *theos* could provide the objective validation of such an order. But the question remains, Is Kohák's version of *theos*, hence of that moral order, continuous with nature? Even if the brackets of the forest reveal God's presence, do they reveal the presence of God who preserves the good and the beautiful from the ravages of time? Does the God revealed in the forest care about the chipmunk's demise, or about my own, in anything like the sense that humans care about each other? I believe not; if the God revealed in the forest does care in this sense, the forest does not show it. To believe otherwise is, Kohák admits, *credo.* But this faith takes us beyond a phenomenology of the forest. In short, Kahák's version of morality and *theos* seems inconsistent with the experience of nature. His nature is Christian. Nature, I think, is not.

Kohák is not unfamiliar with this criticism. In a reply to John Compton, he agreed that the religious sensibility immediately derivable from this experience of nature is that of "natural religion," or "paganism," not that of "historical religions" like Christianity. While historical religion presupposes natural religion, he admits that the step from "the sacred in nature to the sacred in history, is clearly a leap in faith . . ." (Kohák, 1986, p.59). Kohák elsewhere describes the pagan view of nature as a "sacred presence . . . stern and awesome . . . inscrutable, bountiful, self-renewing and infinitely more powerful" than humanity (Kohák, 1994). I am suggesting that, if Kohák's moral sense of *physis* is dependent on his particular *theos*, then his admittedly super-natural Christianity has led him to a moral sense not given by the experience of nature. I must add here that I do not oppose Kohák's Christianity out of an environmentalist critique, as in Compton's judgment that Christian "hierarchical thinking . . . has been responsible, in part, for the death of nature" (Compton, 1986, p.54). I am arguing instead that nature and natural religion are morally inadequate for us, and that Christianity and the modern humanist morality which is in a sense Christianity's progeny, while morally more promising, are contrary to nature.

If the experience of nature is, as Kohák believes, the experience of a relation to a Thou, it is at any rate not a Christian Thou, nor a modern humanist Thou, but a Pagan Thou. Christian love was not meant for nature, Assisi to the contrary notwithstanding. Like Judaism and Islam, it was born not in the forest of New Hampshire – a formidable enough environment from October to April – but in the deserts of Southwest Asia, where nature is sublimely indifferent, where *theos* must either miraculously save us from nature or admit us to an eternal oasis after nature has done with us. The desert religions of the West seem particularly incapable of supporting an environmental ethics. If there is to be a religious defense of the value of nature, it must, I suspect, come from some other source, from Western paganism, Spinozan pantheism, primitive animism, or perhaps Chinese Taoism.

We may admire the traditional hunter–gatherer who prayed to the spirit of the animal as he killed it. This prayer was not solely a gambit to avoid punishment for killing or to ensure good hunting in the future, but a recognition of the place of animal and human in a revered ecology. Some hunter–gatherers, however, also sent old widows out to die and tortured captured enemies. Both acts make sense in a *Weltanschauung* for which suffering is not the ultimate evil. The refusal to abstract the human individual from nature is related to the refusal to abstract the body of the captured enemy from the hated community from which he comes. The individual life is not ultimately important, the individual's pain still less. We may admire a morality which refuses to regard individual longevity, the minimization of pain, and sheer sentience as ultimate values. Both its implications are not easy to retain within any semblance of either Judeo-Christian morality or modern humanism. To be "more natural," in so far as the description makes any sense, means to cease to feel so acutely the value of the living-and-experiencing person, and to cease to prize so highly the individual's life and happiness. If nature is a school, that is what it teaches.

What then of our responsibility to non-human nature? Having argued that there is no morality and no moral value in non-human nature, my discussion leaves two possible bases for responsibility to nature, one secular and one religious. They are *enlightened anthropocentrism* and *natural religion.*

Enlightened anthropocentrism does not revere nature or natural beings in themselves, but cherishes them only *for us.* An enlightened anthropocentrism spares the spotted owl because we do not *want* to live in a world exhausted by our artifice. It is a judgement that human life is better with wild nature than without it, either because of the instrumental or the aesthetic value of wild nature. This perception can then ground a moral *ought* addressed by human beings to each other. Here the ought is hypothetical, that is, it is dependent on what we humans desire for ourselves.

By natural religion I mean reverence for an aesthetically valued nature that transcends us while at the same time being our home. Aesthetic value regarded as objective, that is, as the recognition of a compelling aesthetic quality independent of our perception, can be the basis for *reverence*, the essential religious feeling. In this case the aesthetic value of non-human things is experienced as compulsive, as compelling desire. But note that the moral implications of such reverence are *not* consonant with contemporary morals, including environmental ethics. They do not spell an end to violent use of non-human individuals, since no morality of non-harm or respect for individuals as individuals can be gleaned from nature. A natural religious ethics recognizes the compatibility of destruction and reverence, of slaughter and preservation, of, so to speak, *love in violence*, which makes the living individual a representative whose *life* is less important than its *meaning*, or, what may be the same thing, whose *physical* life is less important than its *metaphorical* life. The ceremonial killing of prey, enemies, and criminals testifies to a moral lesson that is outside the mainstream of contemporary Western morality. How far would we be willing to incorporate such a morality into our social relationships? Not far, I hope.

I conclude with a natural-religious speculation. My difference with Kohák ultimately hangs on how deeply discontinuity or continuity dominate our fundamental intuitions about humanity's place in nature. Kohák's vision is one of *deep continuity*. For him the continuity between my self and nature *holds* my personhood, it obtains at a level at which the complexity of personhood is significant. He is in good company: Jonas and his retrieval of teleological nature; Merleau-Ponty, whose *la chair* is the simultaneously subjective and objective "flesh" of the world; Whitehead's actual occasions or atoms with subjectivity; and Peirce, who neatly summarized his view as "Agapasticism," or cosmic love – for all of them, *kosmos* embraces the human difference. My opposing view must admit to being haunted by a deeper sense of discontinuity, grounded in the belief that morality, tied as it is to the complex construction of a self in social interaction, is a rare and dependent phenomenon. I acknowledge with Kohák that, at the most basic metaphysical level, discontinuity is overcome, but I cannot see that personhood, hence morality, is preserved at this level. Put in aesthetic terms: the sense of divinity appears in two experiences, beauty and sublimity, which respectively generate the reverence of love and the reverence of awe, or to adopt the Nietzschean mythology, a religion of Apollinian harmony or one of Dionysian deindividuation. For Kohák, love of nature's beauty enfolds the human discontinuity and preserves it in its branches; for me nature's beauty is the transient face of its oceanic sublimity, whose waves elevate and then break the anthropic anomaly. Hence at bottom I see, rather than integration, *de-differentiation.*

Perhaps the difference in sensibility is that between Kohák's New Hampshire woodland and my native Rhode Island seashore, between the youthful biosphere of the forest and the hoary indifference of the sea, which is older and deeper than life. Each can be home, but homes of very different kinds. The forest is after all a peculiarly biocentric environment, not only in its volume and variety of life, but in its physical structure, which is itself alive, something that cannot be said of sea, lake, mountain, wetland, or desert, and only in a much lesser sense of grassland. The organic architecture of forests and coral reefs make them the pinnacle of biotic colonization of the inorganic environment. The point is that as a representative of nature, even of nature on Earth, the forest can be misleading.

At any rate, the denial of the enfolding continuity of the forest need not lead back to an existentialist sense of discontinuity. Human meaning is not, as existentialism portrayed it, heroic self-assertion in the midst of silence, divorced from an alien, value-neutral nature. Rather it has its place in the Dionysian sea that tosses it up, only to absorb it back. Sublime nature yields neither a hot existentialism, nor a warm integration, but a cool dissolution. Its contemplation reveals not a present god of love, nor the absent god of whom Dostoyevsky wrote that if there is no God then everything is permitted. Rather, natural sublimity reveals a god *for whom everything is permitted.*

Kohák may well raise a soteriological question at this point. In Michael Tolkin's literally apocalyptic film *The Rapture*, a young woman and her daughter, members of a chiliastic Christian cult, make their Exodus to the Californian desert to await the Second Coming. Near starvation, the woman accedes to the daughter's request and murders her so the girl can reach Heaven more quickly. After the Four Horsemen ride and the world comes to an end, the woman is asked at the gateway to the next life the ultimate question, the answer to which will determine her eternal abode: *Do you love God for giving you life*? Now, since God's gift is existence itself, an undeserved gift that is infinitely better than its alternative, the sensible answer to the question is a grateful yes. Nonetheless, the woman answers in the negative, for she cannot love the God that permitted her to kill her child.

Kohák may ask me the same question: Can you love the sublime God of nature, a God for whom everything is permitted? Is there redemption in such a God, who gives everything, but takes it all back? To this I can only answer: When I know, Erazim, I will be sure to tell you.

Department of Philosophy, Boston University

BIBLIOGRAPHY

Compton, John. 'On the Sense of There Being a Moral Sense of Nature'. *Personalist Forum*. vol. 2 no. 1 1986. pp. 38–55.

Jonas, Hans. *The Imperative of Responsibility: In Search of an Ethics for the Technological Age*. Chicago: University of Chicago, 1984.

Jonas, Hans. *The Phenomenon of Life: Toward a Philosophical Biology*. Chicago: University of Chicago, 1966.

Hobbes, Thomas. *Leviathan*. edited by C.B. Macpherson. Baltimore: Penguin Books, 1968.

Kohák, Erazim. *The Embers and the Stars: A Philosophical Inquiry into the Moral Sense of Nature*. Chicago: University of Chicago, 1986.

Kohák, Erazim. 'Reply' to John Compton. *Personalist Forum*. vol. 2 no. 1 1986. pp. 56–60.

Kohák, Erazim. 'Varieties of Ecological Experience'. American Philosophical Association Eastern Division Meetings. December 1994.

Merleau-Ponty, Maurice. *The Visible and the Invisible.* trans. Alphonso Lingis. Evanston: Northwestern University, 1968.

Nietzsche, Freidrich. *Beyond Good and Evil.* trans. Walter Kaufman. New York: Vintage Books, 1966.

Peirce, Charles Sanders. 'Evolutionary Love'. In Justus Buchler, ed. *Philosophical Writings of Peirce.* New York: Dover, 1955. pp. 361–74.

The Rapture. Written and directed by Michael Tolkin. 1991.

Russell, Bertrand. 'A Free Man's Worship'. In *Mysticism and Logic.* London: Allen and Unwin, 1963.

Whitehead, Alfred North. *Process and Reality: An Essay in Cosmology.* New York: Free Press, 1929.

ERAZIM KOHÁK

"WHOSE NATURE? WHICH MORALITY? A RESPONSE

Professor Cahoone has paid me the highest compliment one philosopher can pay another: he has taken the trouble to understand my position and to stand within it, trying it on for size. It is a rare and precious gift, for only so can philosophy grow. Philosophy, as I understand it, is not an ennobling pastime or a mystic voice from the depth of Beyng but rather the attempt of the species *h. sapiens s.* to make up for a defect in its instinctual endowment. Other species, including those who deliberate, decide and act, as the *canidae* or the other primates, can rely on instinct to provide them with an overall orientation to the world and their calling within it. Humans need to formulate a conscious conception of how the world and their life within it fit together. Philosophy needs to be tested as such.

What I sought to achieve in writing *The Embers* volume – written, incidentally, in the winter of 1976, though not published until 1984, was in part intensely personal, a quest for restoration after one of life's shipwrecks. Hence the stress on the peace of the forest – as Professor Cahoone points out, my personal peace. I had a philosophic purpose as well. I sought first of all, to make my readers sensitive to nature as something worthy of respect and consideration, not simply a store of raw materials. Secondly, I groped for an idiom in which such awareness could be expressed.

Professor Cahoone is, I think, right that speaking of *moral sense* of nature was not a happy choice. Were I to do it over again, I should speak of the moral *significance* instead, stressing that non-human nature needs to be noted, not just used, overlooked, taken for granted. The term *moral* is problematic as well, though I can think of no other. What I wished to say is that non-human nature needs to be treated with respect for its own good as well, with a generous good will. By applying the term *moral significance* to nature I wanted to evoke a sense of the ethical in Aldo Leopold's sense of ethics as voluntary self-restraint for the sake of the whole or the other.

I resorted to the terminology of personalism, close to me both from Bowne and Husserl (notably *Ideen II.iii*), in order to do this. A *person* to me suggests a being worthy of respect, to whom my relation is subject to moral, not merely prudential constraints. In this sense, I found it useful to think of the whole of what is not as an aggregate of inert entities in mechanical motion, nor as an organic complex of living beings but on the model of a *community of persons*,

R.S. Cohen and A.I. Tauber (eds.), Philosophies of Nature: The Human Dimension, 35–37.

thinking of less complex life forms and inanimate objects as persons in a private mode (as Husserl's "sleeping monads") rather than thinking of persons as inanimate objects with some additional properties.

Were I writing today, I might well look for other metaphors. Still, I am convinced that we need to find ways of becoming aware of ourselves as one of many species, distinctive but not privileged. The continuity of humans with other animals and all of nature is to me a very basic experience – we are one species among many, not masters of all we survey or the source of all meaning and value. That I think crucial. We belong.

Today, though, I would stress the way the human species is distinctive – as every species in its way. The distinctiveness of humans appears to me to lie in our freedom – or, if you prefer, in our defective instinctual apparatus. We are the beings whose impact on the earth is not effectively constrained by natural mechanisms. We can compensate for dropping fish stocks by more efficient factory vessels – until there will be no more fish. Both literally and figuratively, humans are the beings who need to impose their own fishing limits if they are not to destroy the earth. Having thrown off (in part) natural restraints, we have assumed the responsibility for imposing self-restraints. Therein I think is the distinctiveness of the human species: we have a moral obligation of responsibility to all other beings, to cherish their kinship and respect their otherness.

To all other beings: at one point, I wrote an essay analyzing our moral obligation to discarded beer cans, the humblest of persons, "Creation's Orphans" [*Personalist Forum*, 1.1 (1985):22–42]. With Hans Jonas, I am convinced that all purposive activity generates value. Life is a value *for* itself – whatever lives, wishes to go on living – and so a value *in* itself, *intrinsically* valuable. Wholly intuitively, I would extend this to all that is. *Humans, not as discontinuous with nature, but as in their freedom distinctive within it, have a moral obligation to cherish and respect all being*. That does not mean they cannot use it, for instance as food, but it does mean they may never use it heedlessly, never destroy it wantomly, never waste it. Waste appears to me as a profound moral wrong.

That is how I would respond to the first part of Professor Cahoone's paper, devoted to a critical reading of my text. To the second half, in which Professor Cahoone proposes his own alternative, basing ecological ethics on aesthetic rather than moral grounds, I could reply only once I had given it the same careful, sympathetic and critical reading he gave my paper. I have not had time to do that, though I want to. Aesthetic categories are both consoling and dangerous: beauty is so easily used to excuse cruelty! Of that, though, more later, perhaps in a year.

Let me close with just one other response to Professor Cahoone. In his aesthetic mode, he suggests our differences may have their roots in the difference of childhood spent on the sun-struck salty beaches of Rhode Island and in the woods of New Hampshire. However, I did not spend my childhood in the woods, but in German occupied Czechoslovakia. My mother was imprisoned in Theresien-kleine Festung, my father in Konzentrationlager Mauthausen, by superhumans invoking Nietzsche – and hailed, though I did not know it at the time, by the *Parteigenosse* Martin Heidegger with the German greeting. One of my early memories is of riding a bicycle after dark, across the footbridge by the sawmill and up the path, taking food to a Jewish family crammed into a crumbling farmhouse, awaiting transport to Auschwitz. It is a memory of the vast inhumanity of humans to humans and to all being. That is an experience which left me with a strong sense of the crucial importance of moral categories. Mine is ultimately not a biocentric or even an ecocentric ethic as much as an *ontocentric* or perhaps an *agathocentric* one – and, yes, in a theocentric cosmos, though that is a topic for another time.

For now, let me just thank Larry Cahoone for a beautiful (his word) reading of my work. When I finally find the time to edit the essays in which I have grappled with these issues since *The Embers and the Stars* appeared, that volume will open with gratitude for this dialogue with him.

Philosophical Faculty, Charles University, Prague/CZ (Boston University, Boston, MA, USA, emeritus)

TIAN YU CAO

MONISM, BUT NOT THROUGH REDUCTIONISM

(I)

It is an undeniable fact that our contemporary conception of nature is, to a large extent, shaped by science. For many educated people, for example, the basic idea of relativity theories and quantum theory, big bang cosmology and cosmic evolution, the origin of life, evolutionary theory, molecular biology and genetics, physiology and neuroscience, computer science and artificial intelligence, and many other widely popularized scientific ideas, have played an important role in their conception of the world. According to this science-based conception, all phenomena, all objects that we have encountered in the world, must have their material embodiment. Nothing is primitively immaterial. Even consciousness, the most immaterial entity, can also be shown to be lodged firmly within the brain, a material organ. When the brain is injured in certain ways, consciousness is impaired. This simple fact clearly indicates that consciousness has no existence independent of brain. In general, according to this conception, whatever phenomena we have encountered in the world, or imaginable in our mind, are, in the last analysis, only different forms and manifestations of the entities, originally present at the very beginning of the universe. That is, they are different manifestations of elementary particles, whose behavior is governed by physical laws described by fundamental physics. Thus, this conception endorses, not only the unity of the world or monism, but also, as some scientists and philosophers would argue, the unity of science.

Of course, nobody would deny that in the course of cosmic evolution, new entities and new phenomena have frequently emerged. But it is arguable that they should be taken only as epistemic emergence rather than ontological emergence. First, the emergent entities and phenomena are always causally connected with the entities from which they have emerged. Second, and this is a more important argument, the laws governing the behavior of the emergent entities are said to be explainable by the more fundamental laws that govern the behavior of the underlying entities. For example, Francis Crick argues that our entire mental life is caused by the behavior of neurons, and all they do is increase or decrease their rates of firing.[1] Steven Weinberg would argue further that the change of firing rates is dictated, and thus explained by physical laws.[2] In this way of reasoning, the monists and reductionists take the occurrence of new entities only as a result of historical accidents, and the evolution of the

R.S. Cohen and A.I. Tauber (eds.), Philosophies of Nature: The Human Dimension, 39–51.

universe only as a pile of frozen accidents. There are different properties and different aspects of entities. But, the reductionist argues, there is no unbridgeable gap which cannot be filled by causal connections.

But this is only one part of the story. The other part of the story is more complicated. No matter how powerful and convincing the scientists arguments for monism and reductionism seem to be, they are not the only decisive factor in the shaping of our conception of nature. For some philosophers, at least, the tradition of metaphysics has also played an important role. Ever since the time of Descartes, if not earlier than that, our conception of nature, concerning the ontological status of minds, has saddled us with dualism of mind and body, or the mental and the material, or the intentional and non-intentional. As is well known, the rise of dualism in the modern period had its roots in the rise of empirical thought. The underlying reason for the cosmic division is scientific accessibility: the physical reality or the bodies are scientifically accessible, while the mental reality of the soul lies outside the scope of empirical investigations.

For this reason, the metaphysical tradition of dualism is not totally metaphysical in origin, but has its roots in the historical specificity of scientific develoment. That is, the incapability of science in exploring mental phenomena. Two implications of this statement immediately come to my mind. First, with the rapid development of neuroscience, there is a real hope to undercut dualism from its foundation. Second, the current incapability of neuroscience may contribute to the fact that some scientifically well-informed philosophers, such as Hilary Putnam[3] and Abner Shimony,[4] and some scientists, such as John Eccles[5] and Roger Penrose,[6] still cannot resist the temptation of dualism.

Contemporary dualism has various shades and tones. There is the crude version of Cartesian type, such as that cherished by Eccles, who believes that there is a separate mental world, the souls, that are attached by God to the unborn fetus at the age of about three weeks.[7] There are sophisticated versions advocated by Shimony and Putnam. Shimony has great confidence in empirical science. He fights vehemently against Putnam and argues for the possibility of naturalizing reason. He believes that the psycho-physical correlations need deep exploration, and that reason can be naturalized. Naturalized, that is, in terms of irreducible mentalistic component that is pervasive in nature at the most fundamental level. More specifically, Shimony believes that the non-physical aspect of nature has its origin in the protomental quantum potential, which is coordinated and intensified by quantum entanglement. And all these are capable of controlled investigations.[8]

Similar ideas have been articulated in a more detailed way by Roger Penrose, who argues forcefully, in his book *Shadows of the Mind*, that no deterministic, rule-based system, that is, neither classical physics, computer science, nor neuroscience, can account for the mind's creative powers and ability to ascertain truth. For Penrose, neuron is a mere shadow of a deeper level of cytoskeletal action, and consciousness is a manifestation of the quantum entangled internal cytoskeletal state and its involvement in the interplay between quantum and classical levels of activity. Although the qualities of consciousness, experience, comprehension, intent and awareness are not taken to be independent spirit or soul, but as aspects of brains, neurons and particles, his aim at reestablishing consciousness as primary in the cosmos deserves the label of aspectual dualism, as Walter Freeman calls it.[9] Yet it is pointed out by neuroscientists and quantum theorists that brains are too large, warm and weak for quantum coherence, and an action potential in neuron activity as a thermal shock would disrupt the quantum coherence rather than amplify it. Some more severe critics even take the quantum approach to consciousness as a category mistake.[10]

Putnam's dualism is of particular interest. For him, on the one hand, the mental is irreducible and enjoys an explanatory priority. For example, perception is taken to be a transition from potential to actual awareness, which is different from the transition of any material from one state to another state. The semantical properties of thought involving meaning and truth, Putnam argues, cannot be reduced or even accounted for in terms of cause-effect, but are intrinsically intentional and shaped in large part by social usage of a speech community. Thus, mind or reason cannot be naturalized in the sense of being exhausted by natural sciences.[11] There is no doubt that his position is greatly strengthened by the current incapability of natural sciences in the understanding of the mind. Putnam also holds that to say that the mind as a system of object-involving capacities is supervenient on the material betrays nothing but the collapse of the materialist monist view, because then the monism itself would have lost all its meaning. Thus, Putnam claims that we simply cannot escape from the Brentano thesis that the mind cannot be reduced to matter and the mind will not go away. (ibid.)

But on the other hand, Putnam also advocates the natural unity of the mind with its constitutive matter, endorses a kind of neo-Aristotelian realism or natural realism, arguing against various forms of representationism, which separate the mind and the world. He says repeatedly in recent years, what we perceive are not our own experiences but things. If I am seeing a chair, then I am not perceiving a mental state of seeing charily, but seeing the chair. According to his neo-Aristotelian position, mind or reason and body are one because form and matter are one. They are not separable. And reason and form are fitted into each other. Thus in a deep sense the problem of monism and dualism for Putnam becomes a pseudo problem. The major concern of Putnam is a resolute rejection of reductionism, because it misses the essential features of the intentional mind.

Putnam's natural realism gives him certain theoretical resources to avoid the most difficult question for various versions of dualism. That is, the question of how it could be possible for the mental and the physical to be in mutual interactions. Yet apart from Aristotelian assertions, Putnam has not provided any mechanism for their interaction.[12]

Scientifically, however, the prospect for monism seems to be more promising. First, the cosmic evolution gives it the most powerful argument for rejecting dualism, except for a kind of pan-mentalism.[13] But pan-mentalism is not very popular because it still has to address the difference between the mentality of the electrons or quantum potentials and the human mentality. And this task is by no means easier than the task of giving an account of the emergence of the mind from matter. Second, the rapid development of neuroscience has revealed many aspects of the secret of the mind. Thus, in terms of methodology, monism seems to be on the side of a progressive research program. In addition, in the present cultural climate of taking reductionism as the embodiment of the scientific spirit, monism could claim a victory if a reductionist solution to the problem of consciousness can be found, either eliminating the mind in terms of artificial intelligence,[14] or giving a causal emergentist explanation of the mind in terms of the binding of neuron firings,[15] or in terms of reentrant excitations in neuron networks.[16]

In any case, one thing is clear: without a naturalistic solution to the problem of consciousness, the debate between dualism and monism cannot be set to rest. I am not stupid enough to venture a solution. But an effort to clarify certain aspects of the debate may be of some interest to some people.

(II)

In order to maintain a materialist monist position, it is necessary to explain the mental in terms of the physical, if the mental cannot be eliminated by being identified with the physical so that its status is reduced to that of folk psychology. Otherwise dualism will always be an alternative option, in addition to the idealist option. With regard to explanation, a consensus among many scientists as well as philosophers is that an explanation is more than establishing invariable correlations, but requires presenting a causal mechanism. By causal mechanism, philosophers usually refer to causal structures, that are produced by causal interactions and propagated through causal processes governed by causal laws. The term cause in turn can be understood in the sense of antecedent, or constitutive.[17] In the latter case, such as explaining the hardness of a diamond in terms of its constituting carbon atoms fitting together neatly, the causal explanation is usually taken to be the paradigmatic case of reductive explanation.[18]

Yet in recent years, a prominent reductionist Steven Weinberg tries very hard to give reductionism some more sophisticated interpretation. For Weinberg the monist, the ultimate aim of science is to reduce the world of physical phenomena to a finite set of equations or principles. For example, Weinberg claims, the bewildering complexities of chemistry and physics should be and have been reduced to two lines of algebraic symbols of quantum mechanics; and the properties of elementary particles can also be understood as mathematical consequences of a fairly simple quantum theory known as the standard model, whose foundation was laid down by quantum field theory. The equations of the standard model deal with fields of force only, particles are just bundles of field energy. According to Weinberg, this simplistic and universalistic style of understanding, known as reductionism, is one of the greatest themes in human intellectual history. Weinberg warns us, however, we should not identify reductionism only with the constitutive causal explanation, which he labeled as petty reductionism. Powerful as it seems to be in many cases,

Weinberg acknowledges three weaknesses of petty reductionism. First, the reduction of Newtonian physics to Einsteinian physics simply cannot be understood in terms of constituents. Second, it does not make any sense to talk about the hardness or temperature or intelligence of individual elementary particles. Third, as was pointed out several decades ago by Heisenberg and many other S-matrix theorists, in the domain of elementary particles, it is impossible to give a precise meaning to statements about particles being composed of other particles. An illuminating example is this. In a sense a proton can be taken to be composed of quarks; yet these quarks in turn can be understood as being made of protons and other particles. Thus the very idea of composition in the domain of elementary particle becomes meaningless.[19]

In contrast with petty reductionism, Weinberg argues, grand reductionism or theory reductionism claims that all of nature is the way it is because of simple universal laws, to which all other scientific laws may in some sense reduced. Weinberg argues that only grand reductionism continues to be worth arguing about. A case can be made, however, against Weinberg's grand reductionism. Usually, physicists take classical physics to be reducible to quantum physics. Otherwise, grand reductionism or theory reductionism would make no sense, and any dream of a final theory would really become a mere dream. However, a universal principle, the principle of Renormalization group flow, which underlies the standard model, was first discovered in the study of critical phenomena in classical statistical mechanics. The principle is applicable both to classical and to quantum system whenever the system possesses no characteristic length and has a statistical continuous limit.[20] In this case, what should we do in the spirit of grand reductionism? Should we reduce statistical mechanics to quantum field theory, or the other way around? Neither direction of reduction is possible. This example has shown that even within the domain of theoretical physics, grand reductionism is often inapplicable and empty.

Weinberg also asks us to notice the distinction between reductionism as a program for scientific research and as a view of nature. While reductionism may or may not be a good guide for a research program, as a world view, however, it certainly provides the necessary insight that there are no autonomous laws that are logically independent of the principles of physics. In Weinberg's opinion, there are arrows flowing into each principle from all the other principles by which it is explained. All arrows are connected and all seem to branch outward from an ultimate law of nature. A concept is deeper than another if it is governed by principles that are closer to the ultimate law. The reductionist program of physics is to search for the common source of all explanation, finding the laws of nature that lie at the starting point of all chains of explanation. Here the conviction of the unity of nature becomes an argument for the unity of science, an old dream of physicalism.

Weinberg acknowledges that there is emergence of life and mind. Yet he insists that the rules they obey are not independent truths, but must follow principles at a deeper level. Apart from historical accidents that by definition cannot be explained, the nervous system evolved to what it is entirely because of physics. Stubborn as Weinberg is as a dedicated reductionist, he realizes that consciousness poses the greatest challenge to reductionism. This is so simply because up until now, nobody knows how to explain consciousness by physical principles.

However, many attempts in the spirit of causal reductionism have been made. Some functionalists follow Putnam in suggesting that the mind is identical with a brain state, in terms of its program rather than its physical terms, which can be realized in different physical compositions, thus transcending the framework of materialism while remaining within it. Functionalism can be rejected for two reasons. First, programs are purely syntactical without content, meaning and interpretation, while the mind has semantics.[21] Another argument against Functionalist is presented by Putnam himself: that is that the mental state is not only compositionally plastic, but also programmatically plastic, meaning a mental state can be realized by different programs.[22]

Another popular approach takes the neuronal processes as the causal basis of our conscious life. Some neurobiologists are searching for elementary components in brain, the reflex, neuron, chemical transmitter molecules, the action potential that release the transmitter, membrane channel that the transmitter opens and the genes that direct the construction of the gates and the transmitter, and the mapping of neuronal connections thus established. Others claim the discovery of feature detector neurons and declare that the secret of consciousness lies in the way in which representations of features combine so as to make representations of objects. Still others try to locate the mental ability in quantum coherent states and take these states to be the physical basis of the mind.[23]

Reductionism as a strategy of explanation and as a world view is resolutely rejected by Putnam. He argues forcefully in general terms that even if the behavior of a system is deducible from the descriptions of its constituents, this deduction is by no means an explanation, because the relevant features in this way of deduction would always be buried in a mass of irrelevant information and cannot be brought out.[24] In a head-on confrontation with Weinberg, Putnam argues that the laws of higher level cannot be deduced from the laws of lower level, because the former needs initial and boundary conditions which, as Weinberg acknowledged, are accidental from the point of view of lower level, but are essential to the description of higher level. For this reason, the laws of higher level are in fact autonomous vis-a-vis laws of lower level. All upper level phenomena and laws have a basis in fundamental physics. Yet they cannot be deduced from the laws of physics. They must be compatible with the laws of physics. But too many things are compatible with physics, thus physics has receded to the background and becomes irrelevant in explaining particular phenomena or laws. Putnam acknowledges that certain features of the lower level may be relevant for the explanation of the higher level phenomena. In this case, however, Putnam argues, these features are specified by the higher level discipline rather than the lower level one.[25]

Physics is important because it is claimed to have discovered the key to objectivity. Yet, Putnam argues, in fact it is a product of some of the same interests that underlie our construction of our lifeworld, and thus has to be constrained and corrected by other human practice, such as studies in history, philosophy and sociology. Physics is chosen for mathematically describing the time development of parameters in a closed system. It is no more than a description of reality in itself than is our everyday talk of colors or desires or values. Economics also reveals a facet of reality that is not revealed by physics. For Putnam, all changes involves physical change, but that does not entail that every fact can be explained by physics. Of course, a mere supervenience claim is undeniable. Yet this very claim betrays the collapse of the reductionist world view, because all meaning and substance of reductionism is lost in this claim.

Based on these anti-reductive consideration, Putnam has developed his own account of the mind, or as he likes to call, cognitive psychology or intentionality. The central theme of Putnam's account is a resolute rejection of representationism. Putnam argues repeatedly that the intentional is not reducible to computer and brain sciences, including Crick's firing as representation, Edelman's neural assemble as the architecture that allows to form inner representation, and Chomsky's inner template as hardwired in the brain by evolution for natural language, semantics or any other conceptual aspects. This kind of reductive pursuit may be of some help, but can never provide any real understanding of the intentional. The fatal defect of representationism lies in the fact that it deals with the link between representations and the external world only causally, but not cognitively, thus has missed an essential feature of the intentional. According to Putnam, the very idea of causation is quite ambiguous. Causation is a relation between events. But the identity of events is always multiple, depending on context and interest. Thus no causal connection between events can take shape without conceptual choices about the division between background conditions and the relevant factors, which are interest dependent and context dependent. That is, the causal linkage between representations and the world is not autonomous. Rather, it is itself dependent on the intentional. In arguing against Jerry Fodor's causal theory of reference, [26] Putnam rightly points out that no reference can be fixed without reflective and normative judgments, which are context and interest dependent and are social in nature.[27]

More specifically, Putnam argues that if the mind can only deal with representations caused by external things, then the mind can never reach out and have access to reality, because the causal connection between representation and the world is of no help in fixing the reference without intentional involvement. There is no need to repeat his famous model theoretical argument about the ambiguity between cat on mat and cherry on tree. But his own suggestion is quite interesting. Opposing any causal explanation of intentionality, Putnam proposes a position he calls natural realism. According to this position, the mind has direct access to reality, because, as a structured system of object-involving ability, the mind involves outside things intentionally and normatively and is extendable. There is no one to one correspondence between brain event and mental event because the mental event is always global. Putnam says, a perception is the exercise of a mental capacity, which is not indentical with the brain events which are necessary for us to have those capacity. More important is the general intelligence which reflects the wholeness of intentionality. Assuming science is reductive in nature and can only deal with causall connections, Putnam endorses the Brentano thesis that intentionality cannot be reduced and will not go away, declares that the mind or reason or intentionality cannot be naturalized, and calls himself a dualist.

But Putnam's dualism is of a very peculiar type. Not only does he accept that the mind is part of nature, supervenient on the physical, he also combines the Brentano thesis with his natural realism, fending off any subjective and relativist interpretation of our knowledge of the external world. Although nature does not have its own language waiting for us to discover and use, and the many ways of talking about nature are created by us, yet, Putnam insists, nature is objective, the truth of the content of our talking is determined by nature.

Still, the thesis that reason cannot be naturalized is quite disturbing in the contemporary minds overwhelmed by the ever-glowing power of science. It seems to me that in a crucial sense, the outcome of the contemporary debate about a naturalizability of epistemology, which is a renewed version of the old monism-dualism debate, depends on the possibility of having a naturalistic explanation of consciousness. In this regard, I find the work of Walter Freeman,[28] a neuroscientist at UC Berkeley, is quite stimulating.

(III)

Reductionism in the study of mind, whether it takes the representationist form of the computer model, the quantum model, or the form of the binding problem, is rejected by both Putnam and Freeman, because it misses the essential feature of the mind: (i) being active and constructive; (ii) being potential, as open possibility toward future; (iii) and being semantical, intentional and social, searching for meaning, truths and reassurance from other minds. Putnam rightly realizes that the structure of the world is constructed by our intentionality, which in turn is determined by the specific practical relationship between us and the environment. Yet his taking position with Aristotle and Wittgenstein in assuming that the mind has direct access to the forms of the external objects, and is able to construct forms which are similar to the forms of objects,[29] is too simplistic at least. It is also inappropriate because, consistent with his rejection of the naturalization of intentionality, it has closed the door for further scientific exploration of intentionality.

Different from Putnam in this regard, Freeman tries to develop a new dynamic model of the objective side of the mental activity, which exhibits many of the essential characteristics of the mind, as I have just mentioned above: (i) active and constructive; (ii) open-ended toward future; (iii) semantical, intentional and social, searching for meaning, truths and reassurance from other minds.[30]

The underlying idea of Freeman's model is dynamical relationship between humans and the environment in which they live. Thus the starting point of his account of the mind is humans' social and practical action in the environment.

First, the metabolic needs of brains for building materials and fuel, which must be searched for and found in the world, dictate that the basic behavior of humans is movement directed into the world, which requires an internal structure of mind for orientation in space and time. This requirement is realized in the natural history by the interactions of neurons with each other to form assemblies, of assemblies to form brains and brains to form societies, with muscles into the shared environment. Second, the transformation of activity of microscopic neurons to populations is described, tentatively, by non-linear dynamics, as the formation of patterns of neurons' chaotic activity. Here, the microscopic neurons contribute to the macroscopic order, and then are enslaved by that order. Third, populations of neurons interact with each other in extended regions of the brain by large bundles and tracts of axons, thus the structures of the patterns are derived from an ever-shifting global consensus. Fourth, and this seems to be the most important part of his story. Based on his experimental discoveries, Freeman tells us that firings go from the pattern of neurons activity into the motor systems that induce search movements, and also go as corollary discharges to all of the sensory cortices to prepare them for the consequences of the intended actions, and to specify the classes of stimuli that are sought. Thus, with the emergent patterns of activity in the brain appears intentional perception.[31]

The constructive nature of Freeman's model derives from his basic finding. The stimulus that has evoked activity from the receptors to the cerebral cortex triggers the state transition in neural population. Then, having set the initial conditions, the sense dependent activity is washed away and vanishes at the central, and is replaced by a new pattern of cortical activity that is created by the brain. That is, the perceptual message sent into the forebrain is the constructions of neural activity, which characteristically lack invariance with respect to the stimuli that triggered them, rather than the traces of stimuli or the residue of a filter or a computational algorithm. Thus the only knowledge we could have of the outside world is what we have made in our own brains. This cortical dynamics is the basis of epistemic solipsism in brain.

At the first stage of perception, the subjective construction is unavoidable and has biological significance. The environment is infinitely rich and complex in information, only a small portion of which forms the basis for action. And that portion is different for every subject. The information is selected for expressing expectancy and attention. The corresponding message is neurologically shaped by the synaptic structure of the neuropil. Thus the response issued from the macroscopic pattern is not a representation of the stimulus, nor is it derivable from it by any filter operations.

More specifically, the macro pattern that enslaves the neurons and determines the global output of the cortex is shaped in small part by micro input, but in large part by the preexisting synaptic network shaped by past experience and by the excitability of the neurons, which is shaped by the interplay of diverse neurohormones and modulators. Even the constant stimulus response relations are mediated by inconsistent patterns of brain activity due to changing perspective and context, which is based on the unity of intentionality.

Thus every perception is influenced by all that has gone in in the past. This contextual dependence of response on the behavioral history and current brain state indicates that the cortical output constitutes the significance, value and meaning of the stimulus. The implication for this constructive nature of perception is that the macroscopic patterns of neuroactivity do not relate to stimulus, but to the meaning of the stimulus. That is, the essence of neural activity is meaning rather than information. In fact meaning is created by neuroactivity that mediates the actualization of intentional structure and shapes meaningful patterns of muscle contractions. And here lies the seed of the unity of intentional structure, which undercuts any representational model of the mind.

Then what is the intentionality structure? According to Freeman, intention is the process of a brain in action having the properties of unity, wholeness and intent of taking in by stretch forth. It is a relation in which a self modifies itself into conformance with aspects of nonself. An intentionality structure is a brain having the capacity to actualize these properties by purposive behavior. Neurologically, intentionality can be described as the on-going dynamic construction of motor patterns with concomitant transmission of discharges to all sensory systems in order to update them in respect to the expected consequences. The unity of the perceptual experience is due to the convergence of perceptual flow. The structure of intentional action extends throughout the forebrain. Each event has its focus in the intentionality structure, and each observed pattern manifests a meaning as the focus of the activity in a region in an evolving landscape provided by the intentionality structure of the subject.

A distinguished feature of intentionality structure in Freeman's theory is that each new state transition in a sensory part initiates the construction of a local pattern that impinges on and modifies the whole intentional structure. Thus each new experience must change all of the existing store in order that a new entry be incorporated and fully deployed in the existing body of experience. This is quite different from the memory of a computer in which new items do not change the old ones. In this theory, the purely mental activity 'thinking' is defined as a process by which some pattern is actualized from the intentional structure into meaning, with which representations are shaped, and deployed into the world. Freeman argues that since the objective aspect of awareness is to be found in the dynamical interplay of motor output and corollary discharge with proprioceptive and exteroceptive feedback, and with repeated update of the limbic mechanisms for orientation of action in time and space, the internal updating and restructuring of its past as the basis for constructing each next step into its future is the essence of the function of each brain.

Supported by experimental analysis and introspection, Freeman further argues that the availability of that structuring for the guidance of action by each individual, uniquely expressed in an evolutionary unfolding, has its subjective aspect in the experience of that individual. All available parts of the forebrain participate, and the entire body of past experience in the form of synaptic modifications and neurohormonal modulations, is brought to bear in varying degree at each moment, waking and sleeping.

According to this constructive theory of the mind, the initial position of every mind must be solipsistic. The reason for this is that the content of mind comes only from its cumulative constructions induced by the sensory milieu, which is unique for each in the common environment, due to differences in perspective and expectancy. Then what is the mechanism for directed modification in both structures of perspective and of expectancy, so that cooperation is possible? How can private experience be shared by others? How can communication and mutual understanding between solipsistic brains be possible? Freeman maintains that there are biological properties by which brains join together in social cooperation. He holds that brains arisen and shaped in evolution are not isolated entities. In fact, channels, codes and agreements that precede the reciprocal mappings of information in dialogues have been created during shared actions. There is chemical mechanism via evolution in support of mammalian reproduction, with which intentional mechanisms is tuned so that sensory cortices can be biased to attractors that classify reproductive signals such as odors. This mechanism can also be used to manipulate other's behavior in order to build channels of communication that are required to overcome the solipsistic gulf in forming a couple. Darwin also notices that the postural and autonomic actions are homologous across mammals in their social interactions. Animals that live in packs and herds have more elaborate systems that are designed to communicate internal states. In addition to homologous signs of emotions and our autonomic responses to them in our feelings and actions, there is language as an effective means for communication. According to Chomsky, sufficiently similar neural structures have been hardwired into human forebrain by evolution in support of communication, which grows in stages during the socialization of children.

Thus, while maintaining that consciousness has its dynamics in individual brains, Freeman shares a position with Putnam that the content and meaning of consciousness can only be derived from social interactions, that any attempt to reify it as a state, process or causal operator within an individual brain has missed the context in which the term derives its operational meaning. That is, as an agent for practical actions into the world, consciousness can only be realized in social cooperation. Without communication no individual brains could understand any meaning. The social character of the intentionality structure sometimes also manifests itself by its rigidity in response to varying stimuli and context, which is often genetically programmed into the survivor according to sociobiological rules for the good of society.

Then what is consciousness? According to Freeman, consciousness is constructed by the intentionality process, and reflects operations by which the entire knowledge store in an intentionality structure is brought instantly into play, As different facets of the same event, the mental event and the neural event are inseparable, and bear no causal relationship, just as the meaning of a book and the printed pages of the same book bear no causal relationship.

The philosophical implications of Freeman's account of intentionality is worth exploring. My immediate feeling is that it gives a natural scientific support to many of Putnam's insights about mind and natural realism, although in some sense it defies his thesis that reason cannot be naturalized. Perhaps Heidegger's concept of temporality can also be analyzed in terms of Freeman's dynamics of intentionality structure.

(IV)

Now let me conclude with two remarks about monism or the unity of nature in face of the bifurcation of matter and mind. First, although the mind emerged from the cosmic evolution of matter cannot be reduced to matter, and the mental, subjective aspect of brain's neuroactivity cannot be reduced to its physical, objective aspect, if the following three conditions can be satisifed, the unity of nature has not been seriously compromised. That is, first, if the objective aspect of the mental activity is scientifically accessible, and can be naturalistically investigated and explained; second, if the content of the mental is individuated by the physical and social environment, and by the practical actions of humans as a physical existence into the physical world; and third, if there are effective interactions between the mental and the physical, then the unity of nature is in no sense in danger. Among contemporary accounts of the mind, Putnam's is quite attractive because it satisfies the last two conditions. But Freeman's account gives a stronger support to monism because it satisfies all three conditions.

Second, the emergence of new, non-primary, yet irreducible ontologies, most important among them is the emergence of the mind, only reveals the openness of nature. That is, it gives more and more surprises and becomes richer and richer in its content. In my opinion, nature's openness should not be interpreted as an indication that nature has become fragmented.

Department of Philosophy, University of Boston

NOTES

[1] Francis Crick, *The Astonishing Hypothesis: The Scientific Search for the Soul* (New York, Schribner's, 1994).
[2] Steven Weinberg, *Reductionism Redux', in New York Review of Books (October 5, 1995), 39–42.*
[3] Hilary Putnam, *Renewing Philosophy* (Harvard University Press, 1992), chapters 1–3; and *Words and Life* (Harvard University Press, 1994), chapters 1–3 and 20–24.
[4] Abner Shimony, 'Reality, causality, and closing the circle', in *Search for A Naturalistic World View* (Cambridge University Press, 1993), vol.1: 21–61.
[5] John C. Eccles, *How the Self Controls its Brain* (Springer, 1994).
[6] Roger Penrose, *Shadow of the Mind* (Oxford University Press, 1994).
[7] See note 5.
[8] See note 4.
[9] Walter Freeman, *Societies of Brains* (Lawrence Erlbaum Associates, 1995).
[10] See note 9; see also John R. Searle, 'The mystery of consciousness', in *New York Review of Books* (November 2, 1995), 60–66; and John G. Taylor's remarks as reported by John Hogan in his 'Can science explain consciousness?' in *Scientific American* (July, 1994), 88–94.
[11] See note 3.
[12] It can be argued that from a purely Aristotelian point of view, interaction might be a non-issue. But I am not totally convinced that this is the reason for Putnam's silence on the mechanism for

mind–body interaction, if we take into account of the fact that Putnam always takes science seriously. I am grateful to Professor Klaus Brinkmann for bringing this point to my attention.

[13] A.N. Whitehead, *Process and Reality* (Macmillan, 1929), *Adventures of Ideas* (Macmillan, 1933), *Modes of Thought* (Macmillan, 1938), Abner Shimony, note 4; Roger Penrose, note 6.

[14] Daniel C. Dennett, *Consciousness Explained* (Little/Brown, 1992).

[15] See note 1.

[16] Gerald Edelman, *The Remembered Present: A Biological Theory of Consciousness* (Basic Books, 1990), *Bright Air, Brilliant Fire: On the Matter of the Mind* (Basic Books, 1992).

[17] Wesley C. Salmon, *Scientific Explanation and the Causal Structure of the World* (Princeton University Press, 1984).

[18] Steven Weinberg, note 2.

[19] Ibid. See also Werner Heisenberg, *Physics and Philosophy (The Revolution in Modern Science)* (Georg Allen and Unwin, 1959); Geoffrey Chew, "S-matrix theory of strong interactions without elementary particles," in *Reviews of Modern Physics, 34*: 394–401.

[20] K.G. Wilson, 'The renormalization group and critical phenomena', *Reviews of Modern Physics, 55*, 583–600.

[21] John R. Searle, 'Minds, brains and programs', *Behavioral and Brain Sciences, 3* (1980): 417–457.

[22] Hilary Putnam, *Representation and Reality* (Bradford, 1988).

[23] For a summary and criticism of these attempts, see note 9.

[24] Hilary Putnam, 'Philosophy and our mental life', in *Mind, Language and Reality* (Cambridge University Press, 1975), 291–303.

[25] See notes 3 and 24.

[26] Jerry Foder, *A Theory of Content* (MIT Press, 1990).

[27] Hilary Putnam, chapter 3 of *Renewing Philosophy* (Harvard University Press, 1992).

[28] See note 9.

[29] Hilary Putnam, chapters 1–3 of *Worlds and Life* (Harvard University Press, 1994).

[30] Walter Freeman, 'Neurodynamics, intentionality and consciousness', (manuscript). I am grateful to Professor Freeman for sending me his most recent work to me when it was still in draft.

[31] Although this intentional perception may still be interpreted as caused, in some sense, by the dual route of firings from the pattern of neurons activity. Yet the appearance of the intentional in this context does provide us a mechanim, with which the intentional emerges from causal processes, thus has features that are absent in causal processes.

MALCOLM DAVID ECKEL

IS THERE A BUDDHIST PHILOSOPHY OF NATURE?*

One of the most common and enduring stereotypes in environmental literature is the idea that Eastern religions promote a sense of harmony between human beings and nature. On the other side of the stereotype stand the religions of the West, promoting the separation of human beings and nature and encouraging acts of domination, exploitation, and control. Roderick Nash gave classic expression to this contrast when he said: "Ancient Eastern cultures are the source of respect for and religious veneration of the natural world" and "In the Far East the man–nature relationship was marked by respect, bordering on love, absent in the West."[1] Y. Murota drew a similar contrast between Japanese attitudes toward nature and the attitudes he felt are operative in the West: "the Japanese view of nature is quite different from that of Westerners . . . For the Japanese nature is an all-pervasive force . . . Nature is at once a blessing and a friend to the Japanese people . . . People in Western cultures, on the other hand, view nature as an object and, often, as an entity set in opposition to mankind."[2]

This contrast between the East and the West owes much of its influence in recent environmental literature to the seminal article by Lynn White, Jr., "The Historical Roots of Our Ecologic Crisis."[3] White depicted the Judeo-Christian tradition as anthropocentric and argued that Judeo-Christian anthropocentrism stripped nature of its sacred status and exposed it to human exploitation and control. While he did not comment at great length about the Eastern traditions, he clearly understood them as the opposite of the traditions of the West.

> The beatniks and hippies, who are the basic revolutionaries of our time, show a sound instinct in their affinity for Zen Buddhism and Hinduism, which conceive of the man–nature relationship as very nearly the mirror image of the Christian view.

White's image of the contrast between East and West was taken up in the same journal seven years later, by the Japanese historian Masao Watanabe.[4] Watanabe associated the Japanese people with "a refined appreciation of the beauty of nature" and said that "the art of living in harmony with nature was considered their wisdom of life." White's image continues to be reflected by some of the best known contemporary writers in the environmental movement. In a recent collection of essays, Gary Snyder, the venerable and respected survivor of Lynn White's "beatniks and hippies," drew a series of graceful

R.S. Cohen and A.I. Tauber (eds.), Philosophies of Nature: The Human Dimension, 53–69.

connections between Henry David Thoreau's concept of the "wild," the Taoist concept of the Tao, and the Buddhist concept of Dharma:

Most of the senses of this second set of definitions [of the wild] come very close to being how the Chinese define the term *Dao*, the *way* of Great Nature: eluding analysis, beyond categories, self-organizing, self-informing, playful, surprising, impermanent, insubstantial, independent, complete, orderly, unmediated, freely manifesting, self-authenticating, self-willed, complex, quite simple. Both empty and real at the same time. In some cases we might call it sacred. It is not far from the Buddhist term *Dharma* with its original sense of forming and firming.[5]

This image of an affirmative Eastern attitude toward nature must have lurked in the minds of the environmental activists and friends of the environment who gathered at Middlebury College in the fall of 1990 to hear the 14th Dalai Lama speak on the topic of "Spirit and Nature." Tibet, like traditional Japan, has been the focus of a certain Western yearning for the East as a place to discover not only a unique sense of wisdom (what one observer called "an intimate and creative relationship with the vast and profound secrets of the human soul") but a wisdom that can insure "the future survival of Earth itself."[6] There was a hush in the Middlebury field house as the Dalai Lama seated himself on the stage and began to speak.[7] It must have been a surprise when he began by saying that he had nothing to offer to those who came expecting to hear about ecology or the environment, and even more surprising when he interpreted the word "nature" as a reference to "the fundamental nature of all reality" and entered into a discourse on the Buddhist concept of Emptiness. To explain the connection between nature and Emptiness, he said: "When talking about the fundamental nature of reality, one could sum up the entire understanding of that nature in a simple verse: 'Form is emptiness and emptiness is form' (*The Heart Sutra*). This simple line sums up the Buddhist understanding of the fundamental nature of reality."[8] And he went on to explain how Tibetan philosophers use logical analysis to develop their view of Emptiness and to pursue what he said was the "expressed aim of Buddhism," namely, the purification and development of the mind.

The Dalai Lama's words were surprising not because he seemed unfriendly toward the "natural" world in the prevailing sense of the word (that is, toward ecosystems of plants, animals, the atmosphere, the ocean, rivers, mountains and so on), but because he so gently and easily shifted attention away from the natural world toward the development of human nature and the purification of the mind. The sense of surprise only became more acute when he began to develop the concept of Emptiness and indicated that it involved a denial of the reality of what he took to be "nature" itself. To say that "Form is Emptiness and Emptiness is Form," in the language of Mahāyāna philosophy, is to say that all things are "empty" of any inherent "nature" or identity.[9] The purification of the mind, which the Dalai Lama called the "expressed aim of Buddhism," comes from stripping away false concepts of the "nature" of things and resting content with their Emptiness. In other words, "nature" (in one possible meaning of the word) may very well be a barrier to overcome in a quest for human development.

What should we make of the gap between the Dalai Lama's words and the conventional image of the Buddhist attitude toward nature? Does the Dalai Lama see something in the Buddhist tradition that others do not? Is the image of Buddhism as an ecologically friendly tradition simply an artifact of the Western imagination? Or is it possible that the Buddhist tradition is a complex combination of ideas and aspirations, some of which are positively disposed toward the environment and some of which are not? If so, is it possible to reconcile the Dalai Lama's approach to the concept of nature with the image of a tradition that seeks to establish harmony between human beings and the natural world? The purpose of this essay is to explore the incongruity in the Dalai Lama's words, to ask where the incongruity comes from, and to ask whether it is possible to identify a "Buddhist philosophy of nature," a philosophy that is genuinely affirmative of what we have come to think of as the "natural" world and, at the same time, true to the complex impulses that shape the Buddhist quest for the purification and development of the mind.

To start with, where do we get the stereotype of Buddhist reverence for the natural world? Masao Watanabe began his account of the Japanese attitude toward nature by telling a story about the nineteenth-century art historian Lafcadio Hearn and the genesis of Western perceptions of Japan. Watanabe said that he read Lafcadio Hearn's account of his first visit to Japan to a group of American students. (It was the trip that led to Hearn's fascination with Japan and to his decision to make Japan his permanent residence.) Watanabe asked his students what they first noticed about Hearn's account of his visit. The answer was Hearn's image of the Japanese love of nature, symbolized in Hearn's story of a Japanese warrior who arranged vases of chrysanthemums to welcome his brother home from a journey. The students' answer then elicited Watanabe's own comments about the sense of natural beauty in Japanese landscape design, flower arrangement, the tea ceremony, poetry, and cuisine.

Watanabe is right to suggest that Western people first approach Japanese views of nature through an aesthetic medium. When Japan opened to the West in the early 1850's, Japanese art flooded into Western markets and had significant effect on the stylistic vision of Western artists as different as James McNeill Whistler and Vincent Van Gogh.[10] There are few more powerful and suggestive icons of the Japanese vision of nature than the gnarled rocks and empty spaces of a Zen garden like the one at Ryoan-ji in Kyoto, and few poets of the natural world can match the grace and intensity that is so evident in the works of the Japanese poet Bashō. It is sometimes said that to grasp the significance of Bashō's poem,

> Old pond –
> Frog jumps in –
> Sound of water!

is to grasp the whole meaning of Buddhism.[11] Certainly, the "meaning" of this poem must have something to do with the condensed appreciation of a single moment in the flow of the natural world, a moment in which the minds of the poet and the reader become absorbed in the natural event itself.

Bashō's poetic appreciation of nature has strong antecedents in Chinese literature, as in the work of the shadowy T'ang Dynasty poet whose identity is known simply by the name Cold Mountain. In the lines of the Cold Mountain poet, the Buddhist "way" takes on a distinctly naturalistic flavor.

> As for me, I delight in the everyday Way,
> Among mist-wrapped vines and rocky caves.
> Here in the wilderness I am completely free,
> With my friends, the white clouds, idling forever.
> There are roads, but they do not reach the world;
> Since I am mindless, who can rouse my thoughts?
> On a bed of stone I sit, alone in the night,
> While the round moon climbs up Cold Mountain.[12]

This verse displays a distinctive sensitivity to the rough, unhewn aspects of nature, to mists, rocks and trees – all the aspects of nature that Gary Snyder associated with Henry David Thoreau's concept of the "wild."[13] But is also expresses important Buddhist values. The lines reflect the traditional theme of the Middle Way, leading from the experience of suffering and ignorance in the world of ordinary people to the wisdom of a solitary and enlightened sage, and they map the contrast between these two realms of experience in a series of standard images. The ordinary world is one of entanglement, obscurity, and darkness, with "mist-wrapped vines," and "idling clouds." The world of enlightenment is one of detachment, coolness, and clarity, where the round moon that symbolizes the Buddha's awareness climbs up Cold Mountain. "Cold Mountain" is not merely the setting for the poems and a reference to the poet's own identity; it also expresses the path a sage has to tread to reach enlightenment and symbolizes enlightenment itself. To combine all of these meanings in a single, concrete image is to suggest that enlightenment involves a sense of fusion between the self and the natural world.

William R. LaFleur has shown that Bashō's poetry is the result of a long process of doctrinal reflection in East Asia about the religious significance of nature.[14] When one of Bashō's predecessors, the poet Saigyō (twelfth century), for example, depicts movement along the road to enlightenment as involving "just a brief stop" to linger in the shade of a willow, he raises a question about the nature of the way itself. Is it better to walk the road like a diligent pilgrim with your eyes fixed firmly on a distant goal or to step off the road and allow your consciousness to merge with some part of the natural world?

> "Just a brief stop"
> I said when stepping off the road
> into a willow's shade
> Where a bubbling stream flows by . . .
> As has time since my "brief stop" began.

Here it is the shade of the willow rather than the pilgrim's road that stops consciousness of the passage of time, and this "stopping" reflects the "cessation" of the Buddha's *nirvāṇa*. But why associate *nirvāṇa* with a willow rather than some other element of the natural world? LaFleur has shown that these lines reflect a complex doctrinal discussion about whether plants in particular can have "Buddha-nature," in other words, whether they can embody the state of enlightenment that the pilgrim is seeking. In China this question was first raised as part of the general discussion of the relationship between Emptiness and ordinary reality. The question then became focused as a specific question about vegetation. Did plants have Buddha-nature? Some Buddhist thinkers found an affirmative answer to this question in the chapter on "Plants" in *The Lotus Sutra*, where it is said that the rain of the Buddha's teaching falls equally on all forms of vegetation, and each plant grows up and is nourished according to its own capacity.[15] In Japan this view evolved into the position represented by Saigyō's "brief stop." The natural world was treated as having special significance as a setting for the experience of enlightenment – enough significance to invite the poet to turn off the path and disappear in the shade of the willow.

Saigyō was not the only one, and his was not the only way, to explore the relationship between the natural world and the experience of enlightenment. Allan G. Grapard has shown that the concept of enlightenment can be mapped onto the physical landscape in even more complex ways.[16] The volcano Futagoyama on the Kunisaki peninsula, for example, was treated as a physical manifestation of the text of *The Lotus Sutra*; its twenty-eight valleys were treated as equivalent to the twenty-eight chapters of *The Lotus Sutra*; and its paths were lined with more than sixty thousand statues representing the total number of ideograms in the text. Here the landscape itself is the text, and the text is the *Dharma*. To walk the paths on the mountain and read its valleys as visual representations of the *Dharma* is to experience the relationship between nature and text, path and goal, and cessation and movement in a way that goes far beyond the simplicity of Saigyō's lines.

More examples could be cited of the relationship between Buddhist values and Japanese appreciation of the natural world, but these should be sufficient to show that Watanabe certainly had reason to say that reverence for nature plays a special part in Japanese culture, including its Buddhist dimension. There also are good reasons to think, however, that this is not the whole picture. In a remarkable article entitled "Concepts of Nature East and West," Stephen R. Kellert has given clear statistical shape to the suspicion that Eastern cultures are just as capable of showing disrespect for nature as their Western counterparts.[17] "In contrast," Kellert says, "to the foregoing descriptions of highly positive Eastern attitudes toward nature, modern Japan and China have been cited for their poor conservation record – including widespread temperate and tropical deforestation, excessive exploitation of wildlife products, indiscriminate and damaging fishing practices, and widespread pollution."[18] Kellert prepared a questionnaire to investigate and compare Japanese and American attitudes toward the natural world. He found that the most common approach to wildlife in both cultures, Japanese and American, was the one that Kellert called "humanistic": both cultures showed "primary interest and strong affection for individual animals such as pets or large wild animals with strong anthropomorphic associations." The percentage of people who held this opinion was 37%

for Japan and 38% for the United States. The second most common attitude in the United States was the "moralistic": 27.5% of the American respondents showed what Kellert called a "primary concern for right and wrong treatment of animals and strong opposition to overexploitation and cruelty toward animals." The second most common attitude in Japan, with 31%, was the attitude that Kellert called "negativistic": a "primary orientation [toward] an active avoidance of animals due to dislike or fear." The third most common Japanese attitude was one that he called "dominionistic" (28%): involving "primary interest in the mastery and control of animals." In other words, more than 50% of Kellert's Japanese respondents feared or disliked animals or were primarily concerned with their mastery or control. Kellert's findings received statistical confirmation from a 1989 survey by the United Nations Environmental Program: the survey found that Japan rated "lowest in environmental concern and awareness" of the 14 countries surveyed.

Kellert pursued his investigation with a series of detailed interviews to elicit explanations of the Japanese attitudes. Many of the people interviewed "indicated that the Japanese tend to place greatest emphasis on the experience and enjoyment of nature in highly structured circumstances." The reasons for this emphasis were diverse but quite revealing. One person referred to "a Japanese love of 'seminature,' somewhat domesticated and tame." Another said that the Japanese "isolate favored environmental features and 'freeze or put walls around them.'" For all of Kellert's informants, the stress fell on the cultural transformation of nature, in which natural elements were refined and abstracted in such a way that they could serve as symbols of harmony, order, and balance. The stress on the cultural transformation of nature rather than nature in its pure, unrefined state has also been noted by Donald Ritchie who has said that "the Japanese attitude to nature is essentially possessive . . . Nature is not natural . . . until the hand of man . . . has properly shaped it."[19]

How can we explain the contradiction between Kellert's findings and the stereotype of the nature-loving Buddhist? How can the Japanese tradition appear to show such deep reverence for nature and yet tolerate or even encourage such pervasive attitudes of cultural domination? One possible explanation is that the Japanese have so thoroughly absorbed a Western preference for the domination and exploitation of nature that the indigenous tradition has simply been overwhelmed in a rush for Western-style economic development. Kellert points out, however, that it is too simplistic to attribute this contradiction merely to the influence of the West. As W. Montgomery Watt noted in his account of alleged external influences on the formation of early Islam, it is difficult for one culture to "influence" another in a deep or significant way unless there are already tendencies or predispositions in the receiving culture that make such influence possible.[20] Could there be predispositions within the Buddhist traditions of Japan that tend to favor this "cultural transformation" of nature? Could Buddhism itself have contributed to such an attitude? The way to explore these questions at the most basic level is to move back to India, the homeland of the Buddhist tradition, and interrogate the tradition in its original setting.

How does the religious literature of India picture the natural world? India is a complex civilization, of course, and it is as complex in its approach to nature as any of the traditions of East Asia or the West, but it does not seem an oversimplification to say that there is a deep and abiding preoccupation in Indian civilization with the distinction between the "human" and the "natural." One of the best sources to use in reflecting on this distinction is the text of the *Bhagavad Gītā*, one of the best known and most beloved of Hindu scriptural texts. The text consists of a dialogue between two figures, the warrior Arjuna who is on the verge of a climactic battle, and his charioteer Kṛṣṇa, who reveals himself to be a manifestation of God. At the beginning of the story, Arjuna shrinks in grief as he contemplates the destruction to be wrought by the battle. Kṛṣṇa counsels him to pick himself up and do his duty as a warrior without feeling fear or grief about the consequences of his actions. The reasoning behind Kṛṣṇa's counsel reflects a fundamental feature of Hindu attitudes toward nature. Kṛṣṇa counsels Arjuna to distinguish between his "soul" (*puruṣa*), which is eternal and cannot die, and his "body," which is mortal, changeable, and destined eventually to be discarded as the soul makes its passage into another life.

> These bodies are said to end, but the embodied self is eternal,
> indestructable, and, immeasurable;
> therefore, you should fight, O Bhārata. (2.18)

If Arjuna knows that his true identity is equated with the "soul" and not the "body," he does not need to be affected by grief or fear.

As the text develops the distinction between "soul" and "body," we find that the "body" is spoken of as *prakṛti*, a concept that is commonly translated as "nature." The distinction between soul and body is a reflection, in the microcosm of the personality, of the distinction in the cosmos at large between the principle of "spirit" and the principle of "nature." What does it mean to say that *prakṛti* is "nature"? The semantic range of the word *prakṛti* might seem at first to be considerably wider than the one that normally is mapped by the English word "nature." *Prakṛti* includes not only the material aspects of the cosmos but also the aspects of the personality called "mind" (*manas*) and "intellect" (*buddhi*). The basic distinction is not between body on one side and mind or spirit on the other, but between the complex of changeable elements in the personality (including body, mind, and intellect) and the eternal, unchangeable soul. The distinction between *puruṣa* and *prakṛti* comes close, however, to the distinction marked by the title of the symposium in which the Dalai Lama gave his Middlebury address: *puruṣa* is "spirit" and *prakṛti* "nature," in the sense that *puruṣa* is conscious, transcendent, and attainable through discipline (*yoga*) or reason while *prakṛti* merely reflects or obscures the consciousness of *puruṣa* and is subject to change and decay. The challenge for human beings in Arjuna's position, caught in the web of confusion spun by the strands of *prakṛti*, is to recognize their true identities as immortal souls and escape the bonds of nature.

In a technical sense, the distinction between *puruṣa* and *prakṛti* belongs to only two of the classic Hindu philosophical traditions, the Sāṃkhya and the Yoga, and these two traditions do not by any means serve as the dominant framework for the interpretation of reality in the Indian tradition. But the distinction has wide influence in Indian culture. When visitors make a journey, for example, to the great ruined temple of Elephanta in Bombay harbor, they travel across the waters of the harbor to a small island, climb a long line of stairs up to the rocky outcropping in the center of the island, then enter a cave where the central shrine has been cut out of the living rock. The journey across the water is a symbolic expression of a journey through the changeable, distracting world of "nature," and entry into the darkness and quiet of the temple represent an approach to the immovable center of "the soul." The religious drama of the journey depends on a basic cultural image of contrast between the world of *prakṛti* and the world of the soul. Even in nondualistic traditions, such as Advaita Vedānta, where the goal is to dissolve the distinction between self and world, the journey of enlightenment is still based on an initial insistence on the "distinction" (*viveka*) between the eternal self and all that is not-self.[21] One can argue with considerable force that the Hindu tradition is driven, even in its nondualistic dimension, by a conviction that eternal things have ultimate value and changeable things do not. "Nature" encompasses the things that change and pass away.

Buddhists do not share the Hindu conviction about the permanence of the individual soul, but they also are suspicious of the difficulties and dangers of the "natural" world. Lambert Schmithausen has noted that, in classical Buddhist sources, Buddhist peasants, townspeople, and even monks preferred the tamed and civilized world of the village and city to the virgin forest or the jungle.[22] The jungle and forest were symbols of death and rebirth (as was the ocean that the worshipper had to cross to reach the temple at Elephanta), and *nirvaṇa*, the cessation of death and rebirth, was represented as a city.[23] Images of Buddhist paradises, when they appear in Indian sources, are generally landscapes in which the "wild" aspects of nature have been thoroughly tamed. With trees laid out in symmetrical grids, rectangular ponds, golden lines, and shiny blue-black surface, the paradise of Sukhāvatī in the Indian *Sukhāvatīvyūha* is more reminiscent of a parking lot than it is of an untamed wilderness.[24] Schmithausen notes quite correctly that a significant number of Buddhist monks chose not to live in cities or towns. In the "hermit strand" of monastic life, one visualized the forest as useful for the practice of meditation. In the forest a monk can avoid the distractions of society and contemplate the impermanence of reality by observing the passage of the seasons. But even here the focus is on the natural world as a locus and a guide for the spiritual transformation of the monk himself, as it was in Grapard's account of the mapping of *The Lotus Sutra* onto the ridges and valleys of Futagoyama.

It is important to be clear that this early strand in the Buddhist tradition is not hostile to nature as such: one does not attempt to dominate or destroy nature (in the form of either animals or plants) in order to seek a human good. But neither is the wild and untamed aspect of nature to be encouraged or cultivated. The natural world functions as a locus and an example of the impermanence and unsatisfactoriness of death and rebirth. The goal to be cultivated is not wilderness in its own right, but a state of awareness in which a practitioner can let go of the "natural" – of all that is impermanent and unsatisfactory – and achieve the sense of peace and freedom that is represented by the state of *nirvaṇa*. One might say that nature is not to be dominated but to be relinquished in order to become free.

In this context the significance of the Dalai Lama's approach to the topic of "spirit and nature" becomes clearer. He was not hostile to nature, but he had other important topics in mind, not the least of which was the purification of the mind itself. When he took up the question of nature in the philosophical style that was appropriate to his own tradition (linking it to the concept of Emptiness), his first step was not unlike Kṛṣṇa's first step in the *Bhagavad Gītā*: he distinguished between the realm of appearance or "nature" and the realm of ultimate reality or Emptiness. He said: "When talking about the fundamental nature of reality, one could sum up the entire understanding of that nature in a simple verse: 'Form is emptiness, and emptiness is form' (*The Heart Sutra*)."[25] The concept of "fundamental nature" might seem to function differently that the concept of *prakṛti* in the *Bhagavad Gītā*, and in a sense it does. It refers to the imagined "essence" or "identity" that a person imposes on reality (the reality of Emptiness) rather than to the distracting and alluring play of "material nature," but it performs the same discriminative function when it comes to the purification of the mind. In the Madhyamaka tradition, out of which the Dalai Lama speaks, the idea of "fundamental nature" (whether it is understood as the Tibetan *ngo bo nyid*, and Sanskrit *svabhāva* or as the Tibetan *rang bzhin* and Sanskrit *prakṛti*) has to be stripped away in order to develop a purified awareness of Emptiness. The term "Emptiness" itself can refer either to the absence of such a "fundamental nature" in all things or to the purified awareness that perceives all things as empty in this way. For the Dalai Lama, the concept of "nature" elicits an image of Emptiness and suggests a practice of purification in which the illusions of "nature" are left behind.

Against this intellectual background, it is not surprising to find that Indian Buddhist literature contains very little of the reverence for the wild and "natural" world that one associates with the tradition of East Asia. Indian poetic accounts of insight or enlightenment often reflect a rhetorical distinction in which the teaching of the Buddha is "greater than" or "in contrast to" the possibilities of the natural world, as in the philosopher Dharmakirti's exploration of the poetic relationship between the Buddha's teaching and the cooling rays of the moon.

> Were not a drop from the Moon of Sages,
> better than a flood of cooling moonlight,
> mixed within the vessel of its thought,
> how would this heart find happiness
> and, though it stood within a cold Himalayan cave,
> how would it endure the unendurable
> fire of separation from its love?[26]

When one puts Dharmakirti's image of the superiority of the "Moon of Sages" (the Buddha) next to a comparable passage by the Chinese poet Li Po (701–762),

> Moonlight in front of my bed –
> I took it for frost on the ground!
> I lift my eyes to watch the mountain moon,
> lower them and dream of home.

one clearly sees the aesthetic and ideological transformation that took place when the wine of the Indian Buddhist tradition was poured into its Chinese bottles. "Nature" in the Indian tradition was a world to be transcended, while in East Asia it took on the capacity to symbolize transcendence itself.

How then should we read the affirmative images of the natural world in the poetry of Saigyō and Bashō? Has the Japanese tradition been so thoroughly infused by Chinese attitudes toward the natural world that it has taken leave entirely from the Indian tradition? Certainly there is a striking contrast between the two traditions, but it is possible to see Indian Buddhism (including the Tibetan tradition of the Dalai Lama) in a way that gives us new eyes for the *Buddhist* dimension of the Japanese poetic tradition. Bashō's "Old pond/frog jumps in/sound of water" can be read as an expression of immersion in the flow of natural processes: a frog jumps into a pond and the mind fuses with the event in a moment of intense perception. But the poem is not, strictly speaking, an expression of the frog or the water in themselves; it is an expression of a moment of perception. The force of the poem lies in the mind of the observer, not to the exclusion of nature, but in the mind's awareness of nature.

When Stephen R. Kellert probed the stereotype of Japanese attitudes toward nature, he found what one of his informants called an "emphasis on the experience and enjoyment of nature in highly structured circumstances." The stress fell less on nature in its raw form than on the cultural transformation of nature: natural elements were refined and abstracted so that they could serve as symbols of harmony, order, or balance. Allan G. Grapard captured the ambiguity and complexity of the same point when he suggested that

> what has been termed "the Japanese love of nature" is actually the "Japanese love of cultural transformations and purification of a world which, if left alone, simply decays." So that the love of culture takes in Japan the form of a love of nature.[27]

Nature may not need to be transformed in an overt, physical fashion to be significant, although the design of a "natural" garden is certainly a refined cultural act, but the significance of the natural setting for the human observer lies principally in the act of perception, and it may be appropriate or even necessary for nature to be fashioned and controlled to make this "natural" mode of perception become clear.

Then is there a "Buddhist" philosophy of nature? If the intention of the question is to identify a simple, unified vision of the sanctity of the natural world, the answer must be no. If anything, there is the opposite. Beneath the evident differences between the Indian and East Asian traditions lies a commitment to the view that human beings work out their fates through the development and purification of their own minds. Riccardo Venturini had something like this in mind, no doubt, when he said that the Buddhist tradition develops its attitude toward nature in the context of an "ecology of the mind" and aims at a "purified" world with man as its steward.[28] Could it be that the Buddhist tradition, which has seemed so promising as a model to escape the destructive consequences of the Western anthropocentric vision of nature, is as much compromised by the flaws of anthropocentrism as its Western counterpart? The question is crucial for understanding the possibility of a Buddhist response to the ecological crisis, and much depends on the meaning of the word "anthropocentrism."

In the summer of 1981, the Dalai Lama gave a series of lectures on Buddhist philosophy in Emerson Hall at Harvard University.[29] At the beginning of the lectures a member of the Harvard community welcomed the Dalai Lama to Emerson Hall by referring to an inscription over the portal of the building: "What is man that thou art mindful of him?" He gave a Tibetan translation of the inscription that related it to one of the key issues of Buddhist philosophy ("What are you referring to when you use the word 'man'?") and said that it seemed particularly appropriate to hear the Dalai Lama's words in a setting where the very issue of human identity had such a rich and controversial history. "What is man that thou art mindful of him?" In the Tibetan and Sanskrit traditions, the word "man" recalls a long controversy about the status of the *pudgala* (commonly translated as "person," but literally "man"). An ancient Indian Buddhist school known as the Personalists (*pudgalavādin*) took the position that a person's identity consisted in a *pudgala* that continued from one moment to the next.[30] This *pudgala* was related ambiguously to the momentary psycho-physical constituents (*skandha*) of the mind and body. The constituents changed at every moment while the *pudgala* continued, and the *pudgala* was neither identical to nor different from the constituents. It seems that the *pudgala* was considered to be something like the "shape" or "configuration" of the personality, so that one could say that a person retained the same "shape" even when all the individual constituents of the personality had changed, perhaps like a car in which all the individual parts have been replaced but which still retains the "shape" of the original car.

The Personalists have long since gone out of existence as an identifiable school, and the controversy about the *pudgala* could be relegated to the status of an obscure historical curiosity if it had not become a symbol for Buddhists of the classic mistake to be avoided when thinking about the nature of the self. One of the most basic themes in Buddhist philosophy is the claim that there is no "self," and by "no-self" is meant at least that there is no continuous *pudgala* that ties together the stream of the personality from one moment to the next. The *pudgalavāda*, the doctine of the "man" or "person" is, as it were, the fundamental Buddhist heresy from which the tradition now chooses to distinguish itself. To ask, "What is man that thou art mindful of him?" or "What are you referring to when you use the word 'man'"? is to probe the foundations of the Buddhist view of the self as its most sensitive point: What is the most basic error that has to be avoided if one is to make progress toward the goal of enlightenment?

Herein lies the paradox of Buddhist "anthropocentrism." The tradition is genuinely concerned with the human achievement of human goals. At a deep historical and conceptual level, the tradition defends an ideal of self-reliance, as in the oft-quoted verse from the *Dhammapada*: "One is one's own Lord (or God or Protector). What other Lord can there be"? But the achievement of self-interest is tied in an equally fundamental way to the decentering of the self. On the intellectual level, the quest for *nirvaṇa* is tied up with a quest for an understanding of "no-self" as both a doctrine and a mode of awareness. On a more practical level, Buddhist discipline is built up of choices, both large and small, that challenge the naïve patterns of self-centeredness from which the fabric of ordinary life is woven. In traditional Buddhist societies in Southeast Asia, Buddhist monks go out each morning to beg their food from lay people, meditating as they go on their "friendliness" or concern for all beings. Lay people prepare the food and enact a model in which their own spiritual benefit is tied to a gesture of renunciation, of giving away the food that sustains the life of a monk. Moral precepts, particularly the prohibition against killing (which is extended not just to human beings but in theory to all sentient beings), cultivate a fundamental respect for life in all its forms. These ideals are realized with greater and lesser degrees of consistency in the Buddhist communities of Southeast Asia, but the theoretical connection between the self-interested decentering of the self and respect for life lies deep in the culture. One could paraphrase Grapard's claim that in Japan the love of culture takes the form of a love of nature by saying that in Buddhist culture at large the cultivation of the self takes the form of a decentering of the self and a concern for a wider network of life.

Steven C. Rockefeller has remarked about the way anthropocentric and utilitarian approaches to environmental ethics take on a more biocentric character when they are combined with a scientific appreciation of ecological interdependence.[31] This conceptual development has its counterpart in the Buddhist tradition as well. To say that one's self-interest is served by realizing and enacting an ideal of no-self is to say that one's own self-interest is best understood by realizing one's location in a network of interdependence or "interdependent co-origination" (*pratītya-samutpāda*). The formulas that express the understanding of no-self in different versions of the Buddhist tradition often equate no-self (or its Mahāyāna counterpart, the doctrine of Emptiness) with interdependent co-origination. A famous verse in Nāgārjuna's root verses of the Madhyamaka school says: "We call interdependent co-origination Emptiness; this is a metaphorical designation, and it is the Middle Path." Other textual sources simply equate interdependent co-origination with the Dharma or with the Buddha himself, as in the common scriptural phrase, "He who sees interdependent co-origination sees the Buddha."

Whether one can interpret the concept of interdependent co-origination as genuinely "biocentric," however, is open to question. If a biocentric approach means recognizing "the intrinsic value of animals, plants, rivers, mountains, and ecosystems rather than simply . . . their utilitarian value of benefit to humans,"[32] then the word "intrinsic" presents a barrier. It seems to suggest precisely the substantial, permanent identity that the ideas of no-self and interdependent co-origination are meant to undermine. But the practical force of an "other-centered" position emerges quite clearly in different kinds of Buddhist meditative traditions. When the Dalai Lama teaches about Buddhist practice he emphasizes the importance of compassion, as is customary in the tradition of the Mahāyāna, and one of his favorite sources for a meditation on compassion is the teaching about the "exchange of self and other" in the 8th chapter of Śāntideva's "Introduction to the Practice of Enlightenment."[33] Imagine, Śāntideva says, that on one side of a divide stands your own needy self and on the other side stand fifty or a hundred needy beings. Whose advantage is best to seek? Should you care just for yourself and cater just to your own limitations and fears? Or should you seek the benefit of the larger group? And what if the larger group is not just fifty or a hundred living beings but all the living beings in the cosmos? Śāntideva says that the answer should be clear. The self's greatest benefit comes from seeking the widest possible benefit for the network of all living beings. Śāntideva's point can be construed as a practical centering of one's concern on others (on the network of *bios* or "life") in order to decenter the self (in the self's own interest).

Here lies another reason why the Dalai Lama was hesitant to address directly the themes and expectations of the Middlebury conference on "Spirit and Nature" and why he shifted attention so gracefully away from the natural world toward the purification of the mind. He was not insensitive to the claims of the natural world, but he felt that there was more important conceptual work to be done before its claims could be made clear. He had to begin with his own understanding of no-self (as expressed in the doctrine of Emptiness) before he could sketch the outline of an ethical response to the natural world, and the response continued to move in the orbit of "interdependence" and "compassion." One moves naturally, as it were, in a series of ever-widening, concentric circles, beginning with the impulse to purify the mind and cultivate one's own sense of self, through the sense of the self's interdependence with a network of all other beings, to a sense of affection and love for all existence. As the circles widen, the center comes under pressure, and the network of existence takes on the appearance of a circle whose center is everywhere and whose circumference is nowhere.

Some of the most forceful and perspicacious Buddhist writing about the environment explores the implication of this basic Buddhist conceptual movement from no-self, to interdependence, to compassion. In his reconstruction of E.F. Schumacher's famous concept of "Buddhist Economics," Stephen Batchelor points out that Buddhist economics has to start from a standpoint of nonduality and Emptiness, and from this point of view the concept of an ethical "center" comes increasingly into question. "In the West we are still caught in a struggle between theocentric and anthropocentric visions, which some Greens now seek to resolve through a notion of biocentrism. Such thoughts are alien to the Buddhist experience of reality which, if anything, has tended to be 'acentric.'"[34] Joanna Macy has charted the same movement from the point of view of the Theravāda tradition, beginning with a sense of "the pathogenic character of the reification of the self," moving on to the concept of interdependence (*paṭicca samuppāda*), and then developing a sense of what might best be called universal "self-interest," in which the world is visualized as one's own body.[35] With the words of Arne Naess and the concept of "deep ecology" in mind, she turns the ethical argument about altruistic motives from one of "duties" rendered by the self to another into an argument about one's own "being." One protects nature in order to protect one's self, and the circle of self encompasses the totality of the natural order.

Certainly the sense of interdependence that is such a crucial part of Buddhist ethical theory gives good reason to be skeptical of any form of "centrism" whether it begins in the *theos*, the *anthropos*, or even more benignly in the *bios*. But do images of the "center" need to be entirely abandoned? Buddhist environmental literature abounds with metaphors of interconnection, from the jeweled "Net of Indra," in which every individual jewel is pictured as reflecting every other, to images of the "web" of existence. But there is another, relatively unexplored body of metaphor that has to do with a sense of "place" or "home." Buddhist sources speak from a very early period about a tradition of pilgrimage in which people visited sites that had been important in the life of the Buddha, "saw" them, and were moved by them. The sites of the Buddha's birth, his enlightenment, his first sermon, and his death were held in special reverence, and traditional sources speak of the throne of the Buddha's enlightenment, under the Bodhi Tree in Bodh Gayā, as the center of the cosmos. Some Mahāyāna texts pass on a tradition that every Buddha, of every era, is enlightened at exactly the same site, and beneath the spot where the Buddha's are enlightened sits a throne that is anchored at the center of the cosmos.[36] If there were a "center" in Buddhist ethical thinking about the environment, perhaps this is where it should be located, at the site where Buddhas attain their enlightenment.

But where is this site? Northern India is one possibility. But the tradition has a distinct aversion to literal conceptions of the Buddha. Embedded in Buddhist tradition is the idea that one finds the Buddha not in his physical form but by understanding the Dharma ("What is there, O Vakkali, in seeing this vile body? He who sees the Dharma sees the Buddha. He who sees the Buddha sees the Dharma.") Then where is the "throne of enlightenment" where one understands the Dharma? One possible answer would be the mind itself. It is in the mind that one understands the nature of Emptiness. But the mind is located in a particular body, and the body is located in a particular place. While Emptiness, in a sense, is everywhere, it is realized only in *this* moment, *this* place, and *this* body. In a fine meditation on "Zen Practice and a Sense of Place," Doug Cochida quotes a reference by the Zen Master Dōgen to the Earth as the "true human body."

The meaning of "true" in "the entire Earth is the true human body" is the actual body. You should know that the entire Earth is not our temporary appearance, but our genuine human body.[37]

The earth is not, as it were, a mere illusion. It is the body of an enlightened sage, and it is as much worthy of reverence as throne of the Buddha.

In his essays on "The Practice of the Wild," Gary Snyder said: "In some cases we might call [nature] sacred."[38] To say only "in some cases" shows an appropriate Buddhist reticence toward attributing sacrality to nature in and of itself. But it is not completely implausible to use the language of "holiness" in speaking of the natural order. The natural world can function as a teacher when one meditates about impermanence. In some strands of the Buddhist tradition it can be thought of as possessing Buddha-nature. But most importantly of all, it is the place that is made holy by the quest for enlightenment. Enlightenment is made present in this body and this earth. To speak of the earth as the throne of enlightenment is a metaphor, of course, and it is not by any means a common metaphor in Buddhist writings. But it is one that resonates deeply with the theistic language of Erazim Kohák, the man to whom this essay is dedicated. Kohák's great meditation on the moral sense of nature, *The Embers and the Stars*, is alive with a sense of the holy or, as Kohák himself says, "the presence of God in the very fact of the world."[39] The Buddhist tradition has problems with the language of classic theism, but a sense of the presence of the holy is hardly unknown in Buddhist experience or imagination. It does not come, however, from the outside, nor is it ready-made. It has to be fashioned and developed by the application of human discipline, imagination, compassion, and awareness. This I take to be the force of the Dalai Lama's Middlebury address, as it is of the tradition more generally. Human beings have to take responsibility themselves for the harmony, the health, and the well being of the setting in which the quest for enlightenment takes place.

Division of Religious Studies, Boston University

NOTES

* Malcolm David Eckel, "Is There a Buddhist Philosphy of Nature?" in Mary Evelyn Tucker and Duncan Ryūken Williams, eds. *Buddhism and Ecology: The Interconnection of Dharma and Deeds*, Cambridge, Mass.: Harvard University Center for the Study of World Religions, 1997.

1 Roderick Frazier Nash, *Wilderness and the American Mind* (New Haven: Yale University Press, 1967): 20–21, 192–193.

2 Y. Murota, 'Culture and Environment in Japan,' *Environmental Management*, *9* (1986): 105–112.

3 Lynn White, Jr., 'The Historical Roots of our Ecologic Crisis,' *Science 155* (1967): 1203–7. Reprinted in *Machina Ex Deo*: *Essays on the Dynamism of Western Culture* (Cambridge, MA: MIT Press, 1968): 75–94.

4 Masao Watanabe, 'The Conception of Nature in Japanese Culture,' *Science 183* (1974): 279–82.

5 Gary Snyder, *The Practice of the Wild* (San Francisco: North Point Press, 1990): 10.

6 The words belong to Richard Gere, the Founding Chairman of Tibet House in New York, and appear in Marylin M. Rhie and Robert A.F. Thurman, eds., *Wisdom and Compassion: The Sacred Art of Tibet* (New York: Harry N Abrams, 1991): 8. The image of Tibet as the "lifeboat of civilization" has been widely remarked upon in Asian studies, notably by Peter Bishop in *The Myth of Shangri-La: Tibet, Travel Writing, and the Western Creation of the Sacred Landscape* (Berkeley: University of California Press, 1989).

[7] The Dalai Lama's speech appears in Steven C. Rockefeller and John C. Elder, eds., *Spirit and Nature: Why the Enviroment Is a Religious Issue* (Boston: Beacon Press, 1992).
[8] Rockefeller and Elder, *Spirit and Nature*: 114.
[9] This formula for the expression of Emptiness comes from the Madhyamaka school of Mahāyāna philosophy, the school within which the Dalai Lama himself speaks. For a more extensive account of this concept and for references to further literature, see Malcolm David Eckel, *To See the Buddha: A Philosopher's Quest for the Meaning of Emptiness* (San Francisco: HarperCollins, 1992; reprint ed., Princeton: Princeton University Press, 1994).
[10] See, for example, Gabriel P. Weisberg et al., eds. *Japonisme: Japanese Influence on French Art 1854–1910* (London: Robert G. Sawyers, 1975).
[11] Robert S. Ellwood and Richard Pilgrim, *Japanese Religion: A Cultural Perspective* (Englewood Cliffs, NJ: Prentice-Hall, 1985): 55.
[12] Burton Watson, trans., *Cold Mountain: 100 Poems by the T'ang Poet Han-shan* (New York: Columbia University Press, 1970): 67.
[13] Gary Snyder has produced some of the most powerful translations of the Cold Mountain poems. See his *Rip Rap and Other Poems* (San Francisco: Grey Fox Press, 1982).
[14] William R. LaFleur, 'Saigyō and the Buddhist Value of Nature,' in *Nature in Asian Traditions of Thought*, ed. J. Baird Callicott and Roger T. Ames (Albany, NY: State University of New York Press, 1989): 183–209.
[15] *The Lotus Sutra*, trans. Burton Watson (New York: Columbia University Press, 1993), chapter 5.
[16] Allan G. Grapard, 'Nature and Culture in Japan,' in *Deep Ecology*, ed. Michael Tobias (San Diego: Avant Books, 1985): 240–55.
[17] Stephen R. Kellert, 'Japanese Perceptions of Wildlife,' *Conservation Biology 5* (1991): 297–308; 'Concepts of Nature East and West,' in *Reinventing Nature? Responses to Postmodern Deconstruction*, ed. Michael E. Soulé and Gary Lease (Washington, DC: Island Press, 1995): 103–121. See also Yi-Fu Tuan, 'Discrepancies Between Environmental Attitude and Behaviour: Examples from Europe and China,' *Canadian Geographer* 12 (1968): 175–91.
[18] Kellert, 'Concepts of Nature East and West': 107.
[19] D. Ritchie, *The Island Sea* (Tokyo: Weatherhill, 1971): 13, quoted in Kellert, 'Concepts of Nature East and West': 115.
[20] W. Montgomery Watt, *Muhammad: Prophet and Statesman* (Oxford: Oxford University Press, 1961).
[21] "Distinction" (*viveka*) is one of the four "qualifications" for the knowledge of Brahman. See Eliot Deutsch, *Advaita Vedānta: A Philosophical Reconstruction* (Honolulu: University of Hawaii Press, 1969), 105.
[22] Lambert Schmithausen, 'Buddhism and Nature,' *Studia Philologica Buddhica: Occasional Paper Series* 7 (Tokyo: The International Institute for Buddhist Studies, 1991).
[23] For sources see Schmithausen: 15. The references to the "city of nirvana" come from texts that are somewhat late. An interesting echo of the metaphor in an early source is a reference to *Suttanipata* 3.109 to nirvana as a level piece of land (*samo bhūmibhāgo*).
[24] *Buddhist Mahāyāna Texts*, trans. Max Müller, Sacred Books of the East, 49 (Oxford: Clarendon Press, 1894; reprint ed., New York: Dover Publications, 1969).
[25] Rockefeller and Elder, *Spirit and Nature*: 114.
[26] Daniel H.H. Ingalls, *An Anthology of Sanskrit Court Poetry* (Cambridge, MA: Harvard University Press, 1965): 438.
[27] Grapard, 'Nature and Culture in Japan': 243.
[28] Riccardo Venturini, 'A Buddhist View on Ecological Balance,' *Dharma World 17* (March–April, 1990): 19–23; quoted in Schmithausen, 'Buddhism and Nature': 17.
[29] The lectures have been published in His Holiness the Dalai Lama of Tibet, Tenzin Gyatso, *The Dalai Lama at Harvard*, trans. and ed. Jeffrey Hopkins (Ithaca: Snow Lion, 1988).
[30] The classic account of the theory of *pudgalavāda* is found in the *Abhidharmakośa*, trans. L. de La Vallée Poussin, *Mélanges chinois et bouddhiques* 16 (1971). A useful English translation of the section from the *Abhidharmakośa* that deals with this theory can be found in Edward Conze, *Buddhist Scriptures* (Baltimore: Penguin Books, 1959): 192–197.
[31] Steven C. Rockefeller, 'Faith and Community in an Ecological Age,' in Rockefeller and Elder, *Spirit and Nature*: 139–171. For further commentary on the issues of "anthropocentrism," see J. Baird Callicott, 'Non-Anthropocentric Value Theory and Environmental Ethics,' *American Philosophical Quarterly 21* (1984).
[32] Rockefeller: 143.

[33] See chapters 8 and 9 of *The Dalai Lama at Harvard.* Śāntideva's own text is available in a number of translations, notably Stephen Batchelor, *A Guide to the Bodhisattva's Way of Life* (Dharamsala: Library of Tibetan Works and Archives, 1979).
[34] Stephen Batchelor, 'Buddhist Economics Reconsidered,' in *Dharma Gaia: A Harvest of Essays in Buddhism and Ecology*, ed. Allan Hunter Badiner (Berkeley: Parallax Press, 1990): 178–182.
[35] Joanna Macy, 'The Greening of the Self,' in Badiner, ed., *Dharma Gaia*: 53–63.
[36] Etienne Lamotte summarizes Mahāyāna traditions about the throne of enlightenment (*bodhimaṇḍa) in The Teaching of Vimalakīrti (Virmalakīrtinirdesa), trans. Sara Boin (London: Pali Text Society, 1976), 94–99.*
[37] Doug Cochida, 'Zen Practice and a Sense of Place,' in Badiner, ed., *Dharma Gaia*: 106–111.
[38] Erazim Kohák, *The Embers and the Stars: A Philosophical Inquiry into the Moral Sense of Nature* (Chicago: University of Chicago Press, 1984): 188.

ALFREDO FERRARIN

ARISTOTELIAN AND NEWTONIAN MODELS IN HEGEL'S PHILOSOPHY OF NATURE[1]

INTRODUCTORY REMARKS

The philosophy of nature has always been considered one of the most controversial and obscure parts of what is usually referred to as Hegel's system. Very often the reader is struck by the awkwardness of certain passages, or by the overall intentions of Hegel, which sometimes are quite hard to make out to begin with. And yet Hegel seems to have devoted many of his efforts to this part. The additions to the philosophy of nature are among the longest, most intricate and exhaustive in the whole *Encyclopaedia*. His knowledge of the contemporary accomplishments in most scientific disciplines was thorough and his discussion thereof quite detailed.

In the past 25 years, since the publication of Petry's translation of and commentary on the philosophy of nature, innumerable contributions to this topic have been published. The whole picture of the philosophy of nature has undergone a substantial change, so that it would be difficult to repeat today, with Croce for example, that the second part of the *Encyclopaedia* should be entirely dismissed as misguided. Hegel cannot be approached as a late nineteenth century historicist who takes his bearings by something like the division between natural sciences and *Geisteswissenschaften* and devotes himself to the latter. For him we do not simply have before us two different models of explanation and approaches to two radically different and mutually exclusive realms. For Hegel nature has a life which cannot be opposed to the life of the mind. It is the idea in its externality. Thus, it is neither alien to reason nor belongs to a realm of objectivity forever separated from man. Rather, this latter position is the view of nature which Hegel sets out to question. Hegel holds that his epoch has separated nature from man and that the main task of the philosophy of nature is to retrieve the substantial unity between the two. He wants to overcome, in other words, the modern conception of nature as the inert and dead opposite of subjectivity.

There is no radical gap between nature and man; nature belongs in us as much as we belong to it. Trying to deny this is an abstract and one-sided approach to both man and nature. Hegel criticizes all understandings of nature which view it as something dead and thereby external to man – whether instrumental to him as in some theological conceptions or as a generalized material mechanism governed by few fundamental laws.

R.S. Cohen and A.I. Tauber (eds.), Philosophies of Nature: The Human Dimension, 71–89.

I think that, in order to have a clearer grasp of the meaning of Hegel's intentions, it is important and fruitful to understand the philosophy of nature against the background of alternative models. In particular I believe that the basic traits of the modern conception of nature as we find them in Newton's *Principia* on the one hand and the Aristotelian conception of *physis* on the other will help us gauge Hegel's standpoint with regard to such issues as the understanding of motion, the notion of organic being and of life.

The purpose of my paper is, first, to show the rationale of Hegel's criticism of Newton and of the mechanical world-view. After an explanation of some of the basic tenets of Hegel's philosophy of nature I will examine the notion of organism or living whole. I will show how Hegel understands his philosophy of nature, at least in its main inspiration, as a revitalization of Aristotelian motives. The interpretation of nature as a living totality and an in-itself concept striving to actualize itself will be the leading thread of Hegel's interpretation and appropriation of Aristotle. Having established that, however, I will then try to spell out the reasons why Hegel misconstrues or cannot render the genuine meaning of a few essential aspects of Aristotle's natural philosophy. In particular, I want to stress some of the fundamental assumptions which Hegel shares with modern physical science and which will be responsible for his misinterpretation of Aristotle.

HEGEL'S CRITICISM OF NEWTON

Let me begin with Hegel's understanding of Newton. From the time of his 1801 *Dissertatio de orbitis planetarum* and through until the last edition of his *Encyclopaedia* appeared the year before his death, Hegel reiterated his criticism of Newton's conception of nature and of the essence of physical laws. For Hegel, Newton confuses physical laws and mathematical laws. Newton was unanimously praised for something which Hegel found both unpalatable and philosophically crude, namely the mechanical world-picture based, as Hegel says, on a Lockean (in the *Lectures on the History of Philosophy* in *Werke* 20, 233) or Baconian (in the *Logic, Werke* 5, 406) empiricism. Further, Newton received all the credit which should be ascribed to Kepler. In fact, Newton's laws are not but the formal mathematical derivation of Kepler's laws. But Hegel goes even further to say that Newton's own contribution is only a superfluous addition to and a misguided interpretation of what Hegel found so speculative about Kepler's laws. So let me have a closer look at the detail of Hegel's criticism. I will not insist on what have been called the ridiculous blunders of Hegel's reading of Newton[2] or try to correct or denounce them. What I am interested to show is the different underlying notion of motion and nature in Hegel and in Newton which I believe is worth spelling out.

In both the *Dissertatio* and in the *Encyclopaedia* (§ 269), in the wake of the Aristotelian tradition which we can still find in Kepler, Hegel distinguishes between heavenly and terrestial bodies and their respective motion.[3] Thereby he overlooks or belittles[4] what we consider the greatest achievement – and the most significant and influential innovation in terms of world-view – of Newton's philosophy. I refer to the reduction of the different – astronomical and mechanical – motions to the fundamental principle of inertia or First Law, and thereby to the substitution of the old rank-ordered and oriented cosmos by the identity of the new uniform laws governing all motion whatsoever of all bodies. As is well known, with Newton all differences in rank between circular and rectilinear motion disappear, and circular motion no more works as a perfect and explicative principle, as it did for the Greeks. It has itself to be accounted for as a constantly accelerated motion retaining an attraction to the center, once inertial rectilinear motion, just like rest, is understood as the natural state of the body. The transition is one from motion as determined by the body's inner nature to a conception of motion regarding masses moving according to mathematical laws in space and time under the influence of forces. "Every body perseveres in its state of rest, or uniform motion in a right line, unless it is compelled to change that state by forces impressed thereon."[5] Thereby motion is not, as for Aristotle, intrinsic to a body's *dynamis*, nature and place, but is determined with respect to time, measurable distances and mass. Accordingly, nature is no longer the inner principle from which the motion of bodies and all change derive, but the spatio-temporal homogeneous realm relative to which bodies in motion occupy determinate positions.

What Hegel understands in this is, according to some authoritative interpreters,[6] rather odd. He criticizes Newton's parallelogram of forces in the decomposition of motion as though Newton had meant to account for the velocity and orbit of a planet by combining two different forces, the centripetal and the centrifugal ones. As we know, instead, Newton only had recourse to the centripetal force.[7] For Newton a centripetal force must be applied to a body to constrain it in circular motion; for Hegel, Newton needs centrifugal force as well. Although in this misunderstanding he was definitely not alone, [8] he does seem to disregard or misunderstand the principle of inertia in that he construes motion as if it were a process which would come to a stop unless a force acted on the moving body. If for Newton we must not explicate motion but only the departure of a body from its uniform rectilinear motion, for Hegel on the contrary, as for Aristotelian and pre-Galilean science, all motion has to be comprehended as a change.[9]

There is something essential in Hegel's criticism, though, which has to be addressed. Hegel thinks that Newton portrayed matter and force as two fixed and dead abstractions external to one another (ENZ § 261). Since Newton refuses to ascribe any active force, including attraction and gravity, to matter,[10] which has only passive resistance to motion, the active principle needed to start motion is found in God's activity. Newton actually confesses his ignorance of the nature and origin of active forces.[11] He claims, just like Galileo before him, that we cannot aspire to explain motion and forces but must limit ourselves to a mathematical description of their action. All we can legitimately say about matter is that it is impenetrable, extended, hard and in particular endowed with an indifference to motion, with a *vis inertiae*, which we can translate freely but, I believe, meaningfully enough, by 'the force of a lack of active force.'

Since Leibniz' *Specimen dynamicum* and Wolff, this absence of force from matter had been widely criticized. For Leibniz the activity of force, far from being an external action changing the motion of a body as for Newton, was inherent in substances and proportional to the magnitude of the body to be moved.

Hegel claims in the *Logic* that Kant's Dynamics in the *Metaphysical Principles of the Science of Nature* has reawakened philosophy of nature for similar reasons.[12] Instead of beginning by positing an indifferent matter and then introducing from without various forces into something alien and dead, Kant conceives of matter as the power of attraction and repulsion. Matter is no longer taken as the abstraction of sensible givenness in perception. It is now an articulate concept.

What is decisive in all this for Hegel is that inertia is an abstraction which considers matter as an indifferent and dead entity. As the same time, Newton does not realize that the matter he treats as dead is not a primary given but itself the result of a universalization from perception (and a bad, intellectualistic universalization at that).[13] Here Hegel thinks he can show why Newton was unclear about the unity of the principles of his philosophy and confused mathematical formalization and physical laws, or, better, physics and philosophy of nature.[14] According to Hegel, Newton never gives a philosophical justification of the principles he makes use of or an account of his choice of them. He claims that the center is a mathematical point and that his forces are of a mathematical nature, but then he treats them as physical and mutually independent (ENZ § 266 R., Petry I 249–50). He takes inertia as an empty and quite abstract postulate of perpetual motion which has no empirical basis. Thereby he thinks gravity away from matter (*ibidem*).

For Newton gravity and attraction are inexplicable because the notion of an *actio in distans* is an unjustified hypothesis[15] and its causality cannot be inferred from phenomena. If for him gravity represents a problem of methodical accessibility, for Hegel it is a phenomenon whose meaning and causality has to be accounted for philosophically just like extension and impenetrability.

Thus, when I said that Hegel finds Newton philosophically crude I meant that he believes that all Newton does is presuppose the validity of the inductive method, set arbitrary definitions as starting points to be uncritically taken for granted, abstract formal notions from what is given as constant in experience, and then proceed to impose in a completely external and arbitrary way a mathematical format to the laws he thinks he has established (cf. *Werke* 6, 99–101). Just as for Kant the possibility of the application of mathematics to motion has to be shown by transcendental philosophy and by the principles of construction of magnitudes, so for Hegel the task of philosophy is that of articulating a logic of measure which must precede, and in fact theoretically comprehends, what empirical sciences do. And if Newton notoriously claimed that he feigned no hypotheses, Hegel not only objects to the quality of the metaphysical assumptions he in fact made,[16] but also to the complete unawareness on Newton's part of a tacit and unacknowledged metaphysics.

It would be easy to show that while by metaphysics or hypotheses Newton meant those of the Cartesian vorteces and full space, Hegel means the adamantine net of presuppositions implicit in everything we say or think. But

Hegel's point is that the thoughtless assumption of metaphysical categories results in an uncritical practice of a necessarily bad – i.e. intellectualistic – metaphysics (ENZ § 38 Remark). It is a bad metaphysics because it cannot make the concrete idea its basis but takes its bearing by one-sided and separate forms of thought rigidly fixed by the understanding.

Coming back to our concern, Hegel thinks that forces have not been brought into matter by God. Instead, they are inherent in matter. In this exclusion of any physical theology or "bad metaphysics" from his consideration and in the supposed understanding of matter *iuxta propria principia*, Hegel thinks he is bringing back to life the old Aristotelian teleological notion of matter and *physis* as an inner principle of change. In the Introduction to the Philosophy of Nature (ENZ, Zusatz before § 245, Petry I, 193) Hegel says that in contradistinction to Wolff we have to retrieve the ancient, that is Aristotelian, philosophy of nature prior to its historical sundering from physics. Thereby we cannot limit ourselves to the question of motion, matter or forces, as modern physics has done, but must move forward to the philosophical question of the essence and meaning of nature as a whole. Echoing Aristotle's *Metaphysics* (1028b 2ff.), Hegel says that the question of the being of nature will "always be asked and never completely answered" (*ibidem*, Petry I, 194). For Hegel the highest view of nature is that which regards it in its proper animation as an end in itself, or, differently stated, as that which does not impose any category external to it. Hegel ascribes this approach to Aristotle (ENZ § 245 Z., Petry I, 196). He then contrasts it with the mechanical frame of modern science, which by remaining external to nature and life begins by applying its most primitive notion – that of death, that is inertia – to matter. Mechanical science cannot de exactly what Aristotle had taught us to look for in nature: a view of it as a living cosmos developing out of itself as an organism.

THE IDEA OF A PHILOSOPHY OF NATURE AND THE ARISTOTELIAN HERITAGE

For Hegel a philosophical understanding of nature requires that we look for a comprehension of what is intelligible and meaningful in it as a whole. In itself nature is the idea in its negated form. In Hegel's language this means that nature represents an existing split between concept and objectivity or an "unsolved contradiction" (ENZ § 248 R). In nature intelligibility and appearance are not reconciled. Nature is not in conformity with its intelligibility. It is the realm of external necessity and chance opposed to its implicit lawful, thereby concept-like, being-in-itself.

Thereby nature is not free, which again in Hegel's language means that it cannot find itself in its other. But the concept is present in a hidden and interior way in nature. Thus, for Hegel the merit of empirical sciences, and of physics in particular, is their attempt at knowing the universality of nature in the form of laws, forces, genera (ENZ § 246), and thereby at overcoming the indifference, externality and split of nature. Its externality to us as investigators, and the split *within* nature between external accidental manifestations and internal intelligibility.

Empirical sciences and philosophy do not differ by way of their more or less close tie with experience, but with respect to their different categories or underlying metaphysical assumptions. In fact, philosophy must acknowledge the results of science. But it immediately transforms its conclusions and critically brings to light its theoretical principles and laws into moments of the concept thanks to its higher awareness and to its vision of the whole. It clarifies them and shows their interrelation. As long as mechanics is the only guidance for our understanding of nature, nature will always remain an unsolved contradiction, because mechanics estranges it from the idea and views it as the corpse of the understanding (ENZ § 247 Z. *Werke* 9: 25, Petry I, 206). Philosophy then must view nature in the light of a different principle: it must regard it as a living or self-articulating whole, that is as the implicit idea. Thus the laws, forces and genera isolated and objectified by natural sciences must become for a thinking consideration of nature a self-determining concept, a self-differentiating universality. But a self-differentiating universality, once again, cannot be found as such in nature. The concept is only hidden in nature, or, also, embodied in its most elementary form in an existing living individual (ENZ § 249).

If the task of the philosophy of nature is that of regarding nature as a whole in its being, then, it must regard all the various forms nature assumes as degrees oriented from what is most elementary towards what is most complex, independent and capable of sustaining itself. For Hegel this is the life of an organism. All this itself amounts to saying that the philosophy of nature is constructed on the basis of its end, the account of the emergence of subjectivity. There is here a conceptual or dialectical teleology which orders sciences, laws and natural scopes to show in them the progress of the idea. But the progress of the idea cannot be taken as a concrete one lying in the natural transition from one species to another as in a great chain of being or *scala naturae*. Since nature "never fails to blur essential limits with intermediate and defective formations, and so to provide instances which qualify every firm distinction," the concepts and prototypes it defines "cannot be drawn from experience, but ha[ve as their] presupposition the independence and worth of notional determination" (ENZ § 250 R., Petry I, 216).

To use Hegel's images, "every drop of water yields an image of the sun" (ENZ § 252 Z., Petry I, 220), but that remains precisely just an inert image standing in need of our interpretation. Or, also, in another metaphor, "in Christ the contradiction is posited and overcome as life, passion and resurrection. Nature is the son of God, not as the son however, but as hardening in otherness – the divine idea as held fast for a moment outside love."[17] One of the implications of this is that a dialectic of the formal moments of the concept is needed to make this hardness fluid and to break the limitation and the mutual externality of the different spheres of nature. Nature is not thereby a process itself but is notionally reconstructed as a progression in complexity. In this idea there is an element which at the same time is inspired by and runs against Aristotle. Like in the principle which Hegel found so speculative in Aristotle's *De Anima*, according to which all forms become matter – or, as Hegel says, inorganic nature – for the subsequent consideration, here we read: "all subsequent stage contains those prior to it, so that the last is the concrete unity of all that have preceded it, and presupposes them as constituting its inorganic nature" (ENZ 252 Z., Petry I, 219). On the other hand, Hegel criticizes all images of a chain of being and all natural metamorphosis and transitions traditionally inspired by Aristotle's *Historia animalium*.

As a consequence, it is the dialectical concept which guides the hierarchical structure of the whole of the philosophy of nature. And the progression will go from what is most external – space, time, motion – to what is most internal, for Hegel the self-relating negativity of an individual organism. Thereby the parts to be dealt with in the philosophy of nature are mechanics, physics and organics. In mechanics matter is indifferent to essential or qualitative determinations and everything is external to itself. In physics processuality is introduced in matter and we first have particular bodies with an immanent determinate form. In organics all determinations are not just specific properties but finally themselves totalities determined qualitatively with regard to one another. Here all elements are posited as finite elements of an ideal nature, that is are subordinated to an overarching and independent whole.

What this entails with regard to the problem we started with, Hegel's confrontation with the scientific revolution as exemplified by Newton's mechanics, is the reduction of the modern scientific approach to nature to one view of which Hegel must show the onesidedness and, at the same time, the pervasive holistic ambitions. I believe that Hegel is very conscious of the problem constituted by a modern scientific world-picture. The positive side of what is so perplexing and awry about the grandiosity of his philosophy of nature is his conviction that unless we thematize the categories used by natural science we would not only have to disregard fundamental questions as to the nature of space, time, motion, substantiality, matter, causality and so on. But also, more importantly, we could always fall prey to a tacit acceptance of the basic underlying assumptions of mathematical physics. The risk is that the form of scientific consciousness actually shapes our very life and the way we ask all fundamental questions, including our understanding of the world and our place in it.[18] In Hegel's reaction to this danger we are bound to see a recurrent temptation we normally try to eschew due to our bad conscience, as Jacob Klein once put it.[19] We believe we – or I, for one – cannot follow in detail the discussion in mathematics or physics, therefore we should forsake all attempt at taking a standpoint on what has long since stopped concerning us as philosophers. This profession of humility is also less dangerous for our narcissism.

For Hegel all this is a typical consequence of modern philosophies of reflection. Science has separated itself from philosophy so that sciences care less and less about their principles and presuppositions and tend to become increasingly empirical and positive, while philosophy becomes more and more abstract and non-scientific. Hegel's precise intention is that of overcoming this pathology. For him philosophy is the paramount case of science and the only comprehensive account of the whole. Therefore the objective is that of articulating the relationship between different levels of scientificity and of bringing back empirical sciences to the scope of the idea.

On this score Hegel is in agreement with Kant,[20] who has shown that nothing whatsoever can be thought without the pure principles of the understanding and that the universality and necessity of physical sciences cannot be found in reality but are a product of pure concepts. If sciences cannot be merely empirical any more, and science of nature itself requires a metaphysics of nature elucidating its principles, for Hegel the task of philosophy is to transform what is familiar and well-known (*bekannt*) from the sciences to a known articulate concept (*erkannt*). The difference between science and philosophy is all in the degree of necessity, that is in encompassing comprehension, and in the assumption or questioning of a given conditional presupposition.

So far I have often used expressions such as organism, or innerly articulated whole. Let me now explain what Hegel means by that. Understanding nature as a living whole, as an organism, comes down to viewing it as a form of subjectivity. For Hegel subjectivity can be natural, and not just spiritual, because he defines this notion more generally as a self-relating process, as a development of itself out of itself or an actualization of a potential state.[21] While doing so he refers back to Aristotle's notion of *physis* as the *archē kinēseōs kai staseōs*, the principle of change immanent in all natural beings. He often praises Kant for revitalizing the notion of organism, but whenever he does so he invariably rushes to add that the limit of Kant's notion is that it is not a cognition of the thing itself but a reflective judgement, while Aristotle had already determined nature more philosophically as internal finality.

The notion of internal finality could not be grasped by modern science. Given nature's internal split, there is always a necessary discrepancy between a law and the objects ruled by it. But these objects, according to the section on objectivity in the *Logic* are themselves conceived by modern science, by and large, as analyzable into distinct parts subject to external forces and devoid of any internal consistency; or, alternatively, as combining according to their inherent affinity or repulsion. A mechanical sort of explanation cannot make sense as a description of a totality. We would have a totality of mutually indifferent substances or of inconsistent objects indifferent to determination and always in need of a mover external to the totality, so that no totality as such could be. Further, no individual stable unity capable of subordinating plurality to itself could exist. This can be exemplified by Newtonian mechanics, in which relations are all external: matter is external to forces as much as the world to its mover, God.

In a chemical explanation, in turn, we first have particular bodies; but if there is a process of combination giving rise to a new product, once the initial properties are lost so is the initial body, which is therefore not independent and capable of reproducing itself. Thereby we cannot give an account of totality as the differentiation and combination among things conceived as independent beings. The consequence in the *Science of Logic* was the necessity to think of reality as a self-differentiating totality; and, in the philosophy of nature, as an organism. In other words the approach is holistic and attempts at a definition of the totality we are dealing with. While a mechanic and a chemical totality are impossible and cannot acccunt for phenomena such as life, a teleological totality can account for both itself and the other inferior forms of totality by subordinating them under itself.

Thus the forms of mechanism or chemism are not merely false. There are more complex scopes in nature, however, for which these explanatory principles are not adequate and sufficient. For example, if we study anything which has to do with life – and we can very well say that the philosophical comprehension of life had always been of crucial importance for Hegel since the time of his early theological writings –, we need a conception of an object as an independent thing not merely subject to external constraint, but deriving its form and constitution from its inner principles. We need a conception of an object as a *Selbstzweck*, an end in itself.

In the *Critique of Judgment*, as I was saying, Kant has made it possible for us again to approach life and organisms in the right manner. Organisms are ends of nature. They are both cause and effect of themselves, in terms of their species through generation and in terms of themselves as individuals insofar as they shape the matter they are composed of.[22] They are organized totalities in which the parts are only possible through the other parts and for the whole. This is the proper mode of being of nature, not as instrumental to something higher but as a regulative form of internal finality.

Thereby, for Hegel, Kant has overcome the old antinomy between necessity and external teleology, between a mechanical view of the totality of nature and a final theological teleology which projects an end outside nature as self-determination. However, Kant interprets this internal teleology as a requisite for our judgment of the intelligibility of living nature, not as knowledge of what life is in itself. In the *Logic*, in the *Lectures on the History of Philosophy* and in the philosophy of nature,[23] Hegel invariably contrasts this approach with Aristotle's internal finality which finds an immanent self-determination in nature. Teleology is not an external principle. We do not need to postulate a connection of objectivity with an extramundane principle of activity to save phenomena. In *Physics* Beta 5 and 9 Aristotle had equated external necessity and chance with matter's resistance to the action of form. External necessity is inherent in matter, but, far from being sufficient to explain change, is defined as the accidental cause which finality uses or, alternatively, finds in its way, while striving to actualize itself. Form uses matter but never actually masters it. Incidentally, this accounts for Aristotle's idea that nature is not to be divinized and occasionally gives birth to monsters.[24]

For Hegel this shows that Aristotle understands nature as an in-itself concept. The concept is defined by Hegel as a purpose which "does not pass over but *preserves* itself, in its operation; i.e. it brings only itself about and is at the *end* what it was in the beginning, or in its originality: what is truly original comes to be only through this self-preservation."[25]

In this view, the organism is a self-sustaining and self-renewing activity of subordinating or subsuming externality under itself. This amounts to the definition of the organism as a negative self-referential unity, which constantly transforms external inorganic nature and assimilates it into its own metabolism or, as Hegel says, in its ideality. Differently stated, its identity is not that of a given form, but that of a rudimentary self, an active shaping of itself and maintaining of its form in relation with externality. Thus the organism is a concrete universal, or the unity of a multiplicity: the process of maintaining its individuality as a realization of its concept or in-itself in objectivity. In this connection we can understand Hegel's treatment of sensation, pain, hunger or desire as a form of contradiction (cf. for example, *Werke* 6, 188). The contradiction is that between the identity of an organism and its alterity. Here we do not have an external relation. The organism can live only insofar as it is a relentless active confrontation with alterity. That is, negativity is constitutive of the organism and does not come to it from without.

In all this Hegel consciously adopts the key points of the Aristotelian philosophy of nature, from the fundamental distinction between all and whole (*pan* and *holon*) to the notion of entelechy, i.e. of *ousia* as a *holon* and as an *aitia* of itself. Hegel translates these two terms by *Ganzes* and *Ursache*, and interprets the organism as a *Selbstzweck*, an end in itself. Thereby he rightly understands Aristotelian natural causality as constitutive of a thing. A natural cause is not, like for modern science, external to its effect, but is the actuality of an *energeia* or the identity of eidos and telos, of formal and final causes. More a ground than a cause, so to speak, the ground of the sublation of the many under the one.

For Hegel Aristotle's *physis* is what has in itself the principle of fulfilling its end and which in its self-identity, while infecting itself with externality, uses it to conform to its concept, that is, to its end (*Werke* 19, 176). Aristotle had distinguished nature from art in *Physics*, Beta 1 as what has the principle of change within itself as opposed to what needs an external agent, and as what can reproduce and perpetuate itself as opposed to what cannot. For Hegel the truth of the organic process in contradistinction to mechanical or physical processes is that it can reproduce itself and contribute to the continuation of the species, while this is impossible for all non-organic, i.e. non independent and self-sustaining – processes. And this is only possible because organic beings are not aggregates of parts but overarching totalities whose particular functions are oriented to the sake of the whole.

Hegel does not mean that everything is to be understood as teleologically structured. For example, when he talks about the earth and he refers to the contemporary earth sciences, in particular to neptunism, plutonism and vulcanism, he treats the earth as the static and petrified foundation for life, its skeleton so to speak. The earth is a dead or always already past processuality, in which time has acquired the solid state of reified space (ENZ § 339 and Z).[26] Once again, internal finality is the principle of a holistic view of nature as a totality of self-determining processes, and it is the only category which can account for both life and what is instrumental to it.

The organics itself shows this inner progression from the least articulate forms of life to the individual living being. For Hegel, plants do not have an internal articulation in discontinuous organs, and their assimilation and reproduction lack a self. That means that, although plants do have a development from themselves to themselves, they do not subordinate their ramifications to their internal unity but repeat themselves indefinitely in their parts.[27] Here is no difference between parts and whole. In plants, neither reproduction nor assimilation are a mediation with otherness (ENZ § 344). In the animal organism, instead, we have three phases or moments: the *Gestalt* or anatomic shape, the relation with the external environment, and the generic process, the relation to an other like itself, e.g. sexual reproduction.

All this hierarchy mirrors the hierarchy that we find in Aristotle's *De anima* stretching from the planets, which have *auxēsis* and uninterrupted food intake, *threptikon*, to the animals which have also locomotion (*hēkata topon kinēsis*), periodic assimilation of food, sensation (*aisthētikon*) and desire (*horexis*) (see, e.g. *De an.* B2, 413a 21 ff.). For Hegel the animal organism is more free than plants. It determines its own relation to place on the basis of its needs and desires; its sensibility as unity of receptivity and activity is the first form of appropriation of the environment. Its desire or lack is its inner split or felt contradiction which pushes the animal to overcome it. But this is only possible insofar as lack or negativity is internal to the organism. In this connection Hegel says that even diseases are a privilege of the animal organism, the negativity internal to it and constituting organisms' finitude as opposed to the negativity coming from without in inorganic beings (ENZ § 371 and Z, Petry III, 194, where the example is that of a stone).

Aristotle spoke of a correlation between matter and form in the beings by nature. In particular, all organs are nothing independent of the *ergon* they perform, so that they are only called organs by homonymy in the case of a dead or impaired organism. An eye which cannot see is no longer an eye but by homonymy (*De an.* B 1, 412b 17 ff.). For Hegel organs are defined by their subordination to the whole, to the living body. The body, in turn, with respect to its functions is the object of the soul's activity.

In Aristotle's biological writings we have a hierarchy of beings corresponding to the combination of matter which defines them. We pass from properties of the four elements such as fluidity, solidity, weight, density etc. to a synthesis of such properties in bodies according to different proportions (*De gen. et corr.* B 7; *De part. anim.* A 1 and B 1–2; *De Gen. anim.* A1). At this stage, unity remains an undifferentiated continuum – flesh, bones, metals, etc. Then we have heterogeneous parts such as organs. These, in turn, are defined by their lack of independence with regard to the functions they perform and to the organism which alone has separate status as living ousia capable of initiating and supporting a change and of actualizing its end (*De part. anim.* B 1, 646a 12–24). In a very striking passage Aristotle adds that nature adjusts organs to their *erga* or functions, not functions to organs (*De part. anim*, IV 12, 694b 13–4). When Hegel comments on Cuvier's comparative anatomy, which for Hegel represents the conception of organisms as systems in opposition to Linnaeus' intellectualistic isolation of characters for the investigator, or when he mentions Lamarck's classification of animals as vertebrates and invertebrates and belonging to water, air and earth, he stresses that Aristotle had already delineated the same hierarchy and understood the organism as a unity in which the parts are moments of a whole.[28]

HEGEL'S MODERNITY

Hegel interprets all stages of the organic life as guided by an internal finality which subordinates inorganic nature to the fulfillment of its purpose. Digestion is no exception. But it is more than just a simple variation on a given theme. it is the very model which Hegel uses for both thought (ENZ § 12 R) and work. Eating is the assimilation of food showing the self's power over externality and the acknowledgement that externality has no for-itself, no independent subsistence. As such it is the self's negation of immediacy and of givenness. In this respect, Hegel speaks of the self's sovereign ingratitude towards what gives it sustenance (ENZ § 381 Z). He also calls food intake or intussusception an infection or a syllogism whereby we mediate ourselves with externality. Thus we transform givenness and idealize it, i.e. give it the higher meaning of our possession (ENZ §§ 363–66).

All organic processes are such mediations. But what this shows is not only externality's lack of for-itself, but also the unconscious and ever active work of idealization on the part of subjectivity. While the formation-process only dealt with the conditions for self-preservation internal to the organism, the assimilation process is the stage of the self's openness to the environment and of its unity with alterity, or the unification of subject and object.

In this process activity is no less important than receptivity. In particular Hegel draws on Spallanzani's research to show that the assimilation would not be possible without the stomach's gastric juices. Digestion would not be thinkable without a reference to the subject's inner constitution and chemical activity which can belabor externality insofar as and to the extent that it produces its own receptivity out of itself.[29]

According to Hegel, Aristotle saw quite clearly that in assimilation the identification between subject and object – or, as in sensation – between perceiver and perceived points to the same activity–within-passivity. The object is assimilated by an active subject predisposed to mediate it (*Werke* 19, 204–7).

The final truth of organic life for Hegel is that the individual living being is inadequate to its genus, in other words it is dominated by it insofar as it lives and dies and its processes are instrumental to the perpetuation of the species (*Gattung*). This is what Hegel means when he says that the existing organism is external to its concept and that its only immortality is that of reproduction (*Werke* 6, 486, as in Diotima's speech, *Symposium* 206e, and in Aristotle's *De anima*, B 4, 415a 26-b1).

Here, I think, we can begin to appreciate, after the similarities and affinities between Aristotle and Hegel, also their fundamental differences. For Hegel, man is not simply his natural life but can rise to coincidence with his species. By thought and action the individual human being can consciously carry out and actualize a spiritual content. Thereby the concept is no more an in-itself but becomes for him. This has the more precise consequences that man can objectify himself in the universal medium of reality and make his individuality universal, part and parcel of the spirit's history. By thus taking nature's independence as an appearance, he can appropriate nature and treat it as a moment of his own ideal life.

In this process spirit appropriates externality and gives it a meaning which, as I said, is only implicit in nature. Thereby the spontaneous or instinctual unconscious teleology active in organisms is made explicit and transformed. It acquires two forms, the theoretical and the practical, in which spirit sublates externality. Reason, reflection and the possibility to inhibit desire and deny natural death are all manifestations of man's consciousness of reality as ideal, that is as negated in its immediacy. But more meaningfully they are an expression of man's superiority over animals and organic life. His control over his instincts, his discipline over his drives, and his satisfaction, depend on his ability to act according to purposes and to temporarily silence his needs. While animals cannot rise to this sovereignty over their naturality, man can disregard it and follow his will. Through work he can transform nature and – this time literally – make it his own. This is where Marx later got his famous comparison between bees and architects. Man can only build a house by rational design, by the mediation of a thought-out project. It is not his immediate nature that pushes him to do so.[30]

The twofold attitude towards nature is all dictated by nature's unsolved contradiction, by its split between intelligibility and accidentality. In the theoretical attitude, at first man finds nature as something alien and subsistent before him into which he tries to gain an insight, to unveil its mysteries. The practical attitude instead is the violent domination of nature, now regarded as a *non ens*. If the former looks for its intelligibility, the latter treats it as mere accidentality. The contradiction will remain unsolved until we overcome the opposition and regard the two opposites as complementary. As usual in Hegel, the separation between theory and practice is overcome at a – however superior – theoretical level. Man's violence and exploitation can only exercise itself on nature's externality; man's theoretical investigation will always differentiate between an essence to grasp and an appearance to discard. The philosophical question about the being of nature purports to bring about precisely a consciousness that if the two levels, in themselves one-sided, are taken as complementary we will stop treating nature as an enemy to colonize or as an externality to unveil. Rather, once nature is understood as a living whole and as an in-itself concept, spirit will comprehend nature as its own inner foundation and at the same time put a limit to the bad infinity of its exploitation of nature through work. This implies that it is spirit itself which has to overcome the opposition between theory and practice. Thereby, it is spirit which has to reconcile nature with itself.

The difference between Aristotle's search for a meaning in nature and Hegel's analogous approach begins to show substantial contours. In both Hegelian theory and practice nature's externality is overcome by man, and likewise is its appearance of givenness overcome. All this is a consequence of the peculiar switch advanced by Hegel in the initial question about the being of nature which he inherits from Aristotle, his definition of it as the idea in its externality. This question, I remind you, distinguishes him from modern science, which stopped asking for the meaning of the being of nature and limited itself to a description of its behavior. For Hegel theory and praxis are different moments of spirit's sublation of externality, while for Aristotle they were separate and independent genera. Aristotle's assimilation (for nutrition, cf. *De an.* B 4, 416a 30 ff.) occurs all within the realm of nature, whose givenness is never put in question. Thus, what I find irreconcilable with Aristotle is the understanding of assimilation, and of teleology in general, as a *sublation of externality*. Hegel has tacitly made a further step: he has set up an opposition between nature and spirit which we cannot find in Aristotle, and he has ascribed to externality the role of a moment in spirit's return to itself.

This is quite consistent with his treatment of the *De anima* in the *Encyclopaedia* and in the *Lectures on the History of Philosophy*. While "psychology" for Aristotle was all internal to natural philosophy, with the exception of the pages on *nous*, Hegel treats it as Aristotle's alleged philosophy of spirit to which he felt so close. Hegel's Philosophy of Spirit, in turn, is subsequent to the philosophy of nature but also opposed to it as to its basis (ENZ § 381). It must be pointed out that the reason by Hegel could make such massive use of the *De anima* was the fact that he found it much more speculative than any ancient or modern work on the subject. Likewise, at the beginning of the philosophy of nature (ENZ § 245 Z) he quotes the verses from the *Antigones* where Sophocles writes that "Wonders are many, and none is more wonderful than man; nothing destined to befall him finds him without resources" (332–33 and 360). In the same sense he could also have quoted the myth of Prometheus as told by Protagoras in the homonymous Platonic dialogue. For Hegel, in sum, the practical attitude towards nature is not typical of modernity but, as an eternal trait of spirit, was already known to the Greeks, who used need and ingenuity to master nature.

What I find disputable in this is not so much an anti-historical lack of any due emphasis on the consideration that only because modern science stopped asking about the being of nature could it objectify it into the realm of externality and begin to subjugate it. Rather, I would like to mention a problem internal to Hegel's system. For us, the main differences between antiquity and modernity are rooted in the respective understandings of nature and of man's place in it. Hegel is free to refuse external history in favor of speculation, to privilege affinities over differences and to isolate what is speculative and a permanent acquisition in philosophy, beyond or possibly against its historical genesis. But this seems to me to run against Hegel's own repeated statement that the progress in the complexity of the idea's thought determinations corresponds to the historical progress in mankind's liberation from externality and to the rise to the absolute level of thought.

Aristotle does not spiritualize man's natural life. The soul is the animation of the body, not its master. The difference between man and other living beings at *De part. anim.* B 10, 656a 5 ff. – that man not only lives but also values life – is based on what falls outside man's body, *nous*, not on a spirituality permeating all forms of man's physical nature, as in Hegel's Anthropology. Accordingly, Aristotle's philosophy of nature, and his "psychology" in particular, can be understood prior to all division and separation from the human world. There is an undeniable difference between this and Hegel's attempt at establishing a continuity or an identity-within-difference between nature and spirit; at overcoming the modern division beteen nature and culture in a renewed understanding of nature's internal finality as the foundation of spirit.

For Aristotle philosophy of nature was a *theōria* of what is given to man and on which man has no influence. As such it was contrasted with art or *technē*. Hegel does not seem to realize that he comes after the fusion of *technē* and nature performed by modern science and that his interpretation of Aristotle is dependent on, and vitiated by, the disregard of this transition. What he disregards is that modern science does not conceive of the investigation of nature as an investigation into the principles of what is given to us in sensation, but begins in open oppositionn to the concepts and methods of ancient science, including the difference in genus between Aristotelian *theōria* and *poiēsis*. Thereby modern science establishes a contrast between a natural prescientific world and the world of objectivity, which has lost all immediate touch with what we see. But it is this latter world that will from now on define what nature is. From Descartes on, mechanics replaces Aristotelian physics because of its objectivity and universality. This is how *technē* and *epistēmē*, for Aristotle "virtues" of two distinct parts of man's rational soul (*Eth.nicom.* Z 2), become two complementary aspects within natural science itself.

Thus, instead of viewing nature as the realm of sensible things in motion, modern natural science treats it as a mechanic and quantitative uniform structure measurable in spatio-temporal terms. Thereby, from Galileo to Newton, it interprets motion and its laws as the motion of geometric bodies in an indifferent space, as the pure abstract structure underlying the motions we perceive. A preliminary split between objectivity and subjectivity in our experience of nature is essential to and presupposed by the new method.

Hegel never mentions the distinction between primary and secondary qualities or experimental method (other than in the *Phenomenology of Spirit*). But this general background is implicitly operative throughout his theory of nature. The way he expounds Aristotle's *Physics* in his *Lectures* bears witness to his own dependence on it. Before passing on to his commentary on internal finality, he begins with the same categories we find at the beginning of his philosophy of nature: space, time and motion (*Werke* 19, 169 and 197–8). He says that Aristotelian motion is the first appearance of substance and that time and space are its moments (*ibid.*, 169).

This beginning can hardly correspond to anything Aristotelian. It is not a Newtonian beginning because space and time are not an absolute framework of motion for Hegel. However, Hegel definitely does assume a modern standpoint, in which sensible experience has already been turned into thought determinations or categories, if only to transform it dialectically, that is to show the inner excitation of its categories. Thus we can begin to realize that Hegel's very idea of nature as an unsolved contradiction is rooted in the very split between intelligibility and appearance which modern physical science has taken so much pain to delineate. Of course, the in-itself concept is not the same as the spatio-temporal lawfulness of mechanics. The concept is not a dead self-same abstraction but nature's inner soul and life. This very concept, which is only for us and can never be for-nature, allows for reason's recognition of itself in nature. Only thus can nature be a moment of the idea. For Aristotle nature is not the in-itself of the concept opposed to its being-for-itself. Aristotle begins by the givenness of the sensible world, not by a contrast between intelligibility and appearance.

Even in his analyses of space, time and motion Hegel draws on a number of Aristotelian notions. For example, I think it can be shown that he takes space as a continuity differentiated only by a proper limit, like the Aristotelian point or *stigmē* (ENZ § 254 Z, Petry I, 224–5), just as time is a continuum separated by the now (*Phys.* IV 10–12). However, in Hegel's philosophy of nature time and space are the necessary presupposition for an investigation of mechanics, that is for the intelligibility of motion. However immortal Aristotle's pages on time are, in the *Lectures* Hegel does not realize that the relation among motion, time and space substantially differs in Aristotelian physics and in modern science.

Time is the number of motion for Aristotle, and thereby allows for a divisibility of space and, in principle, for quantitative and measurable treatment of motion. But motion is first defined without reference to time; in fact, the notion of time is itself articulated in terms of motion. Modern natural science, instead, understands motion in terms of its relation to time as a pure frame of reference; and this model is so successful and pervasive that it guides all modern understanding of time, culminating in Kant's analysis of the relation between causality and the pure form of inner sense. Thus, the reason why the divisibility of time advocated by Aristotle does not give rise to anything like mechanics or uniform mathematical laws of motion is that for Aristotle all physical processes depend on the nature of the substances we are investigating and are not understandable in terms of homogeneous relations binding all bodies equally. And this is the reason why even Aristotelian *phora* or locomotion (one of the three senses of motion, *kinēsis*, in the *Physics*, E 1–2, 225a 1-b 9) differs from the modern understanding of motion.

Hegel also takes motion and matter as correlative notions. Space, as I said, is not an absolute framework but the determinate space of bodies. It is not a Leibnizian relation of order either (ENZ § 254 Z). In its posited form, it is a spatio-temporal self-same place where matter is located. As such it has many properties in common with the Aristotelian *topos*. But it is striking that when Hegel comments on this theme in his *Lectures* he interprets – quite Spinozistically – place as the inner negativity of a thing, whereby its limit is at the same time its negation (*Werke* 19, 184). He does recognize that Aristotle distinguishes high and low. But he goes on to oppose to such natural places the superior dialectical truth of the three dimensions of space.

I believe this is quite meaningful. It shows that when Hegel refuses to treat space and time as a Newtonian ready-made framework before the insertion of bodies and matter in motion, but interprets them as a sort of disposition of matter developing out of itself, he is not going back to an oriented cosmos in which everything tends to its proper place. Hegelian motion or *Bewegung* is not Aristotelian motion, which is qualified by the rest a particular process naturally tends to. For Aristotle, motion is an imperfect process which ends in a clearly identifiable state. When Hegel talks about Aristotelian rest or *stasis*, he translates it as "the possibility of being moved" (ibid., 1982). Thus, he does not see that rest is superior to imperfect motion for Aristotle, because he privileges inner excitation or negativity (which for Aristotle would be potentiality) over rest.

On this score Hegel is more Leibnizian than Aristotelian.[31] This is often true throughout his interpretation of Aristotle. For Aristotle matter always needs an agent to attain its determinate form. As such, it cannot be put on the same level as *physis*, which contains a principle of change within itself. Hegel, instead, universally interprets the inner dynamic negativity of things, substances, spirit itself as a transition from an Aristotelian *dynamis* to its actuality. This is how he can interpret Aristotelian matter as an inner dynamical and quantitative principle,[32] and establish a continuity between matter and form which he sees at work in the Aristotelian pair of *dynamics* and *energeia*. Nothing falls outside *Bewegung* for Hegel. This is so true that he interprets even the *Metaphysics*, which for Aristotle (at least in E 1) dealt with what is not subject to change, as the exposition of the actualization of form, and takes the divine *ousia* as a self-mover.[33]

To conclude, in the philosophy of nature and in the organics in particular Hegel adopts the Aristotelian notion of *physis* as a model of internal finality, of a process in which beginning and end fall within the same subject. Hegel is right to find this incompatible with the mechanical view of nature advanced by modern physical science. But his interpretation of Aristotle's natural philosophy is largely biased by the unacknowledged modernity of his presuppositions.

Department of Philosophy, Boston University

NOTES

1 Quotations will be from G.W.F. Hegel, *Werke in zwanzig Bänden*, Red. E. Moldenhauer and K.M. Michel, Frankfurt a.M. 1969–71. For the second part of the 1830 *Encyclopaedia* (*Werke*, op.cit., vol.IX, abbreviated as ENZ; R = Remark, Z = Zusatz or addition), see *Hegel's Philosophy of Nature*, edited and translated with Introduction and Explanatory notes by M.J. Petry, 3 volumes, London 1970.

2 See, among others, De Gandt (1979), 71 ff., or Shea (in Horstmann-Petry 1986, 33 ff), or Van Lunteren (*ibidem*).

3 See the superiority of circular over rectilinear motion in ENZ § 261 Z.

4 In the addition of ENZ § 270 (Petry I, 272) he does mention the generalization of the law of gravity as a merit of Newton's work. But in the addition to § 264 (Petry I, 246) he expresses scorn for it. I believe we should take Hegel's systematic distinction between finite and absolute or free mechanics as a decisive factor in the evaluation of these two contrasting passages, and therefore consider the former remark inconsistent with it.

[5] I agree with Koyré in preferring Motte's translation over Cajori's. In Latin the sentence reads: "Corpus omne perseverare in statu suo quiescendi vel movendi uniformiter in directum, nisi quatenus a viribus impressis cogitur statum ille mutare." Cf. *Sir Isaac Newton's Mathematical Principles of Natural Philosophy*, translated by A. Motte in 1729, revised by F. Cajori, Berkelely 1946 (from now on: *Principia* Motte-Cajori), p. 13.

[6] Cf. the considerations brought forth by De Gandt and Shea in the works quoted.

[7] *Principia*, Book I, Section 2 ff.

[8] See De Gandt's fifth appendix.

[9] Shea argues for a strong Aristotelianism in Hegel's theory of motion which I believe calls for careful qualification. For the main differences between Aristotelian and Hegelian motion, cf. the conclusion below.

[10] See the Scholium generale added to the second edition of the *Principia* (1713), Bk.III, Motte-Cajori, p. 546.

[11] See the famous letter to Bentley, in *Opera omnia*, ed. by S. Horsley, 5 vols., London 1779–85 (vol.4: 429–42).

[12] In *Werke* 5, 201 ff., and ENZ § 262 R. This holds true despite all the limitations proper to Kant's alleged construction of matter (cf. *Werke* 5, 202 ff.).

[13] The third of the *Regulae philosophandi* (*Principia*, Motte-Cajori, 398 ff.) can be taken as a very good example of what Hegel meant by this universalization of the absolute passivity of matter in perception. Newton writes that extension, impenetrability, hardness, mobility and inertia of bodies are known by our senses. From the parts we conclude to the whole, i.e., we ascribe these properties universally to all bodies whatever, "and this is foundation of all philosophy" (*ibidem*).

[14] Hegel finds Newton's use of the word 'philosophy' quite unjustified but naturally consistent with the British empiricist understanding of philosophy as the knowledge of what is constant and lawful in experience, in all disciplines (cf. ENZ § 7 and remark).

[15] For Newton's understanding of hypothesis, see the second of Koyré's *Newtonian Studies* (1965).

[16] From to the existence of ether to that of atoms and the void, not to mention absolute space, time, God, as abstract and fixed entities. Cf. Newton's *Opticks*, ed. by I.B. Cohen, New York 1952, 369 and 402–3.

[17] ENZ § 247 Z. (my translation). The passage, which is in German more striking than in Petry's translation, reads: "In Christus ist der Widerspruch gesetzt und aufgehoben, als Leben, Leiden und Auferstehen; die Natur ist der Sohn Gottes, aber nicht als der Sohn, sondern als das Verharren im Anderssein, – die göttliche Idee als ausserhalb der Liebe für einen Augenblick festgehalten." Along the same lines § 140 Z.

[18] In this conception I find it particularly meaningful that, to name an example, Hegel links a discussion of Hobbes' political atomism with the scientific world-picture in natural philosophy.

[19] In Klein 1985, p. 2 ff.

[20] See the Introduction to the *Encyclopaedia*, and Illetterati 1995, 304 ff. Hegel's affinity with Kant on this point has been widely recognized. For recent examples in the literature see the essays by Buchdahl, Ihmig and Garrison in Petry 1993. About Kant, see my essay (1995, 155 ff.).

[21] What follows is partly based on my book. For textual evidence and quotation of relevant passages I refer the reader to pp. 66–73, 100–2, 152–3, and 183–5.

[22] *Critique of Judgment* (from now on: KU) in *Kants Werke*, Akademie-Textausgabe, Berlin 1968, Bd. 5, 371–4. See Illetterati 1995, 53 ff.; for an exhaustive and updated bibliography on the organism in Hegel, cf. *ibidem*, 142 fn. and 379–85.

[23] *Werke* 6, 436–400; ENZ § 204 R; *Werke* 19, 173–81; ENZ § 245 Z and § 360 Z.

[24] Likewise, for Hegel contingency and accidentally are ineliminable. They constitute nature's "impotence" (ENZ § 250). Therefore nature should not be divinized (ENZ § 248 R).

[25] ENZ § 204 R, in *The Encyclopaedia Logic*, translated by T.F. Geraets, W.A. Suchting and H.S. Harris, Indianapolis/Cambridge 1991, p. 280.

[26] See T.H. Lever (in Horstmann and Petry 1986).

[27] See Aristotle's *De anima* B 2, 413 b 16 ff., where Aristotle speaks of the "potential multiplicity" constitutive of the soul of plants, whose parts can live on even if they are severed from the original plant.

[28] Cf. ENZ § 370 R, and Aristotle's *Hist. anim.* III, 516b, and II, 505a–b. See Illetterati 1995, Appendix 3.

[29] About Spallanzani and Hegel, see Bodei 1975, 102 ff., and Illetterati 1995, Appendix 2.

[30] Cf. Bodei 1975, 145 ff.

[31] See Bonsiepen 1988 on the importance of Leibniz for Hegel's philosophy of nature.

[32] Hegel is usually very careful not to explicitly superimpose modern categories on Aristotle's philosophy. In fact, he is the first major modern philosopher who studies Aristotle in Greek, so

much so that even Trendelenburg had to acknowledge that. Thus, it is all the more surprising that in a comment on the *Categories* he goes so far as to treat Aristotelian matter (*hylē*) and quantity as synonymous (*Werke* 19, 233).

33 For the textual reason which misled Hegel, cf. my book (1990, 44–8). Let me mention that if the *Metaphysics* clearly distinguishes between the unmoved mover and the first heaven, the eighth book of the *Physics*, on the contrary, seems to identify god with the first moving sphere and speaks of the first mover as moved by itself (256a 4–21).

BIBLIOGRAPHY

R. Bodei, *Sistema ed epoca in Hegel*, Bologna 1975.

W. Bonsiepen, *Hegels Naturphilosophie im Spannungsfeld zwischen antiker und moderner Physik*, in *Philosophie und Poesie*. O. Pöggeler zum 60. Geburtstag, Stuttgart 1988, 349–66.

K. Brinkmann, *Hegel on the Animal Organism*, in Laval théologique et philosophique (forthcoming).

E.A. Burtt, *The Metaphysical Foundations of Modern Science*, Atlantic Highlands, 1952.[2]

F. Chiereghin, *Finalita' e idea della vita. La recezione hegeliana della teleologia di Kant*, in Verifiche XIX 1-2, 1990, 127–229.

R.S. Cohen and M. Wartofski (eds.), *Hegel and the Sciences*, Dordrecht 1984.

F. De Gandt, *Introduction et notes*, in G.W.F. Hegel, *Les Orbites des planètes. Dissertation de 1801*, Paris 1979.

D. Dubarle, *La critique de la mécanique newtonienne dans la philosophie de Hegel*, in *Hegel. L'esprit objectif. L'unité de l'histoire*, Lille 1970, 113–36.

Id., *La nature chez Hegel et Aristotle* in Archives de Philosophie, XXXVIII, 1975, 3–32.

Id., *Preface* to De Gandt, quoted above, 7–23.

A. Ferrarin, *Hegel interprete di Aristotele*, Pisa 1990.

Id., *Construction and Mathematical Schematism. Kant on the Exhibition of a Concept in Intuition*, in Kant-Studien 86, 1995, 131–74.

V. Giacché, *Finalità e soggettività. Forme del finalismo nella Scienza della logica di Hegel*, Genova 1990.

M.L. Gill and J.G. Lennox (eds.), *Self-motion from Aristotle to Newton*, Princeton 1994.

Hegel und die Naturwissenschaften, hrsg. v. M.J. Petry, Stuttgart 1987.

Hegels Philosophie der Natur, hrsg. von R.P. Horstmann und M.J. Petry, Stuttgart 1986.

R.P. Horstmann, *Wahrheit aus dem Begriff. Eine Einführung in Hegel*, Frankfurt a.M. 1990.

V. Hösle, *Hegels System*, vol. 2, Hamburg 1987

L. Illetterati, *Natura e ragione. Sullo sviluppo dell'idea di natura in Hegel*, Padova 1995.

K.H Ilting, *Il concetto hegeliano di filosofia della natura*, in Giornale di Metafisica, 1982, 297–306.

J. Klein, *Lectures and Essays*, ed. by R.B. Williamson and E. Zuckerman, Annapolis 1985.

A. Koyré, *From the Closed World to the Infinite Universe*, Baltimore 1957.

Id., *Newtonian Studies*, Cambridge, MA, 1965.

M.J. Petry (ed.), *Hegel and Newtonianism*, Dordrecht/Boston/London 1993.

V. Verra, *Letture hegeliane*, Bologna 1992.

D. Wandschneider, *Raum, Zeit, Relativität*, Frankfurt a.M. 1982.

LIVIA KOHN

YIN AND YANG: THE NATURAL DIMENSION OF EVIL

INTRODUCTION

Evil comes into the world after the creation of human beings, who are, in most mythologies, at first pure and form an integral part of the primordial unity of natural life. Then, however, something happens, and human beings set themselves apart, rupture the cosmic and communal coherence, by developing conscious knowledge and acting no longer with but vis-à-vis the universe. At this time good and evil first appear, in the human mind and thus in the world.

The decisive event that separates the primordial from the conscious world is, in mythology, often called "the Fall." The Fall describes the transition of primordial humanity to its present state of strife, the change from timelessness to time and history. It signifies the mutation of an entirely harmonious world into one of good and evil, the development of culture from the simple to the complex, the beginning of disharmony and evil in the world (Weizsäcker, 1964, 53). A gradual emergence of more intricate forms of living, it is felt as the irreparable loss of the pristine state of the golden age and often mythically described as a single traumatic event. Thus it can be

> the Fall of the One into the many, the emergence of the physical universe out of a transcendent God, the Fall of the soul into time, the entrapment of an angelic soul into the body of Australopithecus afrarensis, or the Fall of an unconditioned consciousness beyond subject and object into the syntax of thought pounded into forms by each heartbeat. (Thompson 1981, 9–10)

In whichever way the Fall is described, be it the expulsion from the Garden of Eden, the betrayal of a brother, the battle between light and darkness, the loss of unified consciousness, or the derailing of enlightened government (see Mercatante 1978), it is always about cultural progress, the ongoing transformation of the world into more complex forms, the diversification of society into various segments, and the production of more intricate living conditions (Sproul 1979, 24). All these are seen with a sense of great regret, so that, as Douglas Adams describes it in his *Hitchhiker's Guide to the Galaxy*, people came to feel that "they'd all made a big mistake in coming down from the trees in the first place. And some said that even the trees had been a bad move, and that no one should have left the oceans" (cited in Girardot 1985, 77).

R.S. Cohen and A.I. Tauber (eds.), Philosophies of Nature: The Human Dimension, 91–105.

THE FALL IN ANCIENT TAOISM

In ancient China, the dominant mode of creation is by emergence, the dark, abysmal void of chaos giving rise to the Tao as a seed of order in its midst, and the Tao, the unified potentiality of existence, dividing into the two polar forces yin and yang, which stand for female and male, dark and light, winter and summer respectively. These two, in turn, join again into a harmonious combination, named "yin and yang in harmony," and thus give rise to the things and beings of this world.

The world, therefore, begins with chaos, which is gradually and naturally transformed through the growth of the Tao within it, a kernel of being in an ocean of nothingness that is not itself chaos but still close to it, sharing many of its characteristics. As the *Daode jing* (Book of the Tao and Its Virtue) says,

> There is a being, in chaos yet complete;
> It preceded even heaven and earth.
> Silent it is, it never changes;
> It moves around, yet never ends.
> Consider it the mother of all-under-heaven.
> I do not know its name.
> To call it something, I speak of Tao.
> Naming its strength, I call it great.
> (chap. 25; Chan 1963, 152)

Similarly, the *Huainanzi* (Book of the Master of Huainan), a Taoist compendium of about 145 B.C.E., has,

> Heaven and Earth were still shapeless,
> And all was ascending, flying, diving, delving.
> This was Great Brightness.
> Then the Tao began in the obscure void,
> which first produced the cosmos.
> (chap. 3; See Major 1993, 62)

The shapeless or obscure void in these documents is the womb of the Tao. It is unformed, uncreated, vague, obscure, the egg from which all grows. Yet in this process it is neither sacrificed nor destroyed. It is not evil but the primordial matrix of existence; though empty and dark, it is named "great" and "bright." In the midst of this radiant primordiality then, for no specific reason given, is the Tao – silent, alone, eternal, moving; the mother of the universe to be. The Tao, too, gives rise to the world not by active creation or by sacrifice, but by emergence. Again, the *Daode jing*:

> The Tao brought forth the One.
> The One brought forth the Two.
> The Two brought forth the Three.
> The Three brought forth the myriad beings.
> (chap. 42)

Human beings in this system are described as yet another form the combination of yin and yang takes, an integral part of heaven and earth. As the *Huainanzi* says: "Pure energy rose up and became heaven, turbid energy sank down and became earth, and the conjoined energies in the middle became yin and yang [which formed humanity]" (*Huainanzi* 3; Major 1993, 62). People are thus placed in the middle between heaven and earth as the necessary third force, harmoniously combining yin and yang and utterly in accordance with the natural flow of the world. How, then, did they develop the knowledge of good and evil? How, for that matter, could evil ever enter this perfect world?

The answer, given in the ancient Taoist texts, is that evil came with the emergence of consciousness and with a growth in cultural complexity. The *Zhuangzi* (Writings of Master Zhuang), the second major Taoist classic dated to the third century B.C.E., describes the development of discriminating consciousness in four stages. First there is the primordial state of "Chaos Complete," characterized, as the later commentator Guo Xiang (d. 310 C.E.) says, as "the complete forgetfulness of heaven and earth," a mental state of utter unconsciousness, "boundless and free from fetters, going along with beings and fully according with all." Second, there is the stage of "Beings," which means that entities are being recognized but are not yet discriminated or categorized. This happens in the third stage called "Distinctions," when things are clearly delimitated into "this and that" but not yet critically evaluated. Critical evaluation, fourth, is the last and most declined state, called "Right and Wrong" and signalling the "destruction of the Tao" (Knaul 1985, 25).

Supplementary to this, the ancient Taoist understanding finds the key to the Fall in the establishment of Confucian morality. As the *Daode jing* says:

> When the Great Tao declined,
> There were the ideas of benevolence and social responsibility.
> When knowledge and wisdom appeared,
> Great hypocrisy arose.
> When the six family relationships were no longer in harmony,
> Filial piety and the love for children were stressed.
> (chap. 18).

Along the same lines, the *Zhuangzi* holds the reigns of the mythical sage emperors responsible for the decline, each one establishing further delimitations and "differentiating the hearts of people by the sense of kinship" (chap. 14; Watson 1968, 164). The increasing complexity of consciousness goes hand in hand with a more intricate social structure and more sophisticated levels of culture, creating a socially defined vision of evil (Ricoeur 1968, 52; see also Lai 1984). As Norman Girardot has shown, Taoists here speak essentially about the "sociocultural development when men start to classify themselves into particular lineage groups, into a cosmological system of order that is formally reflected in an honorific language of respect, in systems of potlatching prestation designed to gain the return of social prestige, and in discriminating modes of thought" (1985, 84).

Discriminating consciousness, therefore, is the major step that places human beings above the natural cycles and causes the growing complexity of culture and society. Evil in ancient Taoism develops in the human mind and occurs through an active discarding of purity and the growing disharmony of the Tao on earth. It distorts the inherent unity of the universe and causes an alienated state, despite the fact that human beings in their origins were, and to some extent forever are, part of the greater whole.

CHINESE COSMOLOGY

This rather negative vision is reversed in Chinese cosmology as it develops in the Han dynasty (ab. 200 B.C.E.). In the same way as ancient Taoism, cosmology believes that the universe as it was created originally is good and pure and functions to the best of all. It too sees in the Tao the one power underlying the universe, which makes things be what they are, causes the world to come into being and to decay again. In development of classical Taoism, however, Chinese cosmology also sees original unity and harmony in the developed social and cultural complexity on earth and locates its inherent harmony in seasonal and cosmic rhythms. For this reason, more than the evil which arises from discriminating consciousness and the disharmony of the Tao, Chinese cosmology emphasizes the necessity of evil as the complement of good and understands it as a temporary state within the ongoing natural rhythms.

The Tao in Chinese cosmology can best be described as organic order. It is organic in that it is not willful, it is not a conscious or personal creator. It is nature, yet it is more than mere nature, it is also the essence of nature, the inner quality that makes things what they are; governed by the laws of nature, it is also these very laws itself. Yet the Tao is also order because it is rhythmic and predictable in its natural changes. Manifest in the development of the world, the seasons, the directions, and so on, it is predictable and can be analyzed. Its ordered patterns can be described and actively adapted to, they can be used as a model on which to base human behavior and social organization.

This order, aside from the Tao, is described in various technical terms, the most fundamental of which is *Qi*. Qi, translated as cosmic breath, vital energy, ether, pneuma, life force, and so on, is the stuff from which the world is made, the primordial energy of the Tao that lies both at the root of the world and makes up each and every part of it. Both a material and a spiritual force, Qi connects the visible and the invisible, the living and the dead, the created world and the abysmal Tao. Every single living being, every thing that is created and has form, has its particular so-being because of Qi. Qi is the manifest Tao, and it is in deviating from the patterns of one's inherent Qi, in going against the specific Qi of a place or time, in breaking the Qi of a particular social situation that evil and suffering arise. Only through active readjustment of the Qi in its personal, social, and cosmological dimensions can harmony be recovered and full participation in the purity of the Tao be found.

The major division of Qi in the world is into the forces of yin and yang. Originally developed from the observation of the shapes of mountains and valleys, the terms literally mean the sunny (yang) and the shady (yin) side of a hill. From this, even before the Han dynasty, a set of complex associations had developed: bright and dark, light and heavy, strong and weak, above and below, heaven and earth, ruler and minister, male and female, and so on.

Later the polarity was further formalized by subdividing yin and yang into five distinct evolutionary aspects: lesser yang, greater yang, balance of yin and yang, lesser yin, and greater yin. These in turn were associated with five concrete symbols, which originally designed the five fundamental materials of human existence and that were first used in magic and divination (Fung and Bodde 1952, 1:159). Known as the "five phases," they are wood, fire, earth, metal, and water. Sometimes they are also called the "five elements," because three of them (fire, earth, and water) coincide, at least in name, with the elements of ancient India and Greece. However, this appellation is not correct and has recently been firmly rejected by scholars. "Elements" indicates solid building blocks and implies a philosophy of being, of material solidity, that is entirely alien to the Chinese. "Phases," on the contrary, shows the dynamic nature of the Chinese universe, the inherent fluidity of the cosmic system.

Using these five phases as the basic matrix, all existing objects and phenomena were classified in an intricate system of correspondences: colors, seasons, directions, tastes, sensory organs, mountains, rivers, birds, beasts, and many, many more (Fund and Bodde 1952, 1:160). For example

Phase	Direction	Color	Season	Organ	Emotion	Sense
Wood	East	Green	Spring	Liver	Anger	Eyes
Fire	South	Red	Summer	Heart	Joy	Tongue
Earth	Center	Yellow		Spleen	Worry	Lips
Metal	West	White	Fall	Lungs	Sadness	Nose
Water	North	Black	Winter	Kidneys	Fear	Ears

In addition, there was a certain amount of speculation on the exact interaction between the phases, distinguishing two main modes: a productive (water–wood–fire–earth–metal), according to which one phase actively gave rise to the next; and a destructive (earth–wood–metal–fire–water), where the phases overcome or destroy one another. The thinking in each case remained concrete. Water, for example, appears in the world as rain, which in turn makes the trees (wood) grow. This wood next dries and becomes fuel for fire, which turns to ashes and becomes earth. Earth, finally, gives rise to metals in its depth, which on their part attract clouds and thus contribute to the production of rain water – and so the cycle starts again. The same pattern also holds true for the destructive interaction of the phases, and in all cases a highly concrete level of thinking is retained.

The Chinese cosmos thus emerges naturally from the depth of chaos, represented in its inherent order by the Tao, which carries the world through its material aspect of cosmic energy or Qi, divides into the two polar forces yin and yang, and actively manifests in regular cycles as defined in the interaction of the five evolutive phases.

INHERENT AND CREATED EVIL[1]

Evil in this context is understood on two levels: it is first an inherent part of the natural processes, the necessary yin to every yang; it is also "the result of man's temporary distortion of the universal harmony" (Bodde 1981, 287). Evil is thus a natural part of universal life, because without our understanding of evil there can be no appreciation of good, without the shady side of the mountain there cannot be a sunny one, without suffering the toils of life there can be no enjoyment. The world exists in human consciousness through the continued alternation of complementary opposites, the elimination of one of which also means the end of the other. To be at ease and delight in the world, one has to have a certain amount of hardship and toil, suffering and evil are thus necessary aspects of life as a human being.

On the other hand, evil also occurs in the distortion of primordial harmony, as a disharmony with the inherent goodness and purity of the Tao and its manifestation in worldly rhythms. This kind of evil is not necessary and seen as a hostile force, a powerful infringement on the subtlety of the Tao. As such it is, in the Han dynasty, mythologized as demons, a horde of nasty creatures who lurk at the borders of human society, ready to pounce on unsuspecting creatures.

The double definition of evil is also clearly evident in Chinese medicine at the time. Sickness is, on the one hand, the temporary weakness of the body, the response of the physical system to changes in the outside world. As such it is nothing irregular but part of the natural process, a necessary correlate to health and strength, a phase among others in the continuously changing universe. Then again, sickness is understood as caused by active disharmony with the patterns of nature, which in turn rouses the demons and invites serious disorders (Kohn 1995, 128). In other words, by interfering with the patterns of heaven, one causes the body's defenses to weaken. Once the body is weaker, the demons attack and the person gets sick.

The word for physician that develops at the time bears this out: the graph shows a dancing shaman who is armed with a quiver of arrows and a lance (Unschuld 1985, 37). Both a corrector of universal imbalance and a ritual exorcist, the physician is a defender of the integrity of the body as a microcosm and thus quasi-social system. Acupuncture itself, the most typical of Chinese healing methods, may well have developed from this thinking: the needles are miniature spears used to attack the alien forces that have invaded the body (Unschuld 1985, 45).

The understanding of evil in Chinese cosmology is therefore twofold. There is natural or inherent evil that is part of life and which one has to accept and adjust to. To avoid it, people should behave as much as possible in accordance with the seasons and the rhythms of the world, taking the different phases of the universe and the changes in their being as they come. Then there is created evil, caused by an active distortion of primordial harmony which may well lead to a massive attack by demonic forces. This evil has to be fought with all means, using medical cures, repentence, prayers, exorcism, and others. Both types, moreover, are intimately connected, created evil developing only on the basis of inherent evil, often beginning as a resistance to and displeasure with a difficult phase or state, and manifest in the same concrete phenomena – sickness, poverty, pain. To prevent created evil, one therefore must accept inherent evil as part of life, bear one's lot with patience, and never open the doors to the demons.

SOULS OF YIN AND YANG

A later development of this cosmological vision is found in religious Taoism as it grew in the Chinese middle ages (ab. 200–900 C.E.). Here, both the inherent and the created forms of evil are expressed in terms of spiritual entities, souls of yin and yang, on the one hand, and bodily demons and gods, on the other. First, Taoism acknowledges the basic yin–yang makeup of human beings by describing forces of good and evil that naturally reside in the body as its yin and yang souls. Both types of souls are essential for the existence and survival of the individual as they correspond to heaven and earth in the cosmos as well as to the various complementary opposites in the world. Human beings could not live without one or the other, and yet the forces pull people into different directions: the yang souls toward heaven and purity and goodness; the yin souls toward earth and defilement and evil.

A Tang-dynasty Taoist scripture, dated to the ninth century, makes the point quite clear. The *Chu sanshi jiuchong baosheng jing* (Scripture of How to Preserve Life by Expelling the Three Deathbringers and Nine Worms, DZ 871^2); hereafter abbreviated *Baosheng jing*, says

> The three spirit [yang] souls are located beneath the liver. They look like human beings and all wear green robes with yellow inner garments. Every month on the third, thirteenth, and twenty-third, they leave the body in the evening to go wandering about.
>
> At this time, lie down with your feet up, your head supported by a pillow, and your legs stretched out straight. Fold your hands over your heart, close your eyes, and hold your breath. Clap your teeth three times. (1b)

The names of these three wonderful aspects of human existence, who will take their leave and thus allow the body to be vacant for a period on a regular basis, are "Spiritual Guidance," "Inner Radiance," and "Dark Essence" (1a). They look noble and appear human in shape (see Figure 1A), are dressed in courtly garb, and should be supported in their beneficient activities. Certain physical exercises are helpful, as are meditations and special incantations. When the three spirit souls are treated in this manner to bring out the best in them, they will in their turn take care that "there will be no disaster or affliction [to the person], and all evil demons will be subdued. The body at peace, you will attain the Tao. Then there is no more suffering or pain" (2a; see Ishida 1987, 86).

Their immediate yin counterparts, the material or yin souls, on the contrary, are nasty and fundamentally evil creatures, beastly in shape and quite disastrous in their activities. The *Baosheng jing* describes them:

Figure 1A. The three spirit souls.
Source: *Baosheng jing*

The seven material souls consist of the energy of yin and of evil. They are basically demons. They can make a person commit deadly evils, be stingy and greedy, jealous and full of envy. They give people bad dreams and make them clench their teeth, incite them to say "right" when they think "wrong."

In addition, they cause people to lose their vital essence in sexual passion and get dissipated by hankering after luxury and ease. Through them, people will completely lose all original purity and simplicity. (2a)

These souls, far from looking like human beings, are strangely formed devils, having birds' heads, only one leg, tails, abominable outgrowths, and the like (see Figure 1B). Their names are accordingly Corpse Dog, Arrow in Ambush, Bird Darkness, Devouring Robber, Flying Poison, Massive Pollution, and Stinky Lungs. (3ab)

Where the three yang souls, the representatives of heaven, therefore guide people toward greater goodness, the seven yin souls, present as essential material aspects in the human body from birth and necessary for physical survival, pull them toward evil. The various evils incurred, such as defilement, pollution, sin and depravity, in their course lead to suffering through loss of vital energy, illness and death. Evil and suffering here are inherent and natural, a necessary part of life. They can only be alleviated by circumspect action, by the active subduing of one's inner nastiness and nurturing the call of one's higher souls.

GODS AND WORMS

The same twofold structure that is built into the basic makeup of humanity is also found in its link to the gods and spirits, to the cosmos at large. Here, within the human body, we find the representatives of the supernatural administration of the universe, the gods and demons of the wider cosmos that were sent down to help or harm the individual.

Figure 1B. The seven material souls.
Source: *Baosheng jing*.

On the yang side, there are first the Three Ones, pure deities of the Tao who, upon proper meditation, come to reside in the three cinnabar fields, the major energy centers of the human body. They are the immediate manifestation of pure primordial energy, the original material of the Tao, from which all creation springs. At the root of all there is, they represents the divine and utterly pure ground of being and life as such.

As the *Jinque dijun sanyuan zhenyi jing* (Lord Goldtower's Scripture of the Three Primes and Perfect Ones, DZ 253) describes them,

> The Upper One is the Celestial Emperor;
> he resides in the body center [in the head].
> The Middle One is the Cinnabar Sovereign;
> he resides in the Crimson Palace [in the heart].
> The Lower One is the Primordial King;
> he resides in the Yellow Court [in the abdomen].
>
> Together they supervise the twenty-four energies of the body and bring them in accord with the twenty-four deities of Great Tenuity. Through them, these energies combine in the vapor of emptiness and develop into pure spirit.
>
> The Three Ones each have a thousand chariots and ten thousand horsemen; they ride in cloud chariots with feathery canopies . . .
>
> Concentrate firmly on them, and the Three Ones will become visible for you. Once visible, a thousand chariots and ten thousand horsemen will arrive [to escort you on your heavenly journey]. (1a; Kohn 1993, 204)

These three deities reside originally in heaven and represent the primordial state before creation. Once called down from heaven and activated in the adept's body, they control it entirely and raise it to a celestial plane. They are the powerful good forces of the Tao, divine powers of longevity, good fortune, and immortality that can be called upon and utilized toward perfecting one's inherent heavenly nature. They help in all that is good and holy and assist the practitioner in his ascent to the divine.

Their yin counterparts are correspondingly evil: the three worms or deathbringers are a cross between demons and souls, who reside in the head, torso and lower body of the individual. There, assisted by a group of parasites known as the nine worms, do all they can to incite the person to do evil and fall ill. After the death of the host, when the various souls have been sent off to their respective realms of yin and yang, these "noxious parasites" (Seidel 1990, 302) remain with the corpse and gorge themselves on its blood, bones, and muscles. Having partaken of the human body, they are able to assume its former intact shape and appear as ghosts, feasting further on the offerings laid out for the dead. They have thus a vested interest in bringing a person to death quickly and without mercy. Each of the three has a distinct personality and role. As the *Baosheng jing* describes them (see Figure 2A),

The upper deathbringer is called Peng Ju, also known as Shouter. He sits in the head and attacks the cinnabar field in the Niwan Palace [center of the head]. He causes people's heads to be heavy, their eyesight blurred, their tears cold. He makes mucus assemble in their noses and their ears go deaf. Because of him, people's teeth fall out, their mouths rot, and their faces shrink in wrinkles. He deludes people so that they desire carriages and horses and crave for fancy sounds and sights . . .

Figure 2A. The three deathbringers.
Source: *Baosheng jing*.

> The middle deathbringer is called Peng Zhi, also known as Maker. He enjoys deluding people with the five tastes and makes them greedy for the five colors. He lives in the human heart and stomach and attacks the Scarlet Palace [in the heart] together with its Central Heater. He causes people's minds to be confused and forgetful, so that they are full of troubles, dry in saliva and low in energy . . .
>
> The lower deathbringer is called Peng Qiao, also known as Junior. He lives in people's stomachs and legs and attacks the lower parts of the body. He makes energy leak [through the genitals] from the Ocean of Energy [in the abdomen] and thereby invites a multiplicity of ills. Attracting the robbers of human intention, he makes people hanker after women and sex. (7a–8b)

These divine-demonic agents are further assisted by the nine worms who, in contrast, are purely organic in nature. They are the Twist Worm, the Coiling Worm, the White Worm, the Flesh Worm, the Lung Worm, the Stomach Worm, the Cauldron Worm, the Red Worm, and the Scavenger Worm. Classical descriptions include their size (between one and four inches), the harm they do to which part of the body, and the most obvious countermeasures. The *Baosheng jing* also includes their pictures. They look like ordinary worms or insects, some long and slithering, others round with little leg-like extensions, yet others spongy or crab-like (see Figure 2B). They are physical parasites, bacteria that cause illnesses, and serve as the minions of the three deathbringers, who themselves are the managers of evil and suffering.

Measures against them are, first of all, virtuous behavior in accordance with the moral precepts and the continued meditation on the Three Ones, who ideally will replace the three deathbringers in the major energy centers of the body and make the person immortal. For those who lack such saintly virtue and power of concentration, there are more physical countermeasures, such as a vigil on the night, once every two months, when the three have to ascend to the heavens to make their report to the celestial administration. They can only leave the body when the person is asleep and, without proper directives for destruction from above, they will be helpless and disoriented. After seven successful vigils, they will be so emacitated that they die altogether.

For those who cannot stay awake long enough or lack the moral purity required for this vigil, various drugs and medicines as well as specific temporal and spatial taboos keep the three deathbringers from doing too much harm. While these remedies will not eradicate them and people may still die from their activities, they at least hamper them in their evil ways, and one can breathe a little easier[3].

The Taoist vision of created evil is therefore quite parallel to its understanding of inherent evil. Both good and evil forces in the universe can be called upon and followed by the individual human being. Behaving virtuously and meditating with dedication will cause inherent good forces to develop and good deities to descend, thus gradually diminishing the power of inherent and demonic evil. Being egoistic and greedy or committing acts of disharmony, on the contrary, will open the gates wide for the evil spectres and demons to invade and grown strong. They will delude the person's mind and devour his vital energy, leading him further into evil and causing both physical and spiritual suffering. Once a breach has occurred, rituals, repentance, prayers, and

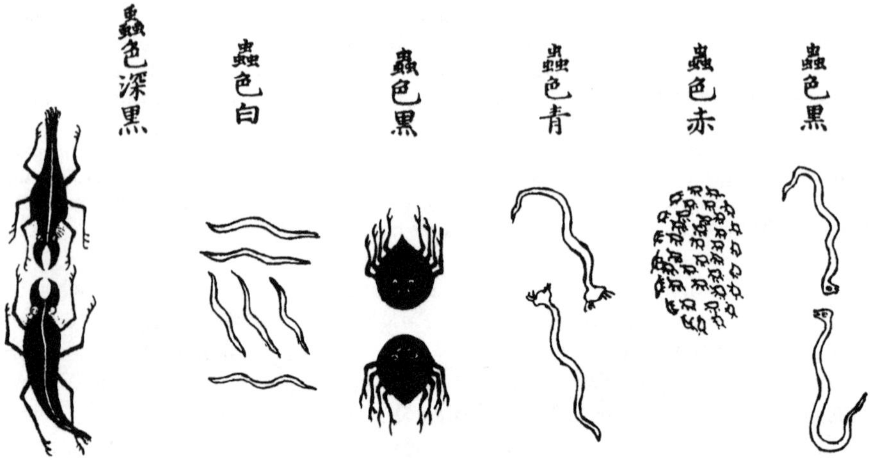

Figure 2B. The nine worms.
Source: *Baosheng jing*.

medicines will help, but ultimately only the full restoration of primordial cosmic harmony and its continued maintainance will guarantee the liberation from all evil.

COSMIC PRECEPTS

In accordance with this vision, religious precepts in medieval Taoism are safeguards not only for moral purity but also against demons and cosmic disaster. The five major precepts, introduced originally from Buddhism – to abstain from killing, stealing, lying, sexual misconduct, and intoxication – in their Taoist reading are thus actively linked with the cosmos at large. As the *Taishang laojun jiejing* (The Precepts of the Highest Venerable Lord, DZ 784) of the sixth century has it,

> The five precepts in heaven are the five planets. As soon as the Tao of heaven loses its precepts, there are natural catastrophes.
>
> The five precepts on earth are the five sacred mountains. As soon as the Tao of earth loses its precepts, the hundred grains can no longer grow.
>
> Among the seasonal patterns, they are the five phases. As soon as the five phases lose their precepts, fire and water fight each other, and metal and wood do each other harm. (11b–12a; Kohn 1994, 203)

Not only generally connected with the five phases, the precepts are also each associated with a particular phase, and through it with a specific direction, bodily organ, and Confucian virtue. Any violation of a given precept, as a result, leads to social conflict and to bodily harm. Thus the text says,

> The precept to abstain from stealing belongs to the north. It embodies the essence of greater yin and presides over the resting and storing of nature. People who steal will receive corresponding calamities in their kidneys . . .
>
> The precept to abstain from lying belongs to the center and to the phase earth; its virtue is faithfulness. People who lie will receive corresponding shame in their spleens. (14b–15a; Kohn 1994, 204–5)

This takes the cosmological vision to a practical, religious level. Each precept is connected with a specific virtue and a particular bodily organ, so that any violation of the precept means a social breach and the immediate harm to the corresponding bodily organ. Evil is again present on two levels – inherent and created. It is inherent in that during certain seasons, some organs are weaker than others and some virtues need to be practiced more assiduously than others. One can only adapt to the natural cycles and do one's best to bear along with them. Yet evil is also created in that it arises through a willful deviation from cosmic harmony, an active breaking of the five Confucian virtues of benevolence, righteousness, ritual formality, wisdom, and faithfulness and/or the key five precepts, originally of Buddhist origin. Such violations of cosmic harmony, unlike the occurrence of inherent phases of evil, immediately reverberate through all levels of the universe, causing cosmic dissonance, social discord, and physical sickness. Created evil, given the choice, is by far the more evil of the two – many human actions are evil, but some are more evil than others.

CONCLUSION

Overall, the Chinese vision of evil reconfirms the basic conviction of an intricate relationship, or even continuity, between humanity and the cosmos. Evil is part of the natural cycle of yin and yang, a phenomenon inherent in life as such, necessary to good as shady is to sunny, down to up, low to high. Without a certain amount of bad fortune and suffering, there would be no appreciation of good luck and benevolence, and life as a whole would be intrinsically boring and, well, it wouldn't really be life any more, would it? Evil on this plane can never be avoided, only alleviated, taken with humor and suffered with patience, but ultimately accepted as one facet within the multiplicity of life as it is.

On the other hand, the Chinese, as first expressed in the writing of classical Taoism, also acknowledge that human beings tend to separate themselves from their cosmic origins and think about and act on, rather than live in, the natural flow of life. They develop a conscious volition, an active will imposed on the world, which constitutes an interference in nature's harmonious cycles – even if done with the best of intentions, maybe even to alleviate inherent evil – and ultimately causes created evil, which is by far the worse of the two. Created evil is not necessary and, unlike inherent evil, will not pass naturally when its phase has reached its peak, will not turn spontaneously around to produce something good in due course. On the contrary, it has the unpleasant characteristic of getting even deeper and darker, of multiplying with amazing speed and intensity, causing an ever more tangled net of disharmony to be woven in and around the world. Created evil cannot and must not be lived with but should be fought at every turn, with all the religious and civil means at society's disposal.

In Taoist mythology, these two aspects of evil are represented in spiritual forces that inhabit the human body. Inherent evil here appears as the material souls of yin quality, seven in number, which are part of the basic makeup of human beings, constituting their active cosmic relation to earth and giving them the tendency toward passions, greed, and sensual excitements. Created evil, on the other hand, is described as the three deathbringers, with their minions, the nine worms, who are heaven-sent demon-parasites that cause people to commit sins and squander their life-force. Having a vested interest in their subject's early demise (as they get to feed on his corpse), they hammer away at him with enthusiasm and make him commit ever more deadly sins. As a result, he gets sucked deeper into the quagmire of evil and dies, forfeiting even his ancestral feeding rights, as the deathbringers, now ghosts in his shape, snatch up his proper offerings.

Still, the two types of evil, however clearly distinguished in the tradition and nicely separated in the analysis, are intimately interrelated. The deathbringers could not do their work if there were not material desires and passionate tendencies inherent in the human being that they could latch on to (Ishida 1987, 84). Both being of yin quality, deathbringers and material souls have a basic nature in common and together make up an important part of human life. As desires, passions, pleasure seeking, and love of excitement are part of the original human constitution, so the violent acquisition of goods and aggressive squandering of life's energies are important sins in the code book of celestial judgement. How could lifespans ever be short, if people did not engage in terrible acts? How could there be a reward of long life and good fortune, if not in the face of created evil?

Reversing the perspective, this also means that one can only eliminate the deathbringers by curbing and controlling the appetites of the material souls, systematically limiting and eventually even destroying the yin aspect of one's being. This, in turn, means that freedom from evil and sin, the attainment of oneness with the Tao or immortality, is an unnatural state that transcends the boundaries of life and can be won only at the great cost of much natural joy. Small wonder, then, that the Chinese have always had some reservations about this goal and continued to give great value to life on earth. Even the Master of the White Stones, a great immortal who, although hundreds of years old, still looked like he was about thirty, responded to the question of why he did not ascend into heaven by saying: "I'm not at all sure I should enjoy myself as much in heaven as I do in this world right here"!

Boston University

NOTES

[1] The following discussion is also contained in my "Disharmonie: Das chinesische Böse." In *Warum gibt es das Leid? Warum gibt es das Böse? Die Antwort von Religionen und Weltanschauungen*, edited by Hermann Kochanek (St.Augustin, 1997).

[2] Texts in the Taoist Canon (*Daozang*, abbreviated DZ) are given according to Schipper 1975.

[3] A detailed study of the three worms, their history and cult both in China and Japan is found in Kohn 1993–95. On their ethical relevance, see Ishida 1987.

REFERENCES

Bodde, Derk. 1981. 'Harmony and Conflict in Chinese Philosophy'. In *Essays on Chinese Civilization*, edited by Charles Le Blanc and Dorothy Borei, 237–96. Princeton: Princeton University Press. Originally published 1953.

Chan, Wing-tsit. 1963. *A Source Book in Chinese Philosophy*. Princeton: Princeton University Press.

Fung Yu-lan and Derk Bodde. 1952. *A History of Chinese Philosophy*. 2 vols. Princeton: Princeton University Press.

Girardot, Norman. 1985. 'Behaving Cosmogonically in Early Taoism.' In *Cosmogony and Ethical Order: New Studies in Comparative Ethics*, edited by Robin W. Lovin and Frank E. Reynolds, 67–97. Chicago: University of Chicago Press.

Ishida Hidemi. 1987. 'Sanshi to shichibaku no ronriteki imi' [The ethical relevance of the three deathbringers and seven material souls.] *Shakai bunka kenkyujo kiyo* 21:73–94.

Knaul, Livia. 1985. 'The Winged Life: Kuo Hsiang's Mystical Philosophy.' *Journal of Chinese Studies* 2.1:17–41.

Kohn, Livia, ed. 1993. *The Taoist Experience: An Anthology*. Albany: State University of New York Press.

Kohn, Livia. 1993–95. Kōshin: A Taoist Cult in Japan.' Parts 1–3. *Japanese Religions* 18.2:113–39, 20.1:34–55, and 20.2:123–42.

Kohn, Livia. 1994. 'The Five Precepts of the Venerable Lord.' *Monumenta Serica* 42:171–215.

Kohn, Livia. 1995. 'Zur Symbolik des Bösen im alten China.' In *Verschiedenheit und Einheit: Studien zum Geist in China und im Abendland: Festschrift für Rolf Trauzettel*, edited by Ingrid Krüssmann, Hans-Georg Müller, Wolfgang Kubin (St. Augustin: Academia Verlag), 113–33.

Lai, Whalen. 1984. 'Symbolism of Evil in China: The K'ung-chia Myth Analyzed.' *History of Religions* 23:316–43.

Major, John S. 1993. *Heaven and Earth in Early Han Thought: Chapters Three, Four, and Five of the Huainanzi*. Albany: State University of New York Press.

Mercatante, Anthony S. 1978. *Good and Evil: Mythology and Folklore*. New York: Harper and Row.

Ricoeur, Paul. 1967. *The Symbolism of Evil*. Boston: Beacon Press.

Schipper, Kristofer. 1975. *Concordance du Tao Tsang: Titres des ouvrages*. Paris: Publications de l'École Francaise d'Extrème-Orient.

Seidel, Anna. 1990a. 'Chronicle of Taoist Studies in the West, 1950–1990.' *Cahiers d'Extrême-Asie* 5:223–347.

Sproul, Barbara C. 1979. *Primal Myths: Creating the World*. San Francisco: Harper and Row.

Thompson, William Irvin. 1981. *The Time Falling Bodies Take to Light: Mythology, Sexuality, and the Origins of Culture*. New York: St. Martin's Press.

Unschuld, Paul. 1985. *Medicine in China: A History of Ideas*. Berkeley: University of California Press.

Watson, Burton. 1968. *The Complete Works of Chuang-tzu*. New York: Columbia University Press.

Weizsäcker, C.F. von. 1964. *The Relevance of Science: Creation and Cosmogony*. London: Collins.

KRZYSZTOF MICHALSKI

HUMAN NATURE AND THE NATURE OF TIME: A NIETZSCHEAN METAPHOR AND ITS CONSEQUENCES

In the first pages of his essay about the advantages and disadvantages of the use of history, Nietzsche introduces the reader into his conceptual world with a group of metaphors, most of which will appear in his later works as well. I will leave open the question whether these metaphors are simply illustrative, helping us to better understand concepts, or whether they not only color concepts, as it were, but give them their ultimate meaning, making them finished conceptions, convincing arguments. In the following pages I will try to demonstrate what Nietzsche could have had in mind when, in the second of his *Untimely Meditations*, he writes about a playing child, grazing cows, an enshrouding cloud, and the lightening flash of light that pierces this cloud.

I have not chosen these metaphors by accident. As you will see in a moment, I believe that these metaphors are an expression of two concepts that are essential to, and perhaps even the most important for Nietzsche's philosophy: the concept of the human condition and the concept of time, as well as their mutual relation to one another. What does the human condition entail? What is time? Or, why is it that the past differs from the future? In what way are these two questions related? Finally, why are these questions important, what are the consequences of providing a particular answer for these questions? Nietzsche, I believe, had these problems in mind when he wrote of children and cattle, clouds and lightening flashes.

"Consider the cattle, grazing as they pass you by: they do not know what is meant by yesterday or today, they leap about, eat, rest [. . .] fettered to the moment and its pleasures or displeasures." (60) Where there is no difference between yesterday and today, between the past and present, there is also no place for memory; these is nothing to be remembered. Cattle do not remember anything. Not because they forgot something that happened to them recently or long ago, but because everything that can happen to them "occurs" in the moment: for the grazing cattle there is no such "ago," recent or long. The "moment" in which cattle live is not, therefore, the "present moment": it is not a present placed between a what was and a what will be, a moment that is preceded by something and after which something will follow.

R.S. Cohen and A.I. Tauber (eds.), Philosophies of Nature: The Human Dimension, 107–119.

For the cow, time as the succession of past, present and future does not exist. And since the time that is a succession of past, present and future is the precondition for memory, the cow does not remember anything.

That makes her happy. For the memory of things past is a "burden," a "chain," a "ghost." Memory is a disturbance of the living moment, a calling back of things that are gone, which infuses it with melancholy, satiety, suffering and conflict. As a consequence, memory does not allow for the carefree abandon in the moment. The cow lives exclusively in the now; she does not bow, as we do, beneath the burden of an "invisible," "dark" yesterday. She is not driven, as we are, to pull on chains that bind it to what was; and she is not, as we are, haunted by the ghosts of the past.

This is why we look at grazing cattle with envy.

But maybe cattle are happy simply because they are stupid (which seems to be a natural assumption). Is their happiness perhaps merely the flip side of their idiocy – the simple undisturbed calm of the moment in which they live, the simplicity of an imbecile?

No doubt this would be so if we were to assume that the perspective of the observer of the grazing cattle corresponds to objective reality, while that of the cattle does not. Or, in other words, if we were to assume that the observer sees the world as it is, while the cattle see a subjectively simplified, distorted image of the world; if we were to assume that temporal succession, the difference between the past, present and future, exists objectively, and that only cows, struck stupid by their forgetfulness, do not see it, and are thus happy.

Is this the case? Is the perspective enjoyed by the onlooker, the one who looks at the cattle, really that of "objective knowledge," while the perspective of the cow is disfigured, misformed?

Let us take a closer look at this "perspective of cattle." First of all, Nietzsche says, not only are cattle unaware of the distinction between yesterday and today, but nor are children. The child, too, plays "in blissful blindness between the hedges of past and future," totally immersed in the moment. When we look at a child playing we are not only envious, as we are in the case of the grazing cattle, but we are moved as well; for in the playing child we have a vision of a lost paradise. For those who look on, the adults, the happiness of the playing child is not something foreign and unreachable, as is the happiness of grazing cattle. Rather, this happiness is something near to them, something that was once theirs, but then one day was lost – maybe necessarily, though for sure irreversibly.

Later on, in *Thus Spoke Zarathustra*, Nietzsche will write the following about the playing child: "The child is innocence and forgetting, a new beginning, a game, a self-propelled wheel, a first movement, a sacred yes." The child does not know what the past is, and can therefore know nothing of guilt. Not because it never committed a crime, but because the concept of guilt has no meaning for somebody who has no past. This is why when children play every move is the first. Each is both beginning and the end, at the same time, as is each and every point on a wheel. The consequence is that the child's playing is at every moment complete and perfect: it lacks nothing, has no goal, strives towards nothing. It is like a turning wheel pushed by no one, going nowhere. The movement of such a wheel is intelligible only in itself: it cannot be understood with reference to anything else. That is why the child playing is the "affirmation," the "sacred yes." This does not simply mean that the child likes to play; the affirmation spoken of here is not an evaluation that would require a distance from its objects, as is the case, for example, when a child says that it likes to play hide and seek. The child "affirms" the playing; the playing itself is the affirmation meant here. Each of its moves is complete, perfect, self-sufficient.

This carefree playing of the child cannot last forever, says Nietzsche on the next page (I am returning to the *Untimely Meditations*). "Yet its play must be disturbed; all too soon it will be called out of its stage of forgetfulness. Then it will learn to understand the phrase "it was": that password which gives conflict, suffering and satiety access to men so as to remind him what his existence fundamentally is – an imperfect tense that can never become a perfect one." Imagine, if you will, a mother calling a child who is playing in the backyard in for lunch. For the child, absorbed in playing, it is always "too soon." But it realizes this only after the interruption of its play by the mother's call. Only then, only when the play is interrupted, do temporal determinations mean anything. Only them do "too soon" and "later," "it was," and "now" mean something. As long as the play lasts, these words are empty, meaningless.

But sooner or later, Nietzsche writes, the playing "must be disturbed." Thus the disturbance of play is not fortuitous, an accident. It is not a mistake that could be corrected. Play itself, out of necessity, leads to this disturbance, just as one necessarily ages, or fruit necessarily ripens. Without this disturbance, playing itself cannot be understood; therefore, according to Nietzsche, the distinction between "it was" and "it will be," the temporal difference, can be understood as the disruption of an originally undifferentiated moment. It is in this sense that the "moment" can be called the "origin" or "foundation" of time. Without such a moment, "time" – the difference between past, present, and future – is unintelligible.

Only here, in step with this disturbance, does human life begin. The temporal differentiation is the condition of the possibility of memory and, therefore, also of conflict, suffering, and satiety, all of which neither the child nor the cow experience, for they are without memory, they know nothing of the difference between "it was" and it "will be." Only when we understand "it was" are we able to fight and suffer, and only somebody capable of conflict and suffering can be called a human being. Here, the human condition comes to the fore, the unfinished temporal character of our existence ("an imperfect tense, that can never become a perfect one"). The loss of paradise, of the carefree, happy playing of the child, of the immersion in the moment of living, is the condition of possibility, the origin of a human life full of suffering, satiety, melancholy and conflict.

If so, then "conflict, suffering and satiety" will never cease, at least as long as we live. "Eternal life" (a life beyond time, not a life which never ends) is an oxymoron. Human life always has a past, thus also a future – the undisturbed happiness of the playing child is beyond its reach.

What does it mean to say that the atemporal "moment," the moment in which grazing cattle and children live, is the "origin" or the foundation of a "human life" stretched between past and future?

Let us return to the metaphor of the playing child. In what sense is the happiness of the child, immersed in the moment of living, out of my reach? Is it out of my reach in the same way that the grazing of cattle, or the lightness of a feather is beyond my grasp? No, that is not what is meant here. As mentioned above, when I see a playing child I am moved. What I see is somehow familiar to me: it is a paradise that I was once in, yet which I lost.

Does that, therefore, mean that the playing of the child is for the me of today a time that has "gone by," one that I remember with nostalgia and affection, as I would the happiness of a former childhood? No. That is also not what is meant here. The temporally undifferentiated moment of playing is not one among many "moments" in time, of a collection of yesterdays, tomorrows, laters, and so on. To contend that the "moment" precedes the succession of time would be meaningless. The comparisons above of the moment with the "origin" of time, of the happiness of the play of the child with a lost paradise, of the disturbance of the atemporal moment with the mother's summons, all could have suggested that the moment precedes temporal succession. This would be nonsense. The happy playing of the child immersed in the moment of living and human life, full of suffering and conflict, are not events, one succeeding the other in time.

If so, then nor could the relation between the "moment" in which the playing child is absorbed and life stretched between past and future be causal. This has nothing to do with nostalgia, with having lost, by growing up, the simple pleasures of childhood. It also has nothing to do with some earlier, cow-like phase of evolution, the traces of which one can discern in later stages.

The tie that is meant here is of a different kind. The "moment" does not pass. It could pass only if it were a fragment of time, a "now" or a "once." Yet the moment is temporally undifferentiated. Just as, for the Christian believer, paradise and its loss is present in every moment of life, giving to this life its only possible meaning, so too is the "moment" present in every moment of time. Without this presence, without this inexhaustible source, the future would not succeed the past, and thus human life would not be possible. The temporally undifferentiated moment, says Nietzsche, is like an atmosphere within which alone life can germinate, and with the destruction of which it must vanish. (64) The child is in this sense "unhistorical," and to the one looking, the one burdened with the past and memory and who is in this sense "historical," the child is an image of a different, non-human, unreachable reality. But what I see is not different and unreachable in the same way as the alienness of another being such as a Martian, an ant, or a feather. Rather, this view, what I see, is a metaphor of an irreducible dimension of my own life, a side of my life without which my specific human character would be lost – namely, that I remember the past, and can thus suffer and fight.

Nietzsche also formulates this thought in another way: memory, he says, is possible only if it is limited; an absolute memory is an absurdity. If we remembered everything we would seal off the only source from which memory flows, we would remove the only condition of its possibility. "Imagine the extremest possible example of a man who did not possess the power of forgetting at all and who was thus condemned to see everywhere a state of becoming: such a man would no longer believe in his own being, would no longer believe in himself, would see everything flowing asunder in moving points and would lose himself in this stream of becoming." (62) An absolute or total recollection would be an exact repetition of what was; or, more precisely, it would be identical to what was. The life of a human being who remembered everything would be the same as that which is remembered. In such a life there would be no place either for the now or for what will be. Absolute memory would have no room for someone "today" who remembers what was "yesterday"; therefore, such a memory would undermine in a single blow both the possibility of life and of itself. Therefore, even though a human being can be distinguished from a child or cattle with reference to the ability to remember, which means, ultimately, the ability to distinguish the past from the future, the contrary ability is equally indispensible: the faculty of blocking memory, or preventing the emergence of the past. Without the latter the former would be unthinkable. Human being needs the capacity to "feel unhistorical," and in this sense to "forget." I do not mean "forgetting" in the usual sense: what is meant here is not that events or names we once remembered escape our recollection against our will; what is meant, rather, is the ability to experience the world as temporally undifferentiated, a sensibility for which the term "it was" has no meaning. What is meant is the sensibility that is expressed by the metaphors of the playing child and the grazing cattle. This obliviousness to the past, which life requires just as much as memory, is the return to the lost paradise of the playing child.

Notice that this obliviousness is spoken of here as an "ability" or "capacity," that is, as a force, not the lack of force (as in the "weakness of memory"). It is something we do, not something done to us; it is an activity, not a passivity; a positive, not a negative determination. This obliviousness, Nietzsche will later write in the *Genealogy of Morals*, "is no mere vis inertia . . .; it is rather an active . . . faculty of repression." (GM57)

But what relation does memory have to this obliviousness? How does the ability to distinguish the past from the future combine with the ability to blur this difference? What is the connection at work between the atemporal moment and temporal succession, where should we look for the principle of its unity, the measures or degrees of both that make up its composite? "To determine this degree, and therewith the boundary at which the past has to be forgotten if it is not to become the gravedigger of the present, one would have to know exactly how great the plastic power of man, a people, a culture is: I mean by the plastic power the capacity to develop out of oneself in one's own way, to transform and incorporate into oneself what is past and foreign, to heal wounds, to replace what has been lost, to recreate broken moulds." (62) This means that the synthesis of memory and obliviousness, of time and the undifferentiated moment, takes place in the life of each human being (or nation, or culture), each time anew. A particular life, with its particular intensity, its particular "force," decides how much we will remember and how much we will forget.

Let us look at the passage from *Untimely Meditations*, the beginning of which I already quoted. "The unhistorical is like an atmosphere within which alone life can germinate and with the destruction of which it must vanish. It is true that only by imposing limits on this unhistorical element, by thinking, reflecting, comparing, distinguishing, drawing conclusions, only through the appearance within that encompassing cloud, of a vivid flash of light – thus only through the power of employing the past for the purposes of life, and of again introducing into history that which has been done and is gone – did man become man: (...) but without the envelope of the unhistorical, man would never have begun." (63-64)

The temporally undifferentiated moment, this "time" of the playing child that does not know the difference between past and future, is the condition of possibility of human life, the atmosphere which surrounds it. But human life begins only with the disturbance of the playing child when the undifferentiated moment is "limited." This does not mean, of course, that the beginning of human life walls in this atmosphere, so that it would only reach a certain point and no further; rather, it "limits" in the same sense that a wave limits the ocean, or a geometrical figure space. It is an internal differentiation. Human being becomes human being when the undifferentiated "moment" differentiates into the past, present, and future.

This differentiation is a "flash of light": only now, with the emergence of difference, is there something that can be as a possible object of knowledge. In the undifferentiated atmosphere, all cows are black. Only now, to be sure, in the aftermath of this flash of light, can the atmosphere which protects human life be called a cloud; only now does it become clear that this atmosphere is not transparent, that nothing can be seen in it. Where there is no light, nothing is as yet dark, nothing is as yet either transparent or opaque. The playing child is unaware, as well, that it does not remember, that it lacks this kind of knowledge.

So understood, the flash of light cannot be gradual. The differentiation is either there or not; even the smallest of differences is a difference. This differentiation is like a first word, or the birth of a child; the world suddenly emerges anew, in one stroke, one indivisible moment. This is why Nietzsche says that the flash of light is a "lightening flash."

How does this differentiation take place? How does the "past" emerge from the cloud of the undifferentiated moment? It takes place only when the past is employed for the purposes of life, when what "is done and over with" becomes "history" again. So it is human life which, thanks to its limited ability to remember, leads the "past" out of the darkness of the undifferentiated "moment." Only the perspective of human life, of human, limited memory, makes room for what is now and what will be – a place for the "future" and the "present." Only in this perspective, to put it in different words, does the difference between past, present, and future have meaning; only here does history "live," "happen" or, to say the same thing, have a future. Otherwise, history dies, ceases to be history.

The past is something I remember, or should remember. The "chains" which limit my possibilities; the "burden" of obligations I have accepted; the "ghosts" of my mistakes and transgressions – the past understood in this way is a part of a living history, the other side of which is the future: the demand of obligations, the promise of possibilities, the inexorability and release of retribution. In this sense, past and future are dimensions of human life, of this "imperfect tense that can never become a perfect one."

Living – which means "thinking, reflecting, comparing distinguishing, drawing conclusions" – opens up a future and, simultaneously, a past. It is human life that is the lightening flash of light that draws what was and what will be out of the darkness. Thus it is human life that introduces the difference between past and future into the undifferentiated, unhistorical "moment," and which in this way limits it, differentiates it into moments such as "yesterday," "today," or "tomorrow." Time does not flow out of itself; it is human life which fashions the difference between past and future. In other words, the past differs from the future in what we do; it is only in our actions that this difference makes sense. The concept of a time without reference to human life is empty, and, on the other hand, human life is essentially this process of distinguishing the past and the future; it is, in its very essence, time. The force of obliviousness and memory that distinguishes the past from the future is the same force that constitutes human life.

Human life, in its essence, Nietzsche contends, is the force that opens the future and the past.

Therefore, this is what I see when I look at the grazing cattle: creatures who know nothing of the difference between "it was" and "it will be," who are, in a manner, comparable to the playing of children, totally immersed in the moment of living (which is not, in this perspective, a "now" crammed between the past and the future), and are thus without memory. And happy. Seeing all this, I become aware who it is that I, the onlooker, am: a creature that goes beyond the now into the future and the past, a creature weighed down with the memory of what was, and who is consequently no longer happy but suffering, fighting, overflowing. Thus what I see is not foreign to me; when I look at the grazing cattle and imagine, in conjunction with what I see, the playing of a child, I am both envious and moved. I hope this interpretation has made clear why this is so: for Nietzsche, this looking on as the cow grazes or the child plays is a metaphor of the fundamental dimension of my human existence. The past and memory, consequently the suffering, conflict and satiety which constitute my humanity, would not be possible without this other kind of sensibility, one that blurs the difference between "it was" and "it will be." In this way the "unhistorical" sensibility is the origin or source of time as the succession of it was, is and will be, and is thus at the same time the "foundation" of what is human.

Time is not, therefore, a "stream" into or out of which I can step at any time, or a "river" which flows regardless of what happens to me. It is a human being, the force of life, that tears away the past from the future, that constitutes the past, present, and future by ripping apart the undifferentiated moment. This force of human life is the ability to remember, thus to suffer and fight – at the same time, it is the ability to forget, the capacity for obliviousness.

If so, then of course it cannot be the case that the scene from which we started, the cattle happily immersed in the lived now, is a metaphor for stupidity. It cannot mean that the point of view of the cow differs from the perspective of the observer as a mistake or an error differs from the truth, or in the same way that a subjective perspective is different from an objective point of view. Temporal succession is not the "objective reality" that we grown-ups recognize but which immature, intellectually undeveloped children and dull witted cows do not. On the other hand, the fact that I long for the simple happiness of childhood does not entail that, when I was a child, I saw the world as it is, maybe even better, more fully than I do as an adult (nor does it follow that the cow is more clever in this respect). The happiness of a child or a cow and the suffering of an adult, the playing child and the conflicts of maturity, the blissful forgetfulness and the ghosts, burdens and chains of memory are two sides of the same, two dimensions of the conditio humana.

Let us look at some consequences of the above. The opposition between "subjective" and "objective" does not help to understand the Nietzschean metaphor interpreted above, the metaphor of a man looking of the grazing cattle and the playing child with envy and affection. Rather the opposite: this metaphor, properly understood, helps us understand the meaning of this opposition, and in a way that we are accustomed to. If the past is only real in its difference with the future, a difference introduced into the world by human life, or if, to put it another way, the "past," as well as the terms "present" and "future," does not mean anything outside of its tie to human life, then the knowledge of the past must seek out such a tie, not try to eliminate it. But, according to Nietzsche, it is exactly such an elimination of any ties with my own life, which is called "objectivity," that is the ideal of knowledge in the age in which we live, an ideal with which we have been inoculated by science, in particular the science called history.

"Objectivity," says Nietzsche, "means a condition in the historian which permits him to observe an event in all its motivations and consequences so purely that is has no effect at all on his own subjectivity." (91) In order to reach such a condition one has to eliminate all emotion, feeling, personal interest with respect to what is being described, so that one becomes totally indifferent to the phenomenon. The historian has to forget himself, has to shed everything that is "subjective"; he must, to use a phrase of Michel Foucault, "mimic death" in order to enter the kingdom of the dead.

In other words, the historian, in trying to realize the ideal of objectivity, understands history only insofar as it is dead. "A historical phenomenon, known clearly and completely, and resolved into a phenomenon of knowledge, is for him who has perceived it, dead: for he has recognized in it the delusion, the injustice, the blind passion, and in general the whole earthly and darkening horizon of this phenomena, and has thereby also understood its power in history. This power has now lost its hold over him, insofar as he is a man of knowledge." Delusion, injustice, and blind passion constitute the historical force of a given phenomenon only insofar as they affect somebody who wants to understand them, only insofar as they matter, or have an effect on somebody. Made into an object of knowledge, described "objectively," they have no such effect; they are, along with the subject who wants to understand them, excluded from what history is: a nexus of effects, to use a phrase of Hans-Georg Gadamer.

I understand the past only insofar as I am in touch with it, only if I am able to include it in the horizon of my life, only if it has an effect on me: that is why the past cannot become exclusively the object of knowledge, and why the knowledge of the past cannot become exclusively "objective."

From this point of view, the question of the objective truth or falsehood of our knowledge of the past loses its dramatic character. This does not mean that it becomes unimportant; but, for sure, it ceases to be the most important question. If "truth" is a "cold, ineffectual knowledge," if it is a character of knowledge and requires "objectivity," the isolation from the coherence of the course of things and human actions, then the question of truth must be amended by an additional consideration of consequences, an evaluation of significance, something that objective knowledge cannot provide. Or, in other words: the question whether the truthfulness of something is important for us or not cannot be answered by objective knowledge alone; we must look somewhere else.

That, of course, does not mean that the concept of "objective truth" is totally useless, or that "truthfulness" of this type is not, under certain conditions, often very important. It means only that if the precondition of the search for truth, of objective knowledge, is my own life, then only with this life, with an evaluation in terms of it, can the importance and thus the meaning of such a knowledge be disclosed. Science, in particular historical science striving to achieve objectivity, requires "supervision and control (Aufsicht und Ueberwachung), a hygiene of life which belongs close beside science." (121)

"Objective" truth acquires a meaning only when inscribed in a life: only then is it a truth that one can enjoy or die for. Torn from this context, truth is ineffectual, a matter of indifference – a human being who succeeds in tearing truth away from this context loses, as a result, its own harmony. "There are very many truths that are a matter of complete indifference: there are problems whose just solution do not even demand an effort, let alone a sacrifice. In this region of indifference and absence of danger man may well succeed in becoming a cold demon of knowledge." (89)

In other words, if the past, placed in relation to the present and the future, is not an endless road which humans once took and will someday abandon, if the past opens up anew in each action of every individual human being, if the past (time) cannot be separated from human life, then an "objective" knowledge about the past is a square circle. Or, rather: knowledge about the past is an abstract knowledge, removed from the context of life, and is thus a derivative knowledge (which does not mean, as I said, useless, that is, cannot be useful under certain conditions). "Life" and "time" are not independent from one another, and to think of their relation as one of subjectivity and objectivity is an illusion. "For how," to quote Nietzsche, "are human actions and the course of things to be distinguished from one another"? (92)

Nietzsche shares the above conviction with authors so different from each other and from him as Karl Marx and Edmund Husserl. Marx believed, as well, that "objective truth" acquires meaning only in the context of human life (which, according to him, consists of an incessant, collective transformation of the world we encounter). Thus, he argued, to search for this meaning we have to use tools other than those employed by a knowledge striving for objectivity. The Husserl of the *Cartesian Meditations* or the *Crisis* also believed that understanding the claim to objective truth (which is constitutive for modern science) requires the disclosure of the motives which lead to it. Therefore, what is needed for such an understanding is a genealogical analysis, something that science, in particular history, cannot provide.

However, contrary to Marx and Husserl, Nietzsche believes that this understanding takes place in each individual life; thus any search for a point of view that goes beyond individual life is necessarily futile. There exists no "subject," no "consciousness" the meaning of which would not be produced each time anew in the life of an individual: there can be neither a universal essence of human species hidden in class consciousness, as Marx would have it, nor Husserl's pure transcendental consciousness that one can discern in various different acts of understanding by phenomenological analysis. Consciousness is a function of my life; it has no sense apart from this life. No consciousness is possible that has an identity independent of somebody's own concrete life. Who I am, and what the world is, is ultimately being decided in what I do and what I neglect; only here is the meaning of the past and of the search for its truth determined.

Life, says Nietzsche, is a "dark, driving power that insatiably thirsts for itself." (76) Life is "dark" because, as the condition of all knowledge, it cannot be fixed as an object of knowledge. Life is a "driving power" because it is not a static structure, a system to which we could refer in order to understand. Rather, life can be understood only through living; it is only in what I do and do not do that life is a reality. As such, life is never complete, never totally determined. And, finally, life "insatiably thirsts for itself" because it is not intelligible from the outside, because every possible external perspective, every order transcendent to life, is an illusion: life does not strive to achieve "something else." Its dynamic is not directed towards a goal, something it lacks and would like to have – there is no success that brings it to rest, to completion.

In other words, it is life which sets up the confines of our understanding. The condition for our understanding of anything is its assimilation into these confines of our living. In Nietzsche's language, it is a question of effect, of force, not argument. My actions, the force or intensity of my life, is what first opens the space for argument: it is the element in which arguments can become significant and convincing.

If this is so, then the hope (shared by Marx and Husserl) of going beyond the horizon of my life, beyond this integration of something into it, is a hopeless illusion. Universal reason, which could justify such a hope, is a fiction. The claim to truth does not build the common space in which the understanding of something different is possible. It is not understanding, the discovery of common ground, that allows me to assimilate a different, foreign-life – on the contrary, this assimilation of something foreign and different is what first allows for its understanding. The assimilation of something different is, therefore, a matter of domination or submission, a question of power and not of arguments. Domination or submission condition and, as a consequence, limit the understanding of something that has been, up to now, foreign, alien. Not every difference that confronts us – another human being, different habits, sudden emotion – is a challenge to discover new, common ground between the other and myself, to revise the understanding I have had of the common up to now. The confrontation between what has up to now been familiar and the new and foreign, the confrontation that forms the horizon in which the lives of each of us takes place, which defines the possibilities of our understanding, is an existential, not an intellectual confrontation. With these thoughts, Nietzsche places himself in the tradition of conservative critics of liberalism, from Joseph de Maistre to Carl Schmitt.

Do Nietzsche's arguments, given above, not render arguments that make a claim to objective truth relative, do they not lead to the worship of force and of de facto success? Not at all, Nietzsche says. (105) "He who has once learned to bend his back and bow his head before the "power of history" at last nods "yes" like a Chinese mechanical doll to every power, whether it is a government or public opinion or a numerical majority, and moves his limbs to the precise rhythm at which any "power" whatever pulls the strings . . ." 106: "In this way you become Devils advocates: you make success, the factual, into your idol, while in reality the factual is always stupid and has at all times resembled a calf rather than a god."

"Fact" is an abstract concept. "Facticity," "past," "what happened" are all, as we already know, what they are only in the perspective of somebody's life, a life that decides their meaning. And this meaning is always being decided over and over again: life is, as we know, necessarily open to the future, it "becomes" rather than "is." Its other dimension, its facticity and pastness can never be completed, finished, closed. If so, then history is not a natural process the necessity of which could be determined once and for all. Each life is a chance for a new beginning: each life can begin history anew by giving it a new sense, by again defining its internal rules, its criteria for success.

If human life constitutes time, if it is only here that the past differs from the future, then history cannot be an "objective" process independent from human life. It cannot be a process that follows "iron laws" to which human beings must adapt, a process that determines in advance what "success" will be.

This Nietzschean criticism of the "iron laws of history" will acquire a new and unexpected importance for the anti-totalitarian thinking of the late Soviet Empire – in particular, in the writings of Erazim Kohák's philosophical friend and my teacher, Jan Patočka.

I am not indifferent either to the "past" or to "facts," whether I am aware of this or not; they appear at all only in the horizon of what is important for me, only in the perspective of a possible usefulness. Moreover, one cannot determine in advance, once and for all, what is important and useful. If it is only my life that can answer the question who I am, if only my life decides what it means to be a Pole, a European, a human being, if there is no subject whose meaning could be determined independently from my own life, then, of course, it does not make sense to speak about the "universal" needs of the human species, or about what is important and what is not for human beings "in general." To live means to reset priorities, to fix what is important and what is not. The "facts" are what they are only in this perspective; every "fact" is for me a "burden," a "ghost" or a "chain." Each effects me in some manner, although I would probably never completely know how and why.

If so, is not the worship of facts a stupidity? And, consequently, could one image a more untenable point of view than relativism? According to Nietzsche, it is the modern ideal of objectivity that leads to relativism, not his belief that objective knowledge presupposes life. For the perspective of an objective, universal reason puts the differences between what is important for my life and what is perhaps important for the life of somebody else on the same level. "Every age is different" – is, says Nietzsche, the message of scientific objectivism – "it does not matter what you are like." (89)

"Force," which enables somebody's life to assimilate something different or foreign; "domination," which first opens up the space for arguments, are not blind, factual, or natural. They always already have a meaning, always already mean something for me. Since the relevant force is the force of a life, it does not make sense to speak of this force independently of living. One cannot measure this force from the outside, as one can measure the capacity an animal or a human being has for physical force. Because my life is always essentially open to the future; because, as we have seen, my life in each of its actions tears away the past from the future, each time in a different way: the sense or the meaning of this force of my life can always be subject to a radical revision. That is why each new life is a chance for a new beginning, a chance for a new answer to the question of who we are, what the world in which we live is.

Let us summarize: if my interpretation of the metaphor in the *Untimely Meditations* of a child's play and the man who looks on with envy and affection is correct, what Nietzsche wanted to do with this metaphor is to show human life as it essentially is: concentrated in the lived moment and, because of this, simultaneously tearing the past from the future. Such a concept of the human condition has far-reaching consequences. If the "past" is a dimension of human life, then it cannot merely be an "object" grasped by pulling it out of its entanglement in subjective life. On the contrary, Nietzsche claims, such an "objectivization" would deny us the only possibility of understanding things past: the "past" can be understood only insofar as it has an effect on what happens now, on life. Not only the past but objective truth in general is intelligible, says Nietzsche, only if it is assimilated into somebody's life, only if it is somebody's truth, only if somebody takes responsibility for it. Furthermore, there is no authority that could relieve me of this responsibility: notions such as "transcendental consciousness," "essence of human species," are for Nietzsche dangerous illusions. I cannot appeal to such authorities to help me understand something foreign, new, different from me. I can rely only on the ability of my life to assimilate, only on its "force."

One can also formulate Nietzsche's idea in the following way: There are differences which cannot be bridged through mutual understanding, differences which only domination and submission can overcome. If so, then the concepts and metaphors that organize my thoughts, my understanding of the world, harbor strategies for conflict, various tactics of confrontation, plans for the uses of force. They should be treated with caution and mistrust, for they do not always mean what they claim to mean. But, on the other hand, this "force" of my life is not blind, it always has a meaning, it always means something for me, and this meaning, or even the question of what counts as a force and what does not, cannot be determined once and for all. Each life, opening new possibilities of meaning and significance, is a new opportunity to answer this question.

Mistrust of concepts, because they conceal mechanisms of domination, and the courage to live on my own responsibility, the courage for a new beginning, is the message that the author of the *Untimely Meditations* wants to convey. This is what he wanted us to learn from the metaphor of the grazing cattle and the playing child.

Institute for Human Sciences, Vienna and Boston University

BIBLIOGRAPHY

Friedrich Nietzsche: 'Untimely Meditations', translated by RJ Hollingdale, Carmbridge University Press, 1983.

Friedrich Nietzsche: 'On the Genealogy of Monet', translated by Walter Kaufman and RJ Hollingdale, Vintage Books, New York, 1967.

ROBERT CUMMINGS NEVILLE

THE CONTINGENCY OF NATURE

I

The thesis of this essay is that there is a primordial apprehension of the contingency of nature in civilized experience.[1] The expressions of this apprehension of contingency have taken many forms, from the mythic to the philosophical and theological. The expressions have different motifs in different cultures, and there is no preferred cultural starting point. My intent here is to explore the truth about this apprehension of contigency, however, and so I shall be arguing for a preferable contemporary way of understanding it, framed in the discourse of Western intellectual traditions as informed by the East and South Asian. In particular, I shall argue that there are two dimensions to the contingency of nature, namely the contingency of natural things within nature, and the contingency of nature as such, and that the apprehension of both is one of the founding definitions of human culture. The first can be called *cosmological contingency*, characteristic of cosmological processes, and the second *ontological contingency*, the contingency of being as such, from *ontos*, one of the Greek words for being. These two senses of contingency are closely related. From a cross-cultural perspective, the notions of both nature and its apprehension are problematic, and some of the discussion to follow treats these problems.

Underlying the discussion is a philosophic supposition that will not be defended here but only illustrated, namely, the hypothesis that reality has characters that are variously discerned by the signs and symbols we have to interpret them.[2] Reality is not created by our signs, as some allege, but engaged by them. Our signs direct our engagements more or less accurately as they interpret reality in important and relevant respects, and insofar as they register and correlate the important and relevant distinctions and features. This is true as well for the signs and symbols of contingency. We need to inquire whether they interpret reality rightly, not whether there is a reality to which they refer. The inquiry might conclude that the signs and symbols of contingency are misleading because everything about nature is necessary, or that they are hopelessly confused and incoherent. But I predict, to the contrary, that we can make sense of the primordial expressions of apprehended contingency and will find that they engage us perspicuously with important elements of reality.

R.S. Cohen and A.I. Tauber (eds.), Philosophies of Nature: The Human Dimension, 121–137.

"Nature" in the modern sense is a notion that has developed swiftly through the evolution of modern science. There is nothing quite like the Western history of the idea of nature in the other great civilizations of South and East Asia.[3] Notwithstanding the diversity of civilizations, we can propose an abstract definition of nature that does indeed have counterparts in all. Let us hypothesize that *a natural thing is anything that is conditioned or caused by something else and that itself can be a condition.* This definition resonates with the Latin roots of the word *nature*, which have to do with being born or arising from something else. The definition is abstract enough to allow for many senses of conditioning or causation, not only within Western cultures but in other cultures as well.

In defining *natural things* I have skirted the problem of defining nature as a whole or nature as such. Whether nature is indeed a whole, a totality, a system, is extremely problematic and is not supposed to be so in all cultures or even consistently in Western scientific culture.[4] A reasonably innocent conception of nature as such can be constructed from the definition of natural things, namely, *that nature is the collection of natural things related as conditioning and conditioned things, such that any two things are connected, however indirectly, by a route of conditions.* This little definition leaves open questions such as tightness of conditioning relations, the uniformity or arbitrariness of orders, whether there is a principle summing up all nature, and whether nature is a whole or totality in contrast to something outside nature. It allows that there might be pockets of order separated by only the vaguest and most trivial connections.

This vague definition of nature holds distant promise for explicating the two dimensions of nature's contingency, because the mutual conditioning of natural things is the realm of cosmological contingency, and the question of why there are any natural things at all, why there is any cosmos with one thing causing another, addresses the ontological dimension of contingency.

That distant promise must be approached step by step, and the first step is to ask whether there are any boundaries to nature. That question sounds peculiar to us, because we assume that any boundaries are natural boundaries and what lies on the other side must be more nature.[5] The question of boundaries made much sense in the ancient world, however, when there was deep concern to distinguish nature from the supernatural on the one hand and the unnatural on the other.[6] The supernatural was conceived by the mythopoeic imagination as impacting otherwise ordinary events of nature and human society but also as being expressive of powers and kinds of action that had no imaginable limits. When the supernatural was personified in gods, the gods were imagined on the surface as super-intense people or animals but were imagined on the inside as not bound by the kinds of things that limit human or animal behavior. Marduk, for instance, could be depicted graphically as a matricidal warrior but one whose action creates the entire space for life and action in the natural senses. The unnatural, on the other hand, was conceived as not the intensification of the natural but the evisceration of it, the realm of the dead, of the shades, of those things impacting ordinary life through witchcraft and demonry. Because both the supernatural and the unnatural condition ordinary affairs, they seem to fit within the preliminary defintion of the natural.

Does nature have closure, however, against the supernatural and the unnatural? By this, I mean to ask whether there is a characteristic or trait that distinguishes all things in nature from the unnatural or supernatural. This is like closure in the mathematical sense in which it is said, for instance, that the set of even numbers is infinite but has closure over against all the odd numbers. In mythopoeic thought, the natural, the supernatural and the unnatural had permeable borders, with intrusions and overflows. When the Israelites left Egypt their magicians, Moses and Aaron, proved more able than the Egyptian magicians at calling down the supernatural, and Israel's God triumphed over the other gods in Egypt and Canaan. As the Israelites' religion developed toward monotheism, however, God became more distant from affairs and real kings took over the management of the realm from the judges who acted as God's surrogates.[7] By the time of Saul and David, witchcraft for contacting the dead was dangerous and forbidden, though practised.[8] By the time of the prophets of the exilic period, in the sixth century B.C.E., the world was sharply distinguished from the creator by being wholly dependent and created.[9] Any fudging of createdness as the definition of closure is idolatry, they thought, leading to witchcraft and magic which were condemned.[10] In that same century, the pre-Socratic philosophers and the playrights broke the hold of mythopoeic thought in Greece. Gautama the Buddha and Mahavira did the same in India. So did Confucius and Laozi in China. This was the pivot of the Axial Age, as Jaspers called it, and one of its main features was the placing of closure on the natural world over against the supernatural and unnatural.[11] Instead of the mythopoeic commerce and mingling among the three realms, the natural world became the human home and relations to the others, particularly to the supernatural, were transformed into issues of transcendence.

Without a clearer and popularly accepted definition of what createdness consists in, however, it was difficult for those traditions sharply to maintain the separation of the natural and supernatural. Nature in Jewish, Christian, and Muslim popular cultures, as well as popular East Asian and South Asian cultures, continued to be filled with gods, angels, and devils, and religion and magic were continually tempted to treat the divine as manipulable. Because the supernatural or divine had come to be defined in such sharp contrast with the natural world, and yet people had no clear definition of the closure of nature, the popular cultural reality became thick with intermediaries to the supernatural: angels, postumously effective saints and ancestors, djinns, incarnations of the divine, avatars, Bodhisattvas, shamanistic totems, Daoist gods and immortals.

With the rise of mathematical science in the West, however, those intermediaries were relegated to superstition by people shaped by the new culture. This was because it became possible to give a new definition of the closure of nature, namely, that it consists in whatever is measurable.[12] Put crudely, if something could not be found and measured, at least in principle, it was conceived to be a fiction. And if it could be found and measured, at least in principle, it was fitted into the system of nature, sparking a vast extension of nature's image. The mind-blowing expansion of nature's bounds occured early in modernity with the naturalization of the starry heavens and the explanation of previously occult biological processes by microscopically revealed mechanisms. Cosmic big bangs and dancing quarks express the continuing naturalization of the supernatural into our own time.

Spurred by the cultural ideal of quantifiable measurability, the unnatural in modern culture was conceived not as diminishment or deviance but as absolute nothingness. The supernatural was conceived not as the realm of especially powerful beings but as the Infinite or the Immense, "immense" meaning literally "unmeasurable."

Subsequent developments of Western modernity have damped the initial enthusiasm for universal quantifiable measurability. Too many important things, such as mind and value, are not measurable in that sense. But that definition of closure itself can be generalized to include the mental and valuational things that have resisted objective quantification. We can say that nature consists of all things that are *determinate.*[13] Determinateness is the universal trait of nature and the natural world. If something cannot be measured in a quantifiable way, that must be because it has some determinate character that makes it so unmeasurable. To be determinate is to have an identity different from, and therefore relative to, other identities. Determinateness is what makes things related to each other and also different, and so measurable relative to one another in that larger sense.

II

My hypothesis now is that nature consists of all the things that are determinate, and that determinateness is to be understood in part in terms of things being conditioned by and conditioning each other. I want to explicate and defend this hypothesis and finally return at the end to consider how it makes sense of the pervasive senses of cosmological and ontological contingency expressed in many cultures.

Determinateness needs first to be explicated on an abstract metaphysical level. To be determinate is to be determinate with respect to something else, to be *this* rather than *that*. A determinate thing thus needs to have features serving two kinds of function. On the one hand it needs conditional features coming from the other thing that condition it and with respect to which it is determinate. These features express the senses in which a thing is caused, or environed, or shaped by other things, according to any conception of conditioning whatsoever. On the other hand it needs essential features that integrate the influences of other things into the determinate identity of the thing itself. The essential features give the thing its position, or substantiality. Without them there could not be any conditional features because there would be nothing to conditions, there would be only the other things as potential conditions. Yet the essential features do not constitute anything determinate or substantial by themselves: their nature is to bring the conditional features to determinateness as a thing in relation to the conditioning things. The essential features need the conditional features just as much as the conditional features need the essential ones. Their harmony is the determinate thing, and without their harmony, neither is determinate over against the other.

Atomistic philosophies attempt to get along with only essential features, suppressing the conditional ones. They assume that a thing can have an internal determinate nature and enter into relations with other things in wholly external ways like atoms in the void. Atomisms have great difficulty accounting for why any relations are important to the internal natures of things. They have to assume that things internally are unaffected by their environment, and this usually collapses to the assumption that atomic things are internally homogeneous and, in the end, not different from one another in character. But if things are wholly homogeneous internally, and not determinate with respect to one another, they are indistinguishable from nothing and have no internal determinateness at all. Things thus can be defined not internally but only in terms of their external relations, such as position, direction, velocity, and mass for moving other things. The problem for atomists then becomes that of giving an account of external relations where things function as mere markers on an extensional field. Either the field is a special thing, internally determinate with places and geometries of movement, contradicting the atomistic hypothesis, or it is a fiction. If it is a fiction, atomism loses all determinateness, both internal and external. Atomism ought to recognize that external conditions must be harmonized internally to make things determinate with regard to one another in whatever field they might be together, which is to say that things have conditional features.

Idealisms by contrast suppress essential features in favor of conditional features, defining determinateness in terms of relations. Determinate things are treated as terms of relations, and their internal natures are analyzed as relations with subordinate terms. But idealisms fail to account for the difference between a possible relation that in fact relates no real things, and an actual relation among real things. An actual thing cannot be a mere term in a relation without some essential features to determine it over against the functional character of being a term in the relation. The functional character by itself is only a possibility. Without standing of its own, a thing is reduced to its relations and hence cannot actually function as a term over against other terms. Moreover, the relations themselves are determinate only as terms in higher relations, which themselves suppose higher relations to the point of an absolute relation, which is itself indeterminate and hence incapable of relating any subordinate relations. As F.H. Bradley argued, the idealism of relations supposes a mystical absolute with no determinateness whatsoever.[14] If idealism is to "save the appearance" of really different related things in the world, it needs to appeal to something essential in things.

This brief survey of the superiority of the hypothesis of conditional and essential features to atomism and idealism is schematic and not exhaustive. I hope it will be made plausible by the use to which it is put in the following discussion.

The hypothesis is that to be determinate is to be a harmony of conditional and essential features, and thus in connection with other determinate things. A thing must be determinate in some respects to have any identity whatsoever. Otherwise it is the same as nothing. But a thing need not be determinate in all respects.[15]

The next step in the argument is to make that metaphysical hypothesis cosmologically specific with regard to the flow of time. Cosmological contingency has to do with the conditioning of things within time. The hypothesis about temporality is that time has three determinately different modes, past, present, and future, and that all three must be acknowledged in interconnection in order to account for the causal passage of time.

As determinate, the modes of time have essential and conditional features.[16] The essential features of the past give it actual fixity and complete determinateness. The past is what it is and does not change in actual structure or value except insofar as more things become past with the passage of time. The essential features of the future have to do with formal possibility. Relative to the processes ongoing in the present, certain outcomes are possible and others not, and these are formally connected providing a field for all things in process. The essential features of the present have to do with spontaneity and decisiveness involved in actualizing possibilities. The present is when actual states of affairs emerge from mere possibilities. The essential features of the three modes of time cannot be expressed without reference to the conditional features, and in fact my exposition just now cheated.

The conditional features the past receives from the future are the determinate forms that get actualized and the values these structures bear. The conditional features the past receives from the present are the accretions of new actual states of affairs; the actuality itself is the product of present actualization. The conditional features the future receives from the past are the actual structures and lines of causal force that give definiteness to pure form. Without the actual diversity of things to integrate, the future as pure form would not be determinate. The future consists of formed patterns because of the determinate character of the past. The conditional features the future receives from the date by date movement of the present are the continual kaleidoscopic shifting of those patterns as different actual decisions are made. With each present set of actualizations, the future's formal structure changes. The conditional features the present receives from the past are the actual things that function in the present as potentials for integration in emerging actualities. This is causation in the most ordinary sense: the past provides the actual stuff for new actualization, and that stuff has structures that must be taken into account in any new emergence. In certain respects the past might not allow alternatives for subsequent actualizations, and in these respects, deterministic laws prevail; in other respects, the past might allow of several alternatives for emerging things. The conditional features the present receives from the future are the formal possibilities for integrating actual structures that are potentials for new actualization. The present is peculiarly creative in that its essential features integrate both given actual conditions and formal possibilities for combination into new determinate actualities. As soon as the present actualizes something, that actual something is past and thus fixed.

Each of the temporal modes has exercised a powerful and often exclusive effect on the imagination of temporality itself. Modern deterministic physics, for instance, imagines all time as if it were modeled on the fixed past, with determinate structures stretched out on a time line. On this image, the place of the present is arbitrary, and one can imagine the direction of time to be reversed. As Bergson complained, this is a spatialization of time. Existentialists, on the other hand, take their image of time from the present, emphasizing decisiveness and the intentionality of consciousness. Process philosophy also emphasizes the present as alone real, with the past significant only as entering into present actualization and the future as real only as anticipation; process philosophy rightly expresses the creativity in the present, pointing out that the action of actualization is not from the past, which cannot change, but an essential constituent of the present as the emerging of actuality. The future is the paradigm of time itself for those who see it only as a changing pattern of forms, one thing after another. Perhaps this is the dominant popular image of time in our day when things seem not to add up, only to pass, and when we look to the future for things to be different far more than we look to the past for the values that have been actualized or to the present for the sites of our responsibilities. But none of these images of time adequately represents the experienced flow of time in which possibilities are selected among alternatives, locating present action and building an actual world where decisions count.

The real flow of time involves the interaction of all three modes of time. At any one present date, things are being decided that change the future and add to the past. Thus together with the change that consists in present actualization there is the change that consists in the unfolding of past actuality and the change that consists in the shifting of formal future possibilities. The togetherness of these three kinds of change is not temporal togetherness. Temporal things are in time, structured by the temporal dimensions. They have a future, move from date to date of present actualization, and lay behind them an actual past. Temporal things do not exist only in the present, anticipating a future that does not exist yet and remembering a past that no longer exists. Of course the future and the past do not exist in the present save through anticipation and memory. But they exist in their own temporal modes as possibilities and actual fact. The future possibilities cannot be unreal because their existence is only as possibilities anyway; they are what they are irrespective of whether they are actualized, although their patterns depend on what has been actualized. The past cannot change any more, and hence cannot cease to exist. The flow of time in temporal things consists of the dynamism in which the date of the present advances so as to move possibilities to either definite actualization or exclusion, building up a history of achieved character and structure. The human feeling of time flowing in the present includes the apprehension that the future has its formal patterns or limitations regardless of whether we rightly anticipate them, and also the apprehension that the past has an actuality that measures our memory rather than the other way around. The present alone is an abstract part of the larger temporal reality that includes the actuality of what has happened and the formal limitations of what yet might.

The picture of nature emerging with this hypothesis is far more complicated than might be conveyed by the introductory remark that nature consists of things being conditioned by and conditioning one another. The introductory language might suggest only that things are conditioned by the past lines of causation that lead up to them, or by their environment. Now we can note that at every date of their temporal existence things are conditioned by future possibilities, by past actual things providing their potentials for actualization, and by the random and directed decisions taking place in that date as present. But because no temporal thing exists at only one date, but rather through a temporal stretch of dates, each things is conditioned by the shifts in its possibilities through time, by the ever changing cumulative character of its past, and by the diverse acts of creativity in each of its dates when those dates are in the mode of present time. The complicated dynamism of temporality means that every one of its dates when it is present has a somewhat open future, a past providing its potentials, and a decisiveness of actualization. Moreover, every one of its dates has a myriad of open forms structuring it as future possibility relative to each prior date, and also a fixed structure as part of an actual past that grows with the movement of the present. A dynamic temporal thing is never fully existent at a given temporal moment, although of course that is all the thing is at that moment. What is left out of a given moment, say of present existence, are the thing's future as that will have been actualized later and the past when those past moments were future or present. Any given moment is an abstraction from the whole of the temporal thing's reality.

Precisely because of the dynamism of time's flow, constantly adding to the past, shifting the future, and making decision after decision, a temporal thing is fully real only in eternity. Eternity is not a happy notion in modernity and it needs explication here. Eternity is not static because there is no time when the three temporal modes are not changing. Rather, the eternity in which time flows is fully dynamic in at least the ways mentioned here, and things are in eternity precisely because they are temporal. This is because the relations among the temporal modes are eternal, not temporal. The past is not before or after the present or future, but together they define temporal relations for temporal things.[17] The mutual conditioning of natural things is at least as complicated as described here, and surely far more complicated.

Yet another step must be taken to acknowledge diverse kinds of temporal things. I have spoken as if temporal things were all of one sort, and probably most people had in mind individuals or Aristotelian substances. But in fact there are many crucial determinate distinctions between kinds of things. Human discrimination usually focuses on items picked out of backgrounds.[18] Sometimes these can be more or less autonomous organisms in an environment. Other things are so intertwined that the language of ecosystems is best for their description. Some things do indeed maintain a fairly continuous character through significant changes in environment. Other things such as human beings build a tight identity over time by extraordinarily sensitive adaptations to a variety of things, many at a great distance. There are aggregates of things such as sandpiles whose members' internal relations with one another are relatively insignificant; there are organic environments in which any change anywhere resonates with changes in everything else. Human societies exhibit vast mixtures of organic individuals living lifetimes, natural environments slowly changing, social institutions that exist in the human enactment of meaningful behaviors, and languages that provide meaning and change with use. Whereas any thing in any of these senses involves all of the kinds of conditioning of past, present, and future temporality, at every date of its existence, each kind of thing has different kinds of temporal conditioning relative to itself and to those things that are its conditions and which in turn conditions. A temporal thing is in eternity in the sense that its whole identify includes each of its dates in all its future states toward which it might be responsible, each of its dates in its proper present of decisiveness, and each of its dates as given determinate actuality.

III

Up until now I have been fleshing out the hypothesis about cosmological contingency. A natural thing is contingent upon its entire shifting array of temporal conditions, inclusive of all of its dates in their eternal togetherness. What about ontological contingency? I shall make two approaches to this question, one from the cosmology of temporal relations and the other from the metaphysics of conditional and essential features sketched earlier.

With regard to the cosmology of temporal relations, eternity was the name given to the context in which the temporal modes are together, because unlike things in time they are not together temporally. Eternity is also the context in which temporal things have their full identify, in which each of their dates is future, present, and also past, eternally connected so that they are together temporally yet lived dynamically with a moving succession of presents. What can this eternity be? It is the ultimate condition for the existence of temporal things and their temporal conditionings. What can provide that condition?

Eternity cannot be a static form because that cannot be the context for either existential actualization or the solidity of actual fact. Eternity cannot be a fully determinate whole, as absolute idealists might suggest, because that cannot be the context for the dynamics of shifting future possibilities nor existential actualizations. Eternity cannot be a temporal, present God, merely remembering the past and anticipating the future, however perfectly, because that cannot be the context including the reality of the past to which memory ought to be true and the future that is so sensitive to change.

Rather, I submit that the only plausible hypothesis is that eternity is the act that creates everything determinate, everything variously determinate in all of their dates as past, present, and future, and as mutually conditioning one another in whatever ways they do. The determinate things have their ontological existence as the interrelated termini of the creative act. They are created to be together in the ways they cosmologically are, with the regularities and decisive spontaneities of their temporal unfolding, in one act. The act is not in time but creates time with its dynamism by creating the temporal things in their shifting natural conditioning relations. The act is not anything determinate apart from creating, and its determinate nature in only that of the determinate world. The act does not proceed from any determinate potentials, because all determinate things are created. So it can be said, in the language of Western theology, to be the act of creation *ex nihilo*. In the West, this has been identified with God, although not all Western conceptions of God would accept the characterization of creation *ex nihilo*. I shall discuss South and East Asian variants shortly.

The argument I have given here is not coercive because it merely eliminates some obvious candidates for eternity and then says that I can think of only one more possibility, the act of creation of all temporal reality. Furthermore, the vast complexity of natural temporal conditions described makes it hard to get the argument in mind. But in the discussion of temporality, the very distinctness and connection of the temporal modes and hence of the ways different temporal structures provide conditions for temporal things was described in terms of essential and conditional features. Because of the essential features the temporal modes cannot be reduced to one another; because of the conditional features each is defined in part by its relations to the others; the determinateness of past, present, and future come from each being the harmony of its essential and conditional features. From this, the argument for creation *ex nihilo* as the context for ontological contingency can now be restated in terms of the metaphysics of essential and conditional features.

The conditional features constitute a complicated set of cosmological, indeed temporal and spatial, togetherness. But the essential features of each thing are not contained in the other things to which it is related by conditional features. Rather, the essential features are what give it its own existence, allowing it its own being so as to be able to enter into conditioning relations with other things. The other things each have their own essential features by the same token. The mutual togetherness of the conditions does not include the togetherness of the essential features of the several harmonies together. Yet the things could not have their mutually conditioning conditional features if those features were not harmonized with the essential features in each case. Therefore, there is a deeper, ontological, context in which things are together with their essential as well as their conditional features. This deeper ontological context of mutual relevance is what allows things to condition one another. What can be the ontological context that allows harmonies of essential and conditional features to be together without swallowing each within the others? What is the context that allows the harmonies to be, which they could not if they were not separate in essential features and mutually implicated in their conditional ones?

The ontological context of mutual relevance could not be a common property such as *being*, because then all the other properties constituting things, all their essential and conditional features, would be outside of or in addition to the context that is supposed to contain them. Nor could the ontological context be some kind of determinate container such as the Absolute, because then there would have to be a yet deeper ontological context to relate determinate harmonies to the determinate container. The context itself cannot be determinate without an infinite bad regress. Yet if it is merely indeterminate, it does not contextualize anything. I propose, therefore, that the context is the creative act that is nothing without acting and which results in the world of determinate harmonies. The determinate harmonies with their essential features are together as created together, and their conditional features specify their temporal relations with one another. The sense in which things are in an ontological context of mutual relevance, which they must be in order to be determinate through conditioning one another and having essential features, is the sense in which they are created together. Togetherness in the creative act is eternal because it makes possible the temporal relations described above. In their temporal relations, things are together as earlier, later, and as contemporaries, and this is possible only because they are also together eternally as creatures. Such is ontological contingency.

The great abstractness of this hypothesis is a philosophical virtue even if a rhetorical vice. For the thesis that all determinate being is contingent upon an act of ontological creation is compatible with any determinate world whatsoever that science and the other forms of cognition might discover. The universe might begin temporally with a big bang, or it might be continuous through endless years as was believed in the steady state theory until recently. Human beings might be the highest form of intelligence, or there might be improvements. The running of the world might be practically deterministic, or it might involve both chance and responsible freedom. Whatever is determinate, including determinate indeterminacies such as openness to the future, can be created *ex nihilo*, and it is up to whatever forms of inquiry recommend themselves to ascertain what the determinate world is. My hypothesis opens inquiry to radically empirical questioning, at the same time that it provides at least the abstract beginnings of a reconstruction of the world's religion's apprehensions of ontological contingency.

Because the theological character emerging from this hypothesis is hardly tolerable to the modern Western sensibility, I need to address the peculiarity of the idea of an ontological act of creation.

The oddest element of the creation *ex nihilo* hypothesis about ontological contingency is that it describes, points to, and names an act. The hypothesis claims that the way to understand how there can be a world of nature with many different but connected things is by identifying an act that makes them be as determinate things. This is different from our customary intellectual strategy, namely, to understand things by identifying the principles or structures from which they derive. More concretely, our customary strategy is to explain things by reference to a being, or set of beings, that has a determinate nature, such that the world to be explained follows from that ontologically foundational nature. This strategy is illustrated in common understandings of God according to which the world's goodness, rationality, or purpose derive from the divine nature. Science no less than theology employs the strategy of explanation by principle and by appeal to the nature of foundational beings when it cites the non-obvious implications of mathematical structures and the natures of elementary particles. But these strategies, whatever their merit in orienting us to certain commonsense matters or for explaining the regularities science discovers, do not address ontological matters at all. Stephen Hawking remarked, "The usual approach of science of constructing a mathematical model cannot answer the questions of why there should be a universe for the model to describe."[19] The question is more radical that he knew, because we can also ask why there should be mathematical determinateness so that models are conceivable.[20]

Put bluntly, curiosity about why there is a world can be answered only by reference to an act or acts that make there be a world, with whatever character it has. An appeal to first principles or to the nature of basic beings is unhelpful if the question is why or how there are principles and beings at all. Even within ordinary experience we appeal to acts of making more than the European Enlightenment story about reason leads us to believe. In moral situations, we want to know who is responsible for the way things are and whose responsibility it is to act in the matter. When we ask why people do what they do, we presuppose an understanding of the principles and possibilities of action, including the possible motives; but the moral force of the question is why the people adopt the motive they do when they could have acted out of other motives, why they choose to have the moral character that comes from their choice of action, why they make one thing happen rather than another when the decision is up to them. Scanning any ordinary situation and taking up an orientation to it, we look to all the myriad choice points that need to be observed to follow what is going on, intentional choice points of human actors and random choice points of chance collisions among causal processes. In ordinary experience we do not limit our considerations to principles of possibilities or to the natures of established actuality but look also to the pulse of existential decisions or actualizations. Understanding by reference to decision points is the analogue for noting ontological creation, the result of which is the entire realm of determinate things, in all temporal modes, *sub specie aeternitatis.*

Critics will point out that the idea of an ontological creative act is wholly discontinuous with all senses of causation, conditioning, or creation within temporal existence. Yes and no. Yes, all finite acts of creation are conditioned by actual antecedents that must be taken into account and also by formal possibilities, and ontological creation produces both of those kinds of conditions rather than being conditioned by them. No, in every present moment of finite actualization some new reality is produced that was not contained in the antecedent actual conditions. If that were not so, there would be no difference between the situation of the actual conditions without the new actualization and the situation in which the new actualization has added something to what before was actual. Formal possibilities conceived next to old actualities cannot provide any new actuality. The actuality in the actualization is all the old actuality plus something that makes the actualization an addition. Human beings experience this in every single intentional act for which responsibility can be taken. The ontological creative act is like every act of finite actualization except that everything in its outcome is new and nothing is a potential deriving from a previous actualization. There are countless finite analogues to the ontological creative act, and we can say precisely what is not analogous – antecedent determinations that condition the creative act. Ontological creativity as expressed within time is involved in every precise moment of spontaneity, in any respect in which something new is added to the inherited actual conditions. Moreover, every one of those past conditions, when it enjoyed its own present, had ontological creativity in its novelty; each past actuality had its own past when it was present, but then its past too involved ontological creativity insofar as something novel happened when it, earlier, had been present. So in the whole temporal stretch of things, any actual change involves some novelty, and every novelty is, was, or will be of the ontological act. No actual condition fails to be analyzable into its own expressions of novelty in its present moments and the actual conditions for those moments, which themselves derive from inherited acts of novel creation. Actual existence is not the whole of reality, of course, because there is also possibility, which I have argued is the character of the future and therefore an integral part of time's flow. The pure form constituting possibilities as such is created itself in the eternal ontological act in connection with the shifting novel moments of the present and the accretions of the past. In sum, the ontological creative act is like every finite act of present creativity except that it creates everything new and inherits nothing.

The ontological creative act is radically asymmetrical, as a finite creative act is partially asymmetrical, namely, in adding something new to the actualized past. Ontological creativity is the making of something, starting from nothing (not just the past) and ending with the entire world of determinate things (not a mere addition). The explanatory reference to the ontological act has three dimensions: it requires an icon, an index, and a conventional symbol or name. Treating these three modes of reference shall bring me to the end of my argument.

The icon of the ontological creative act is, of course, the theory that says the reality is like what the theory describes. The ontological creative act must be understood in terms of three notions, the act of making, the source of the act, and the product. None can be understood without the other, and in this sense the three terms are symmetrical. Yet their connection is such that they assert the asymmetry of creation. The source of the act is indeterminate, because everything determinate is in need of being created. Moreover, only something indeterminate escapes the requirement for a further ontological ground of mutual relevance between itself and the determinate world. Without the act of creation, the indeterminate ground is nothing, absolute nothingness; it is not even the ground of anything and is wholly inconceivable. In fact, without the act of creation there would be no determinate conceivability at all. But there is the creative act, and therefore the ground is indeed the source of the world. The world itself consists of all determinate things, nature in the most inclusive sense whatever that might turn out to be. The act itself is determinate only in the sense that it is the creation of the determinate world. The act has no determinate extensionality of its own, no medium that constitutes a nature apart from its product, because all determinateness is in the product; it is simply an act of making. Taken together, the indeterminate source, the act, and the determinate product are the ontological creation of the world and of the creator. Theisms, starting with metaphors of gods as super intensifications of human agents, point out that ontological creation is the voluntaristic making of the nature of God which is itself dependent on the divine creative will. Duns Scotus was a voluntarist in this sense in contrast to Thomas Aquinas who claimed that God creates according to a pre-existent nature, albeit one that is too simple to be determinate! Just as a moral agent gives himself or herself the morally relevant parts of personal identity, so God makes the divine nature in the ontological creative act. The icon of source, act, and determinate product describes the ontological situation of contingency.

The indexical reference to the ontological act is to point our attention toward the act, like finding out who the moral agents are in a complex situation, or where the important causal happenings take place. The entire dialectical argument I have presented here is a single indexical reference, starting with a general historical assertion of pervasive senses of ontological contingency, directing the issue of contingency to determinateness as the mark of closure for nature, giving a metaphysical analysis of determinateness, and a cosmological analysis of temporal conditioning, pointing out the eternal character of the context in which determinateness and temporality are possible, and then asking for what might be that eternal context. That question indexically points us toward what makes the workd in the most basic sense of making.

Such a dialectical index is appropriate for a philosopher such as myself and, I suppose, for anyone still awake in the audience. Many other things function as indices for contingency and its ground for most people. According to the iconic theory, there ought to be three main classes of indexical experience. The determinate world is itself contingent, and any special marker of that, such as death, an intimation of one's morality, a natural disaster, or an observed cosmic singularity can trigger the look toward the creative act. For many people the creative dynamism of the act itself, never separate from its determinate work, is an indexical pointer of attention. Gerard Manly Hopkins' haecceities shimmer creativity like shaken foil, and most of us become nature-romantics at least temporarily when confronted with a sublime sunset or sea-swell. Then for the mystics the experiences of penetrating deeper and deeper into the basic elements of the created world, until the abyss of nothingness yawns like the fire whose flames come from nothing, point to the source sourcing, that on which all depends, nothing in itself, everything for us.

V

The conventional references to the ontological creative act on which all nature is contingent are to be found, of course, in the cultural references to such contingency in the world's great civilizations. My conclusion here will be a brief review of some of the standard ancient motifs of East and South Asian civilizations, showing how they mark contingency with different symbols. I trust that the language of my argument so far has called up the parallel Western ideas, such as God's creation of the world so eloquently expressed in the Hebrew Bible and presupposed in the New Testament, Plato's claims that the form of the Good creates the world and that soul is the primitive source of motion, and Aristotle's explanation of the contingency of all motion on the imitation of perfection as self-sufficiency. This review of non-Western motifs comes last rather than at the beginning of my argument because I hope the philosophical analysis above can be seen to resonate with these symbols, finding in them concrete specializations of the abstract notions. The symbols themselves are generally familiar.

From China I cite two basic motifs. The first and most obvious is the opening of the Daodejing:

> The Tao (Way) that can be told of is not the eternal Tao;
> The name that can be named is not the eternal name.
> The Nameless is the origin of Heaven and Earth;
> The Named is the mother of all things.
> Therefore let there always be non-being so we may see their subtlety,
> And let there always be being so that we may see their outcome.
> The two are the same,
> But after they are produced, they have different names.[21]

The Confucian tradition picks up the more general Chinese sense of causal process as the subtle interweavings of yin and yang and gives that process an ontological interpretation. The eleventh century Neo-Confucian, Chou Tuni-i wrote:

> The Ultimate of Non-Being and also the Great Ultimate! The Great Ultimate through movement generates Yang. When its activity reaches its limit, it becomes tranquil. Through tranquility the Great Ultimate generates yin. When tranquility reaches its limit, activity begins again. So movement and tranquility alternate and become the root of each other, giving rise to the distinction of yin and yang, and the two modes are thus established.[22]

From the Indian tradition there are many accounts of the creation of the world with both mythical expressions and highly philosophical expressions of the contingency of not only matter but appearance, as in Advaita Vedant. But I shall quote part of the ancient hymn of creation from the Rg Veda:

> Non-being then existed not nor being:
> There was no air, nor sky that is beyond it.
> What was concealed? Wherein? In whose protection?
> And was there deep unfathomable water?
> Death then existed not nor life immortal;
> Of neither night nor day was any token.
> By its inherent force the One breathed windless:
> No other thing than that beyond existed . . . .
> Desire entered the One in the beginning:

It was the earliest seed, of thought the product.
The sages searched in their hearts with wisdom,
Found out the bond of being in non-being . . .
Who knows for certain? Who shall declare it?
Whence was it born, and whence came this creation?
The gods were born after this world's creation:
Then who can know from whence it has arisen?
None knoweth whence creation has arisen;
And whether he has or has not produced it:
He who surveys it in the highest heaven,
He only know, or haply he may know not.[23] (R&M 23-24)

I accede at last to the claim philosophers should resist until the end, namely, that nature's existence is a mystery. Why it is a mystery should be clear by now: nature rests on a creative act which has no nature, no necessity, and no existence save in the standing forth of determinate being from nothingness. The alternative to the world is very serious nothingness, and the mystery is the act that makes the difference. Given the primitive character of the conception of making something new, which applies throughout nature in many limited ways, we stand in awe before the Ultimate Making, and the Maker made in the Making.

Department of Philosophy, Boston University

NOTES

[1] Indeed perhaps in all human experience although the discussion here is limited to its civilized expressions.
[2] The extended defense of this thesis is complicated, but I have attempted it in print. Part of the defense is to show how symbolic representation is itself a part of nature. For this we need not only a theory of signs, such as that provided by Peirce, but a theory of nature as well. Then we need an analysis of some important kinds of symbols, such as religious ones, that purport to engage "ontological" dimensions of reality as well as garden variety things. My trilogy, *Axiology of Thinking*, attempts the theory of signs and nature; it consists in *Reconstruction of Thinking* (Albany: SUNY Press, 1981), *Recovery of the Measure* (Albany: SUNY Press, 1987), and *Normative Cultures* (Albany: SUNY Press, 1995). For a theory of religious symbols and their ontological reach, see my *The Truth of Broken Symbols* (Albany: SUNY Press, 1996).
[3] By "West" here I mean Western Asia with its European appendage and American colonies, and include the religious traditions of Judaism, Christianity, and Islam, each of which has been called a civilization in its own right.
[4] The criticism of the idea of totality in nature in Kant's Antinomies (*Critique of Pure Reason*, B 455-462). I have discussed it at some length regarding Hegel and Whitehead's approach to nature in *The Highroad around Modernism* (Albany: SUNY Press, 1992), chapter 5.
[5] This was the nub of Kant's argument about totality.
[6] Many of the elements of nature discussed here, in the ancient conceptions, are discussed compendiously by Richard Sorabji in *Time, Creation, and the Continuum: Theories in Antiquity and the Early Middle Ages* (Ithaca: Cornell University Press, 1993).
[7] See the book of *Judges* in the Bible.
[8] See *1 Samuel* in the Bible.
[9] See *Isaiah* 40–66; *Ezekiel*; *Job*.
[10] On the secularizing of God's creation, see Tikva Frymer-Kensky's *In the Wake of the Goddesses: Women, Culture, and the Biblical Transformation of Pagan Myth* (New York: Free Press, 1992).
[11] See Karl Jaspers' *Way to Wisdom*, translated by Ralph Manheim (New Haven: Yale University Press, 1954).

[12] The ideal of early modern science was to quantify measurement, as in Descartes' analytical geometry. See my discussion in *Reconstruction of Thinking*, chapter 1.
[13] The discussion to follow summarizes in part my analysis of determinateness in *God the Creator* (Chicago: University of Chicago Press, 1968; reprint with a new preface, Albany: SUNY Press, 1992).
[14] See Bradley's *Appearance and Reality* (Second edition; Oxford: Clarendon Press, 1897).
[15] There might be many things with respect to which it is only partially determinate, or indeterminate. It might be, for instance, that the determinate character of the present is such that several different responses to future possibilities can be made; with respect to the future, the present is somewhat open or indeterminate, awaiting subsequent deciding events.
[16] This analysis summarizes that in *Recovery of the Measure*, chapters 9–10.
[17] Our late modern age has more difficulty than the ancient Western world in imagining dynamic eternity as the context for emersion in the temporality of life. But that is because of an impoverishment in the conception of time as much as it is a collapse of eternalistic religious sensibilities.
[18] See the analysis of perception and appearance in *Reconstruction of Thinking*, part 2.
[19] See Stephen Hawking, *A Brief History of Time* (New York: Bantam, 1988), p. 174.
[20] The argument above urged that things are determinate only by being related to other things, and that they and the other things must have essential features that are together on a more basic level than their conditioning relations. Thus even the laws of logic and mathematics are ontologically contingent even if we cannot imagine alternatives – indeed, imagination itself is contingent.
[21] Translated by Wing-tsit Chan in his edited *Source Book in Chinese Philosophy* (Princeton: Princeton University Press, 1963), p. 139.
[22] Chan, p. 463.
[23] Translated by A.A. Macdonell in *A Source Book in Indian Philosophy* edited by Sarvepalli Radhakrishna and Charles A. Moore (Princeton: Princeton University Press, 1957), pp. 23–24.

ALAN M. OLSON

THEOLOGICAL REFLECTIONS ON THE NATURE OF NATURE: REVOLUTION, REFORMATION, RESTORATION

"The God that meets me in nature is the God of wrath" (Luther)

In his contribution to Habermas's *Stichworte zur 'geistigen Situation der Zeit'* (1979), Jürgen Moltmann provides a retrospective analysis of reformation, revolution, and liberation within the context of German theology. While Molmann's lamentations are largely political, the deeper and more fundamental philosophical and theological problem of *nature* emerges through a closer reading.

> Of all European nations, Germany is the country of missed and miscarried revolutions of liberation. The German consciousness emerged out of the Reformation, whose partial success went hand in hand with the altogether unsuccessful German Peasant Revolt in 1524. Ever since Luther's decision against Thomas Müntzer and the peasants, the theme of 'Reformation and Revolution' has always been present in the German consciousness, although constantly suppressed. Since 1524 all routes in Protestant Germany from Reformation to Revolution have been obstructed by fundamental anxieties, warnings, and the condemnation of heretics; here threatens, so it has been claimed, *fanaticism, the politization of the Gospel, hubris, insurrection against God, the Spirit of the Antichrist [and so forth]* . . . Confronted with the choice between revolution and dictatorship, the Germans have always opted for the dictator. Sins from above are more forgivable than sins from below, and violence from above is more bearable than violence from below, for the one is [identified with] order and the other [with] chaos. This precedent was already apparent in Luther's pastoral letter to the peasants. It is for this reason that the French Revolution did not penetrate the German states. It is for this same reason that the bourgeois revolution of 1848 miscarried when the Prussian power of order destroyed democratic freedom. Even the decisions of 1919 and 1933 are to be seen against this historical background. It is owing to this sixteenth-century reformation without revolution that in the nineteenth century the French Revolution led with us only to a *revolution in thought*. Its impulses were absorbed by German inwardness and preserved in a pure form by intellectuals through an *inner emigration.*[1]

Filled with obvious longing for "what might have been," Moltmann's provocative words also give rise to the perplexing problem of natural theology. By his choice of spatial metaphors, Moltmann implies that the Germans (especially Protestant Germans) have a traditional aversion for nature or the "lower-world" because it is the source of "chaos." As a consequence, the Germans consistently have taken refuge in the "upper-world" of eidetic or "rational order" – however potentially oppressive or demonic that order might

R.S. Cohen and A.I. Tauber (eds.), Philosophies of Nature: The Human Dimension, 139–149.

be. Obviously sympathetic to the Marxian inversion objective and subjective Spirit, Moltmann implies that the defect in German political consciousness originates in the inability to come to terms with the pre-reflective, pre-linguistic given, that is, nature or the *Lebenswelt*. It follows, then, beginning with Luther, that the great lacuna in German thinking regarding nature and natural theology has been the source of dire social and political consequences. Post-Reformation accomplishments in the neotic world of German Idealism must be viewed as coming at the expense of a constructively sympathetic knowledge of the natural order – from which, Moltmann, like Rousseau, seems to suggest a greater sense of humanity should emerge. But since the Uranic dimension consistently and fatefully overpowers the Chthonic in German thinking, since reason negates or sublates feeling, the Germans, according to Moltmann, have excluded themselves from the liberating potential that comes through a deeper sense of nature, the Germans have somehow been fated to their social and political woes. In making this charge, Moltmann echoes the famous observation of Heinrich Heine who asserted, out of Jean-Paul, that the Germans, in order to compensate for the British empire of the sea, and the French empire of land, were left to conquer the only remaining region, "the empire of the air."

But is it reasonable to attribute so much to an apparent defect (if it be such) in theology and philosophy? In other words, does the neglect of a sympathetically constructive theology of nature lead necessarily and inexorably to the kinds of social and political deformations Moltmann enumerates? After all, Heidegger too was highly critical of the conceptual drift of philosophy and theology away from what might be termed a primordial sense of nature and what is immediately given in experience and into what he considered to be the empty and barren categories of onto-theology. Or is it the case, given the will to rational order, that when a repressed sense of nature (as in Freud) breaks out from under the bondage of rationality, it leads to catastrophe?

Like Heidegger, Erazim Kohák, in *The Embers and the Stars*, speaks at length of the dialectic of *techne* and *poiesis*, and how *dichten* overtakes *denken* with the coming of the night. But he also warns us of the dangers of substantializing, in some fundamental ontological way, the terms of the dialectic – whether the dualistic properties of the light, on the one hand, or the welcoming monism of the night, on the other. Unlike Heidegger, Kohák calls our attention to an "intermediate" level of experience and Being between the alternatives of *techne* and *poiesis*. This "intermediate *metaxy*," so to speak, is the originating ground of the deeper "moral sense of nature" to which we obviously need to become attuned.[2]

I think that Kohák is correct in this assertion, and it is this intermediate region that I would here explore in order to make a case for the mediating position that more accurately reflects a philosophical theology of nature informed by the German philosophical and theological context. After sketching out some of the theological aspects of "the nature of nature" in the first part of my essay, I go on to suggest, in the second part, that if, during the late 18th and early 19th centuries, *revolution* comes to mean the *revolt against nature*, and if *restoration* means the return of *natural law*, the only way to reconcile this polarity is by way of what Hegel calls "the reformation of consciousness as conscience (*Die Umgestaltung vom Bewusstsein also Gewissen*)." As Hegel asserts emphatically at the end of his *Lectures on the Philosophy of History*, "there can be no revolution without reformation." This notion, I believe, that is implicit in Kohák's reformulation of a "moral sense of nature."

LUTHERANS AND CATHOLICS ON N ATURE

Whether as a signifying category or that which is signified, whether *natura naturans* or *natural naturata*, nature has had a checkered history in Christian theology generally. One might even be so bold as to suggest that the history of institutional Christianity, to a considerable degree, has turned on violent disagreements regarding the legitimacy of natural theology. Moltmann is correct, therefore, in suggesting that the turning point regarding the conception of nature, and the turning point for Western or Latin Rite Christianity generally,[3] comes with the Protestant Reformation as evidenced by Luther's uncompromising assertion, against scholastic realism or essentialism, that "the God who meets me in nature is the God of wrath." Luther and his orthodox Protestant successors[4] held that no saving or *salutary* knowledge of the Divine is possible through the *mirror of nature*, as it were. This anti-natural theology polemic probably reinforced what has long been perceived by Christian theologians, whether accurately or inaccurately, as the anti-nature stance of the Bible.[5] This anti-nature stance is usually located in the struggle of the Hebrews with the fertility cults of ancient Canaan, beginning with the "cursing of the ground" (Genesis 3) and Moses's destruction of the Golden Calf at the foot of Mt. Sinai (Exodus 32). Such events dramatize, they argue, the incommensurability between "naturalistic" or "idolatrous" representations of the Divine, on the one hand, and devotion to the revealed Moral Law, on the other. The anikonic prohibition of idolatry, as evidenced by the command in the First Table of the Law, "Thou shalt not make unto the Lord thy God any graven image or any likeness of any thing that is in heaven above, or that is in the earty beneath, or that is in the water under the earth,"[6] is precisely what Kant, in his *Critique of Judgment*, counts amoung "the most profound words in the Bible" – "most profound" because the Divine essence or noumenon remains necessarily hidden and is cognizable only through the Moral Law.

Now the Catholic conception of nature, especially after Thomas Aquinas and the Christian appropriation of Aristotle, is more accommodating with respect to the prospect of a natural theology – at least on the surface. As such, the spirit of the Catholic position seems closer to Saint Paul's use of the "dark" or the "foggy" glass metaphor through which we try to make sense out of the world. Thomas makes repeated use of Saint Paul's *Corinthian* image of "seeing through a glass darkly" throughout the *Summa Theologica* and the *Summa Contra Gentiles* where he speaks of natural law and the natural or innate knowledge of God.

Certainly a relatively non-problematical *Übergang* between Plato and Aristotle, Augustine and himself, is implicit in Thomas's famous proposition that

Revelation [grace] does not negate nature but fulfills it. And the spatialized transcendental passage between the temporal and the eternal is never so confident or exhilarating as in Bonaventure's *Itinerarium mentis ad deum* where he confidently delineates the presence of the Divine in the books of nature, the self, and revelation. As in the case of his 13th century contemporary, Dante Alegheri, guided by the almost blessed Virgil through the *Inferno* and the *Purgatorio*, and by the unblemished and wholly blessed Beatrice, into the mystic rose of *Paradiso*, Bonaventure passes ecstatically, like the Buddha, through the "triple world" of *ratio, intellectus and intelligentia* into the Enlightened serenification and *Gelassenheit* of *affective rationality*.

But it is a long way from the erotic Mediterranean to the brackish Baltic, and the topography of natural theology changes dramatically along the way. Like Nietzsche's "earth loosed from its sun," the prospects of natural theology seem to become progressively colder as one moves towards the Nordkapp. To be sure, Luther maintained that a working or functional knowledge of nature was possible; and it was the purpose of the *usus primus legis* or the "first use of the law" to elucidate precisely this necessity. Luther believed that the created order, though corrupted by the Fall, retained patterns of order and regularity sufficient to the development of what might now be viewed as an instrumental conception of nature and even the basis for a "soft" natural law theory based on the so-called "orders of creation."[7] No doubt this impersonal, quasi-nominalistic view of nature greatly facilitated the rise of empirical science, for it provided the de-spiritualized vision of nature that technology presupposes. Certainly the nominalistic conditioning of the *via moderna* accounts for the more hospitable climate that Kepler enjoyed in 17th century northern Europen religious culture as contrast to the stormy climate betwixt Galileo and the Holy Sea in Southern Europe. In other words, because Luther and his reforming contemporaries did not believe that a metaphysical knowledge of nature was possible, and since the Protestants, as a consequence, did not really have a soteriological stake in nature, there could be no hierarchical encroachments on the Divine through its scientific or empirical investigation.[8] Luther's skepticism regarding the ontological claims of Thomistic Realism or Essentialism, and his deprecating comments on Aristotle, might also be viewed as playing a critical role in preparing the way for the transcendentalized neo-Kantian spheres or containers of reality that emerge during the late 19th and early 20th centuries, one for physical–empirical reality, and the other for values.

But Luther is also suspicious (in a Rorty-like way!) of theologies constructed at the other end of the experiential spectrum – in this case, the "spiritual" side of the mirror symbolized by the "sanctified epistemic subject," so to speak.[9] In the history of dogma, this skepticism is aimed at the *usus triplex legis* or the "third use of the law" – the "law of the regenerate." A century after Luther, in sharp reaction to Lutheran scholasticism, Pietists were highly sympathetic to the "third use of the law," the law of sanctification. They believed that while nature is corrupt and remains corrupt, it nevertheless is possible for regenerated individuals (those who have experienced their hearts "strangely warmed," in Wesley's phase) to improve their moral condition and to strive not only towards the perfection of themselves but also the perfection of the created order as a whole.[10]

But there were also those who, like Kant, Hegel, and Albrecht Ritschl, were deeply suspicious and critical of Pietism – not only on the notion that it invited all manner of religious fanaticism, but also because they believed that the law of sanctification might reopen the door for a return of the "ascetical virtues" of the Roman Church and its substantialistic sacramental system – chief among which was the hierarchical distinction between the "religious" orders and the laity, the sacred and the secular.[11] Thus on balance, I believe that it is safe to conclude that Luther himself would have been critical of such utopian, perfectionistic modes of thinking – even though the social, economic and political conditions of Europe were vastly different in the 19th and 20th centuries than they were in the 16th. For as his "Treatise on Vocation" clearly indicates, Luther firmly believed that all "callings" or *Berufen* have dignity and worth provided that the individual recognizes the eschatological provisionality of life in this world, and the fact that no vocations provide natural advantage with respect to one'e eternal destiny – since, according to Saint Paul in his *Epistle to the Romans*, "all have sinned and fallen short of the glory of God" – be that individual be Pope or Environmentalist, Braumeister, Butcher, Baker or Candlestick Maker!

What remains distinct and determinative, then, in German Protestant and especially in Lutheran theology, is the legacy of intense focus on the meaning of the "middle ground" occupied by the *usus duplex legis*. The second use of the law has to do with the mediating ground between the reality of a fallen nature, on the one hand, and the prospect of a regenerate nature, on the other. Focussed upon the consciousness of the sinner, this middle ground is the locus of the origination of the notions of subject and object, as Kant, Fichte, Schelling and especially Hegel were to show in great detail. The critical question, of course, is whether this mediating ground actually mediates in a unique and effective way, and whether this position can bear the theological and philosophical freight intended, or whether, on the other hand and as Moltmann suggests, this middle ground provides the basis for "the inner emigration" and a kind of moral paralysis, especially in the areas of political philosophy and social praxis.[12]

Situated between the *usus primus legis* and *usus triplex legis*, the *usus duplex legis* or "second use of the law" is the "rock" Luther repeatedly refers to in terms of the doctrine of justification (*Rechtfertigung*). Whatever one makes of the particulars of this doctrine (especially in the christological aspect upon which it utterly depends), one cannot underestimate the subsequent influence of this doctrine on the growth and development of German thinking. German Idealism might be viewed, in fact, as the progressive secularization and demythologization of this doctrine (e.g., "legitimation theory"). Whatever the case, there is, in post-Reformation thinking, a progressive movement away from the work of Christ (especially as represented in the forensic or substitutionary sense of the "perfect sacrifice" in Latin rite Christianity) and a movement towards the speculative consideration of the *mind of Christ* with respect to the transcendental properties and moral attributes of conscience and consciousness (*Gewissen und Bewusstsein*). For transcendental-existential-moral nexus of "self-consciousness" is where laws regarding the meaning of nature and God impinge directly on my self-being, and it is here where the focus of German Idealism is the most intense. This intensity clearly begins with Luther.[13]

The primary function of second use of the law, in Luther, is entirely clear, namely, that "the law always accuses" (*lex semper accusit*). As such, he holds that the primary function of the law is not to yield metaphysical knowledge regarding particulars in the "great chain of Being," nor does it provide special information regarding the prospective social programs of sanctified souls. Its function rather is to bring one to the knowledge of one's ignorance of God, that is, to the knowledge of one's non-knowledge of God (as in Cusanus and Johannes Tauler) and the exigent need for grace and forgiveness. For it simply is not true, according to Luther, that my ignorance of God is caused by the ignorance of myself, and it is not true that my ignorance of myself is caused by my ignorance of nature and natural theology. It is rather the case that this cycle of ignorance, the cycle my *cor incorvatum*, is a vicious circle consisting of the fact that I persist in a state of sin no matter how much I know, and that I willfully persist "in this body of flesh," as Saint Paul put it, because my *Wilkür* is stronger than my *Wille*, to use Kant's famous distinctions regarding the problematical relationship of the executive and the legislative will.[14]

If one happens to be raised in the rhetorical immediacy of this religious tradition, that is, brought up in the unadulterated, non-revised Lutheran tradition prior to the egalitarian revisionism that has so dramatically altered these traditions (and all mainstream religious traditions, for that matter (during the past couple of decades), it is impossible to underestimate the psychological force and power of Luther's assertion that "the God that meets me in nature is the God of wrath"! For in this tradition it is customary to begin one's worship with the confession of a sinful and corrupt nature as evidenced by the words: "I confess unto Thee that I am *by nature* sinful and unclean, and that I have sinned against Thee by thought, word and deed, wherefore I flee for refuge to Thine infinite mercy, seeking and imploring Thy grace . . ."[15]

In Luther, at least, the message regarding the prospects of natural theology is entirely clear: Like Job, he recognizes that the God one meets in the double mirror of nature and mind is the *Deus Absconditus* and not the *Deus Revelatus*. Obviously, this pessimistic view finds an echo in Thomas Hobbe's observation regarding the "original state of nature as nasty, brutish, and short," and one can certainly find the echo of Luther in Hegel's comments that "the original state of nature is savagery," and that "the truth of nature is that nature has no truth."[16] To have saving knowledge of the Divine one must, therefore, look for another kind of God – to the *Deus Revealatus* or the "God of History" and to what JCK von Hoffman later referred to as the processes of *Heilsgeschichte* that provide determinate evidence regarding God's saving activities, whether for nature or for ourselves. Left to ourselves, we are helpless, and if we think otherwise, we deceive ourselves.[17]

It is scarcely surprising that many *upbeat* Enlightenment and Post-Enlightenment thinkers had difficulties with these "hard sayings," nor is it surprising that they turned from the dogmatism of orthodox biblical theology to an investigation of historical consciousness itself as the primary clue to the meaning of Being. Indeed, preoccupation with the meaning of *historicity (Geschichtlichkeit)* became, from Schleiermacher to Dilthey, Bultmann and Jaspers, the late 19th century philosophical analog of *redemption* and this motif has continued far into the 20th century. As Gadamer observes in *Truth and Method*, "this consciousness no longer is concerned with any particular historical event or events that have saving power; saving power now rather lies in understanding historical consciousness itself." No doubt, this is what Moltmann, like Kierkegaard, views as symptomatic of the *inner emigration* of the German philosophical and theological consciousness. For with this subjective emigration, historical consciousness has forgotten, in many instances, the originating conditions of the emigration – conditions that were not aesthetic but moral in nature.

Hegel is a major exception to this pattern. For in Hegel – especially in the *Encyclopedia* form of his *Logic* and in the *Philosophy of Right* with which it is contemporary, the antecedent theological formulations come back in the form of objective spirit, subjective spirit, and absolute spirit. These parallels are to be found, I would argue,[18] in Hegel's reconstituted three-fold use of the law (as articulated above) as objective, subjective, and absolute spirit; a reformulation that not only includes the classical and medieval antecedents of this understanding (with which Hegel obviously was familiar), namely, created nature, redeemed nature, and sanctified nature, but also the medieval *threefold* conception of faith as *notitia, assensus* and *fiducia*. These reformulated conceptions, I think, have a direct bearing on the meaning of the terms *revolution, reformation, and restoration* and their relation to nature.

REVOLUTION, REFORMATION, RESTORATION

Following the social and political upheavals of the late 18th century, the meaning of nature is transformed by the symbolical terms *revolution* and *restoration*. The term *reformation*, which I would here consider as the mediating term, is progressively forgotten. As such, the fate of *reformation*, as a meaningful political category, is akin to the fate of the *soul* as a meaningful anthropological category. For as philosophy abandoned the *tricotomist* view of the self, which prevailed from the classical through the medieval periods of philosophy, for the *dichotomist* view that emerges after Descartes, what remains is the incompatible, if not incommensurable, mind – body split between mentalism and physicalism, subject and object, and the difficult problem of establishing a relation between them in the absence of a viable mediating category.

A similar situation obtains in the case of *revolution* and *restoration* when the mediating term, *reformation*, is abandoned. This problem is deeply embedded in the German language itself – which is why, perhaps, neo-Latinate French and English words are used inorder to convey meanings that are not self-evident in ordinary German. For example, the modern term *revolution* suggests a dramatically new change in the order of things – something that is *unnatural*, so to speak, otherwise it would not be a revolution. *Restoration*, on the other hand, denotes a counter-revolutionary *return* to the order of things, that is, a return to what is naturally and even divinely given.

While these English terms all have the same Latin participial prefix, viz., *re*, German cognates are formed across the spectrum of the prepositional/pronomonal prefix "*Um*" and the adverbial prefix *wieder*. Taken literally, *revolution*, in English, does not really signify the commencement of a wholly different order of things, but suggests rather a different *spin* on things which remain the same after all.[19] The German words *Umdrehung* and especially *Umwälzhung* have similar connotations – as in the various "turnings" of dancers on a solid floor. The cosmological meaning of *starry heavens above*, even in Kant, provides the outer foundation for "turnings of the planets," just as *mögliche Existenz* is grounded within the inner foundation of the Moral Law. Given these meanings, *restoration*, following political upheaval, seems a natural possibility as the historical re-turn to the way things have always been. The possibilities of temporal transcendence as "re-turn" are deeply embedded in terms like *Wiederherstellung* and *Wiedereinsetzung* and *Wiedererstattung* – all of which have to do with "putting back into place" by way of a *Wiederholung* or a temporal repetition.

What makes the German terms significantly different from their English equivalents is the adverbial, temporal prefix, *wieder*, which, in the case of *restoration*, places the accent on time or history, whereas the spatial or circular prepositional prefix *Um*, in the case of "revolution," places emphasis on space or nature. Perhaps in his attempt to move from the former to the latter, Heidegger provides his unique, anti-revolutionary, and quite Catholic nuance to *Umkehr* – namely, "the turning that is a re-turn to what dwells." In other words, the Heideggerian *Umkehr* or "turning that is a re-turn" seems to be predicted on precisely the same confluence of categories – the leading edge of signification, in this instance, being natural or spatial rather than temporal and historical, Catholic rather than Protestant. The point is that there seems to be nothing linguistically given in these terms upon which to base the radically new meanings attributed to them in post-Enlightenment, post-Revolutionary European conceptions of nature.

This confused semantical state-of-affairs may be reflected, we are told, in the discomfiture of the Crown Prince of Prussia when he read Hegel's *Philosophy of Right*, since Hegel's non-foundational approach to the legitimacy of monarchy makes no mention of natural or divine right. What Hegel does consider, however, is the power of *Reformation* – precisely that which, as the product of subjective spirit, stands betwixt Revolution and Restoration as the mediating center of consciousness qua conscience. It may be helpful to note that the ordinary German terms for the Latin, *reformatio*, have various prefixes – whether *Umgestaltung*, *Verbesserung*, or *wieder-formieren*, all suggesting that reformation entails a change for the better, and that such new-formations

(*Neuesformieren*) must encompasss the spatial *and* the temporal, the natural *and* the historical, the communal *and* the individual, to be lasting and effective models of transcendence.

This horizon of meaning is implicit, it seems to me, in Hegel's emphatic assertion: "It is a false principle that the fetters which bind Right and Freedom can be broken without the emancipation of conscience. There can be no Revolution without Reformation."[20] To understand the weight of this assertion, one must understand what Hegel means by "the emancipation of conscience" (*die Befreiung des Gewissens*). *Befreiung* means "emancipation" or "liberation" as its root *frei* suggests, that is, a "freeing up" of human nature from the bondage of ignorance. The rhetorical forms of freedom, needless to say, are intense nowadays, and most expressions are political, such as "empowerment" and "emancipatory praxis." In both instances, these slogans make plain the need to be freed from oppression or from those conditions deemed "unnatural" and therefore oppressive. Indeed, the former term, "empowerment," is presently undergoing transmutation into a noun as though it is some kind of *thing* that can be legislated and delivered as an object.

Most people, perhaps, are sympathetic and responsive to these notions since all of us feel a little oppressed from time to time, believing that we could use a little more "power" and that, moreover, we have a natural "right" to this. In Hegel's use, however, *Befreiung* also has the sense of "satisfaction" embedded within it. It is precisely this nuance (and its Paulinean nuance) that makes it more than just a political or psychological term, and certainly something more than is given by *nature.* If we examine the word that *Befreiung* modifies, namely *Gewissen* or "conscience," we note that it is a word carefully chosen. Hegel could have used the term "mind" (*Gemüth* or *Geist* or *Bewusstsein*), in which case we would have a "free spirit" or a "satisified mind."[21] But Hegel is deliberate in the choice of *Gewissen*, and by this choice he points to the moral essence of consciousness and to a "certainty" (*Gewissheit*) that is possible only by way of the continuous integration of the intellectual, moral and religious elements of spiritual rebirth and conversion. And this, of course, was precisely the intent of Luther's reformational conception of *Rechtfertigung*. As such, it might be argued that *reformation* stands between the spatial-natural and the temporal-historical as the absolutely essential mediating category of possible Existenz. Apart from this necessary mediation, the intellectual and moral confusions of foundationalism or anti-foundationalism are inevitable.[22]

What makes Hegel's assertion difficult, of course, is the obvious fact that there are no easy formulae for making it work. In many ways, this is what informs Moltmann's lament. Truly enlightened emancipation is the project of possible Existenz, and the project of possible Existenz is the project of understanding that mysterious place, according to Goethe, "where nature and history coincide." That place of coincidence, according to Erazim Kohák, is "the human heart . . . between the *embers and the stars*," where our "moral sense of nature" is born.

Department of Philosophy, Boston University

NOTES

[1] Jürgen Moltmann, "Theology in German Today" in *Contemporary German Perspectives on 'The Spiritual Situation of the Age'* ed. Jürgen Habermas, trans. Andrew Buchwalter (MIT, 1984), p. 187. German Edition, *Stichworte zur 'Geistigen Situation der Zeit'* (Suhrkamp, 1979).
[2] Erazim Kohák, *The Embers and the Stars: A Philosophical Inquiry into the Moral Sense of Nature* (Chicago, 1984), pp. 32, 49, 66–68. The metaphoric title of this work, and its meditative character, conveys an accurate sense of the spirit informing Kant's understanding of "the starry heavens above and the moral law within" as something more than a purely formal transcendental deduction.
[3] The permanent separation of the Eastern (Greek) and the Western (Latin) Church in 1054, beginning with the Iconic Controversy in the 9th century and continuing to the present day, originated in conflicting conceptions of nature and the divine. This is another aspect of the "civilizational fault line," to use Huntington's phrase, so evident in the religo-ethnic conflict in the former Yugoslavia.
[4] Martin Chemnitz in his *Loci Theologici* (1591) says that "Our natural knowledge of God either amounts to nothing, or is imperfect, or languid" – *amounts to nothing* in the sense that it has no saving power, *imperfect* because it knows nothing of the "inner" or moral nature of the Law, and *languid* in the sense that it is "feeble and easily shaken by doubts." Chemnitz cites as an example the *Tusculan Disputations* where Cicero tells Antony "to read diligently Plato's treatise concerning the immortality of the soul. Nothing will be left to you to desire." "This I have done," Antony replies, "and somehow or other as long as I read, I assent, but whenever I lay down the volume and begin to reflect concerning the immortality of the soul, all my assent glides away" *(Loci, I, 20)*. So also Johannes Quenstedt, in his *Theologia Didactico-Polemica (1685)*, asserts that while one can make a case for the natural knowledge of God (*aletheia* and *aletheian tou theou*, Romans 1:18, 25) as partly *hemphotos* or "innate, implanted and constitutional," and partly *epikatatos* or "acquired by reasoning and research," both are imperfect with respect (a) "to their object, this object being either altogether unknown or inadequately known," and (b) "to its subject, this subject being the inability to recognize God with sufficient constancy . . . in consequence of congenital corruption" (I, 253).
[5] Jewish scholars have long pointed out that such perceptions are mistaken or at least distorted by the Augustinian notion of *concupiscence*, a notion that substantializes the famous line from the Song of David (Psalm 51) that "in sin did my mother conceive me" as distinct from "unto sin." The latter contains more of an Aristotelian notion of *potency*, and this may be a more accurate reading. Whatever the case, a thousand years of celibate tradition could not fail to color the views of the Reformers, even though they had broken the pattern during the Reformation.
[6] Exodus 20:4 (King James Version). Luther's rendering is instructive: *Du sollst dir kein Bildnis noch irgendein Gleichnis machen, weder von dem, was oben von Himmel, noch von dem, was unten auf Erden, noch von dem, was im Wasser unter der Erde ist.*
[7] Luther lived on the cusp of what R.G. Collingwood accurately describes, I think, as the transition from the "organic" conception of nature as a "thinking substance" in the classical and early medieval worlds, to the "mechanistic" view that emerged in the Renaissance and continued through the Enlightenment. Both are "analogical," he argues, the analog being the consciousness of the machine in the Renaissance and Enlightenment; hence the movement from Plotinian emanationism, in the former, to deism, in the latter. See R.G. Collingwood, *The Idea of Nature* (Oxford, 1945), p. 5f.
[8] A notable exception, of course, is the 17th century speculative theosophist, Jacob Boehme – who, of course, was condemned by the Lutheran establishment as a heretic!
[9] Of course, this is the general basis of Richard Rorty's skeptical argument in *Philosophy and the Mirror of Nature* (Princeton, 1984).
[10] It is possible, I think, to establish a line from this view to the rather more optimistic holistic environmental movements of the present day – movements driven, on the one hand, by egalitarian conceptions of right that include even species-beings other than humans, and, on the other hand, by a deeply felt, intuitive devotion to nature as an in-and-for-itself.
[11] I've developed this at some length in *Hegel and the Spirit: Philosphy as Pneumatology* (Princeton, 1992), pointing out that while Pietistic excess was repugnant to both Kant and Hegel, the essence of Pietism, especially from the standpoint of social and political philosophy, was deeply appreciated and deemed an intrinsic part of the civil society.
[12] Luther characterizes this dialectic as something akin to "two foxes tied together by their tails," the only thing being different about them is that "their heads are pointing in opposite directions." Both foxes, as in Isaiah Berlin's famous characterization, are Pelagian in the sense of claiming

essential knowledge regarding the nature of nature "by their own wits," so to speak; on the one hand, the Thomists, on the other, the Anabaptists, "and while they outwardly profess to be the great enemies of one another, inwardly they think, teach, and defend one and the same thing . . .," viz., self-salvation. See *Commentary on Saint Paul's Epistle to the Galatians (1531)*, Middleton Text (1575) prepared by Philip S. Watson (London: James Clarke & Co., 1953), p. 19.

13 I might add that this "intensity" also contributes to what remains, for many, the "inscrutable" dimension of German Idealism (especially certain Anglo-American thinkers), and that this inscrutability remains such apart from an appreciation of the religioius and theological antecedents of 18th and 19th century German philosophy.

14 On a personal note, I might mention that, many years ago, I discussed with Professor Kohák the prospect of doing a dissertation on Erasmus's *libro arbitrio* and Luther's *servo arbitrio*, developing this historical problematic within the context of Ricoeur's *Le Volontaire et l'involontaire*. "Now that is a dissertation I'd like to advise"! he replied. I decided instead to work on the problem of Transcendence in the philosophy of the unabashed voluntarist, Karl Jaspers, and Professor Kohák, thankfully, still served as one of my advisors.

15 What precisely "nature" means in this instance is unclear, for there is nothing in its formulation that approximates the notion of *phusis* (especially as Kohák develops it in *The Embers and the Stars*. In popular usage, this formula seems to be a fall-back to the substantialistic language of the scholastics, whose discourse was informed, in the main, by the christological controversies and the language of "nature" considered as *hypostasis* – that is, as "fixed." Calvinism, of course, reinforces this use through its notion of "total depravity." What it has meant to the millions of individuals who have "confessed" their "sinful and unclean" nature down through the centuries is anyone's guess – although it may be safe to speculate that for Lutherian and Catholic alike, it connotes the Augustinian sense of *concupiscence*, or at least it used to. As we discussed this paper, Professor Kohák related the humorous story of two children, one Lutheran and the other Anglican, the former saying to the latter: "You are just an occasional offender, but I am a natural sinner"!

16 See Hegel, *VPRel.*, *Enz.*, *passim*.

17 In his critique of liberalism, Karl Barth made frequent use of the late-19th, early-20th century comic-liberal figure, Baron von Munchausen, who 'was forever trying to pull himself out of the quagmire of contingency and sin by pulling upwards on his handlebar mustache'!

18 See *my Hegel and the Spirit: Philosophy as Pneumatology* (Princeton, 1992).

19 A current example of this might be adduced in the cynical contemporary term *spinmaster*, that is, the publicist who, for their clients, puts a different "spin" on things which, otherwise, remain very much the same, viz., "politics as usual."

20 Revolutionary conceptions of Liberalism, understood as an abstract principle, are "falsche" because "die Fesseln des Rechts und der Freiheit ohne die Befreiung des Gewissens abgestreift werden, daß eine Revolution ohne Reformation sein könne." *Lectures on the Philosophy of History*, trans. J. Sibree (New York: Prometheus, 1991), p. 453; Suhrkamp, XII, p. 535.

21 See Robert Pippin, *Hegel's Idealism: The Satisfactions of Self-Consciousness* (Cambridge, 1988). Pippin is entirely correct in suggesting, as he does extremely well in this book, that one of Hegel's chief concerns was to broaden or existentialize (and historicize) the implications of the transcendental turn in Kent.

22 See Tom Rockmore, *Hegel's Circular Epistemology* (Indiana, 1986).

STANLEY ROSEN

REMARKS ON HUMAN NATURE IN PLATO

The word *nature* is ambiguous and has undergone various shifts in meaning during its long history. Perhaps the most important of these shifts is that which distinguishes the predominantly classical sense of *phusis* or *natura* as divine from the typical modern conception of nature as *res extensa* or in other words as extension or body in contrast with mind, soul, or spirit. In all epochs going back to classical antiquity, nature has been defined by way of contrast with or opposition to an antonym. The most important of which I am aware are *nomos* or custom, *techne* or art, soul, intellect, spirit and history. Each of these antonyms is itself ambiguous and admits of various senses. On the whole, however, it does seem possible to detect a common theme in the diverse understandings of the term "nature." To borrow an expression from Aristotle, nature is that which happens always or for the most part. This is so whether we think of nature as divine or lifeless. Custom, art, and history are all changing.

The distinction between regularity and change can also be applied to the soul, intellect, or spirit in such a way as to reveal yet another dimension in the ambiguity of our conception of nature. Soul, intellect, and spirit could each be regarded as the nature of the human being even though we distinguish between natural and non-natural operations or productions of soul, intellect, or spirit. One of the greatest difficulties in the history of philosophy is precisely to determine which acts, if any, express the nature of the human being. In my talk today I am going to restrict myself for the most part to a discussion of what it might mean to speak of human nature in the dialogues of Plato; needless to say, I cannot exhaust even this restricted topic in a lecture. I will try to give a general characterization of the problem of human nature as it is presented by Plato, a characterization based upon a sampling of the most important evidence. To do even this in my opinion is a difficult and perhaps even questionable procedure, since it requires me to take passages out of their dramatic context, and to attribute general doctrines to Plato as though he had written treatises explaining discursively and in detail his understanding of human nature. Nevertheless, I will try to say something useful on the topic.

Speaking very generally, there are three broad positions that we might take on the question whether human beings have a common nature. The first two are obvious enough; we may either affirm or deny the proposition in question. The third position is much subtler and accordingly more difficult to articulate; it is

R.S. Cohen and A.I. Tauber (eds.), Philosophies of Nature: The Human Dimension, 151–162.

expressed by Nietzsche's Zarathustra in the assertion that man is the not yet established or ascertained animal: *festgestellte* can also be translated literally as "securely positioned." In other words: man is an animal, not a god or a machine, and an animal that has a position or is partly put together in a way that can be ascertained but not precisely defined. Human beings can become not merely who they are, as Zarathustra says in partial citation of Pindar (he omits Pindar's qualifying *mathōn*, "by understanding"); they can also become other than they are. But "other" is circumscribed; man can become a superman (*Übermensch*); but a "super" man is still some sort of man.

What in fact do the two extreme answers to our question mean? I suppose that they must mean something like this: either human being has a fixed nature or it has no nature and so no fixity whatsoever. Now it seems obvious that both of these views, if taken literally, are absurd. I cannot think of anything about which we could state with certitude that it has a fixed nature, if this last expression means "fixed in all respects." But even allowing for the distinction between nature and chance, there is an indeterminate range in human activity that rules out any reference to our kind as fixed in nature. To take a simple example, those who believe in human nature normally speak about natural ends or laws that guide or govern our acts. But it is obvious that we can deviate from these ends and disobey these laws. Perhaps we ought not to do so, but we can. Conversely, it is absurd to say that human beings exhibit no fixity whatsoever, since this would mean that we have no defining properties, and so that it makes no sense to speak about "human being" at all. I will not attempt to provide a detailed refutation of what is manfestly absurd. Instead, at least for the purposes of this talk, I am going to assume that all sensible or defensible theses about human being, when they are formulated with moderate care, are versions of the third view. I restate this third view as follows: human being has a nature, but it is partly open. By "partly open" I mean that human being is not entirely fixed; it is part of our nature to be varying or adding to as well as subtracting from what we choose to be. In speaking about Plato, I need to make explicit a corollary to the thesis of partial openness, a corollary that states that human nature is partly divided against itself. Human being is stamped by what Hegel calls an "innere Zerrissenheit," a tear or rip in our internal fabric which cannot be entirely stitched together by our own efforts.

Now there cannot be any doubt that Plato also adhered to a view of *innere Zerrisenheit* in human nature. This can be illustrated very easily by citing the doctrine of the soul in the *Republic*. Socrates speaks there of the three parts of the soul, intelligence, spiritedness, and desire. He makes it explicit that desire and intellect are in conflict, and further that the intellect is incapable of controlling desire directly, but can do so only through the mediation of spiritedness. Socrates refers to the soul as part master and part slave "by nature" (IV.444b4-5). Human nature is thus divided against itself, and there is no natural guarantee that intelligence will triumph. This triumph requires the rule of philosophy, or in other words, art.

Before I go any further, let me reply to a possible objection. Is there not a difference between granting that human beings do not always live in accord with their nature and the thesis that human nature is intrinsically open? One might hold, in other words, that whereas human beings may disregard what is right by nature, this does not change the fact that it is both possible and better for them to live in accord with nature. To this I reply as follows. It may or may not be true that there is a way of human life that is best by nature. But it is certainly true that we must be persuaded to live in accord with whatever philosophers claim to be natural for us. There is a critical zone in which nature does not enforce her ends or laws. Since we are free to disobey these laws (but not, of course, all natural laws), we are open to a life in opposition to nature. Furthermore, there is an important sense in which we are compelled to live in opposition to nature, a sense that is guaranteed by nature herself. This view is to be found in Plato as well as in modern thinkers like Hobbes and Hegel.

In short, I exclude as absurd the thesis that human nature is non-existent or that we are simply and completely "open" as though we constituted an ontological void or nullity. Equally absurd, however, is the thesis that we are capable of living "in accord with nature" by the straightforward expression of our natural constitution. To say that we are partly open is also to say that we are partly closed. As to the question whether what is now partly closed might be subsequently opened, or conversely whether what is now open might be subsequently closed, I reply that there are no certain limits to the fixity of human nature. But in saying this, I do not mean that we are in total ignorance concerning what it is to be human. I mean that this question is finally not scientific or metaphysical but rather belongs to the domain of common sense and sound judgment. We have to evaluate the results of science and the claims of metaphysics on the basis of what we actually are, always or for the most part. This includes deciding whether we ought to be other than we are. But as soon as we become something that is always or for the most part other than what we human beings presently are and have been, then we shall cease to be human beings, and with us, philosophy will also cease.

In my remarks to this point I have tried to sketch the general terrain within which to locate the Platonic view of human nature. I now turn more precisely to Plato. And first I need a technical distinction. Plato speaks of nature in two broad senses. In the first sense, that which exists by nature is not man-made. The most dramatic instance is the cosmos, which the dialogues attribute, whether metaphorically or literally, to a divine artificer. In the second sense, the dialogues speak of "nature" as the power or essence of something, whether that thing is itself natural or artifactual. I want to illustrate this second sense with a crucial example, namely, the *polis* or city-state. As is well-known, Aristotle includes among his three or four definitions of human being the assertion that man is by nature the political animal. What is not so well-known is the fact that no such assertion occurs in the Platonic dialogues. Let us make a preliminary assessment of what this means.

Again I start with a broad distinction. The root of the Greek word *phusis* signifies originally both "grow" and "showing forth" or "emergence into the domain of visibility or sunlight." The cosmos is visible but one could not say that it grows; plants grow, and in fact the Greek word for plants is *ta phuta*.

Parenthetically, I note that in a striking passage in Book Ten of the *Republic* (597d5), Socrates uses the word *phutourgos* or "gardener" to refer to the god who grows the ideas or paradigms of the entities of genesis. But animals also grow, and thus so too do human beings, whereas one would not normally speak of the "growth" of human artifacts except in a metaphorical sense, or when referring to the results of breeding plants or animals, and so to a modification of living nature. *Phusis* is clearly associated with the notion of life; since this life is not man-made, it is normally attributed to the activity of the gods; and indeed, *phusis* was popularly regarded by the ancients as a goddess. In sum, let us say that the natural is initially and primarily divine life; derivatively, it is everything that is present to human perception and cognition as given to us by the gods. Third, there are deeds and speeches which can be called "natural" even though they are neither simply given nor do they grow forth from the given without human action. Cities do not grow on trees or bushes, but they may be regarded as natural in the sense that human nature can be fulfilled only through political or communal life.

Aristotle's famous assertion, previously cited, that the natural is what occurs always or for the most part, is a kind of secularized version of this broader Greek view. The human changes continually, as does life itself. But the ways in which human things change, like the ways of life and genesis altogether, do not change. This is not enough for an exhaustive account of Aristotle's doctrine of nature, but it will do for our present purposes. "Nature" is orderly change; the ways of change are fixed. This means, of course, that they are not arbitrary or spontaneous, or dependent upon the inscrutable will of the gods. What is fixed in the sense of regular or orderly is knowable. We can *count* on this order, an expression that brings out very nicely the important role played by mathematics on the one hand and what Aristotle would call the human or political implications of natural order on the other. Nature is not only orderly but it is also authoritative and reliable. As one could also express this, nature is both beautiful and good.

When Aristotle says that man is by nature the political animal, he certainly means to say that human beings must bring cities into existence, but more importantly that the political telos of human nature, while of course depending upon practical intelligence, is accessible to us through practice. I call to your attention a crucial implication of this view. Philosophy, and above all theory, is not ncessary for the coming into existence of cities that are adequate, whatever their faults, to the expression of our natural political telos. In other words, closely connected with the thesis that man is by nature the political animal is the Aristotelian tripartition of theory, practice, and production. The perfection of theory, or the *bios theoretikos*, is not the same as the perfection of practice or the *bios praktikos*. It is not merely that cities come into existence without philosophers; what is also the case is that the perfection of the political component of human nature does not depend upon philosophy. It is therefore also true that for Aristotle, the satisfaction of the political telos does not depend upon the achievement of some extreme, unnatural form of perfection. This constellation of doctrines is quite foreign to, and was certainly intended as a criticism of, Plato.

Whereas Plato of course speaks of "the nature" of the city, just as he does of the nature of man, he does so to designate its power or essence: the way in which it must be, in order to fulfil the function of a city. The city is not, however, "natural" in the Aristotelian sense that it is the orderly consequence of man's political nature. On the contrary, Socrates speaks on more than one occasion of human beings as sick by nature, as having disordered souls that require philosophical therapy or psychiatry, as incapable of living just or good lives unless they are healed or governed by philosopher-kings. For this reason his usual account of the city is that of an artifact. This is evident in the terminology of the *Republic* and in the central paradigm of the *Statesman*, to mention only the two most authoritative sources for Plato's political views.

Aristotle would no doubt agree with Plato's view of the defects of human nature, but not at all with the diagnosis of a sickness that can be cured only by a combination of violence and philosophy. Aristotle's *Ethics* and *Politics* are intended as expressions of political excellence that is accessible by nature. Plato's *Republic*, even if we do not take it literally as a prescription for actual political revolution, nevertheless and even more on this supposition expresses the impossibility of attaining to political excellence by what Aristotle would call natural means.

In the *Gorgias*, Socrates speaks of justice as the medicine of the soul (478d6-7). Whereas health is the order of the body, justice and temperance are the order of the soul (504c5-d4). In an earlier passage, Gorgias, the great rhetorician, describes how his brother, a physician, engaged him to persuade a patient to take a bitter medicine (456b6ff). Socrates subsequently says that the art of rhetoric in the usual sense stands to medicine as does the art of the pastry chef to that of the gymnastics trainer (465c1 ff). But Socrates claims to be virtually the only Athenian who practices the true art of politics, and in so doing, compares himself to a physician (521d6ff, e3ff). The genuine art of politics is in other words the medicine of the soul; philosophy is required to heal injustice. Art is required in order to rectify the natural disorder of the human being. This series of passages should be connected to the discussion of genuine or philosophical rhetoric in the *Phaedrus*, where Socrates refers to rhetoric as the medicine of the soul (270b1-c2). The rhetoric of Gorgias is akin to that of the pastry cook, but this does not mean that rhetoric is superfluous to politics. What we require is the rhetoric of Socrates or Plato himself, or genuine rather than specious psychiatry.

The intrinsic sickness or psychic disorder of the human being is also central to the political discussion in the *Republic*. I have already made this general point with respect to the parts of the soul. It can be made with equal ease with respect to the city. What Socrates calls the "true" and "healthy" city (372e6-8) is almost immediately rejected because of Glaucon's complaint that its citizens are restricted to a vegetarian diet and have no couches on which to recline while eating (372e6-8). Socrates does not mention it, but we may note that in this true city, which is fit for pigs but not human beings, there is no philosophy. Socrates is forced to start again in the construction of what will be the just or "beautiful" city (VII.527c2) in which philosophers rule. He calls this city "luxurious" (*truphōsan*: II.372e3) and "fevered" (*phlegmainousan*: e6-8). The natural disorder or sickness of the human soul is thus represented by Glaucon, the extremely erotic youth (III.403a4-5; V.474d4) who is also very courageous (II.357a3). Glaucon is the interlocutor in the most philosophical passages of the dialogue and it is he who encourages Socrates to speak about topics which he is reluctant to discuss, including the noble lie (III.414c11) and the nature of the good (VI.509c3).

Glaucon's erotic nature is a perfect image of the sickness intrinsic to the human being. On the one hand, eros is an essential component of the philosophical nature, as we know not only from the *Symposium* and *Phaedrus* but also from the discussion of the philosopher in the *Republic* (for a detailed treatment of this point, see my "The Role of Eros in Plato's *Republic*" in *The Quarrel between Philosophy and Poetry*). On the other hand, the *Republic* makes explicit the political danger of eros by identifying it as a tyrant and madman that drives out the useful desires such as shame and temperance (IX.573b6ff). Eros *turannos* (573d3) is obviously unjust, and hence constitutes the sickness of the soul. The intrinsic ambiguity of eros is not overcome by the claim that the philosophical eros differs in nature from that of the body; this is not correct.

As is obvious throughout the dialogues, the human eros must be transformed or deflected from its corporeal objects to a desire for the Platonic Ideas; but it is the same eros in all cases. In the *Symposium*, it is not Socrates who distinguishes between a noble and a base eros but the refined pederast Pausanias. That the philosopher, as hyper-erotic and even manic, is an extremely dangerous type, quite unstable in its youth, is made plain in the *Republic* at VI.491e1 ff, where Socrates explains how the best natures are the most harmed by a bad education, and that unless he is saved by good luck, even the well-educated philosopher will be destroyed by the wrong soil and nurture.

The philosopher is a physician of the soul (cf. VI.489c1) who must moderate or shape the fundamentally erotic nature of the human being by an elaborate political rhetoric that features the telling of medicinal lies (II. 382c6ff; III. 389b2ff, 414b8ff; V. 459c2-d1) and the comprehensive regulation of education from the cradle to the grave in the case of the guardians. This of course raises the question of who will properly train the potential philosophers. Socrates' reply is explicit if not entirely unambiguous. The fate of the potential philosophers depends upon good luck, as for example in his own case, namely, the *daimonion* or demonic voice that saved him from his own divided nature. As for his successors, they will be trained by Socrates, or rather by Plato, who assumes the prototypical figure of the philosophical prophet and lawgiver.

I note that at III. 399e5-7, Socrates claims that the musical and gymnastic education they have designed amounts to a purging of the luxury of the city. He does not say that it has purged the fevers of the city. But we do not need to argue from silence. The elaborate steps taken to control sexual reproduction among the guardians, including rigged lotteries to select mates, direct lies, and the destruction of offspring of uneugenic intercourse (V. 459a1-460c5) will eventually fail, thus leading to the destruction of the city (VIII. 546a2ff). But even while the city exists, it will contain injustice. Socrates says (III. 410a2-4) that the physicians of this city will put to death those who are sick in the soul. More dramatically still, the largest class in the city, consisting of workers of all kinds, is not subjected to the education, restrictions on breeding, or supervision of occupations that stamp the souls of the guardians. It is obvious that the largest portion of the city is thus characterized by an intrinsic or natural injustice that cannot be removed by philosophy but only regulated by the police. Nor should one forget the philosophers themselves, who would prefer to philosophize but who are compelled to rule. It would take too long to analyze this complex and controversial issue; suffice it to say that there is legitimate reason for supposing that, in compelling them to choose an inferior over a superior life, the city is doing an injustice to its best citizens. Finally, I remind you that the first act of the philosopher kings will be to send everyone over the age of ten into the countryside (VII. 540e5), an obvious euphemism for something very close to what is today called "ethnic cleansing."

It is true that, according to Socrates, the beautiful city has been built "in accord with nature" (IV. 428e9). But it is equally true that the human being is divided against itself by nature, as is most obvious from the ambiguity of eros, the principle of philosophy and tyranny. Plato's point is no different from that of Machiavelli in the *Prince*: justice depends upon initial injustice; nor will the latter ever be eradicated from human affairs. It can only be controlled by medicinal lies, rhetoric, and purges, and these are to be administered by the most virtuous and hence the most just citizens. There is no equivalent in Plato to the Stoic doctrine of living harmoniously in accord with nature, a doctrine that is so castigated by Nietzsche. Parenthetically, I note that in the Romantic period, of which the twentieth century, for better and for worse, is a distorted continuation, there comes into view a doctrine of living in accord with nature understood as the countryside in opposition to the city. I suppose that Rousseau is as responsible for this doctrine as anyone; it is clear, however, that this is not a political but an anti-political teaching, and that it too points to the underlying notion of a divided human nature. Rousseau was both the citizen of Geneva and the solitary promenader. This of course is not the Platonic view; in the *Phaedrus*, Socrates makes it clear that philosophy belongs to the city rather than to the countryside or the natural landscape, to mention only the decisive point. If it is true in a general sense that Plato accepts the view of human being as by nature political, the fact remains that human nature is divided against itself. What Socrates calls "the healthy and just character" of the philosophical nature, which hates the lie in the soul and avoids the chorus of evils that follow the many nonphilosophers (VI. 490b8-c6), must regulate the erotic madness of the many by its own erotic madness. What is the tyranny of intemperance in the many is to be transformed into the tyranny of sobriety in the few.

In sum: according to Plato, the human being must be modified by philosophical *techne*, and this includes philosophical rhetoric; the human being is sick by nature and must be healed by philosophical psychiatry, in order to live in a propertly constructed city. The citizen is a work of art, not of course a perfect work of art but one in which the inner contradiction of human nature can be contained within a harmony induced by the right kind of *paideia* for as long as possible. I can now refine my thesis about Plato as follows. The city as such is natural in the sense that it comes into existence because the individual human being is not autarchic but has need of others (II. 369b5ff). But there is no perfection of *praxis* or political virtue by nature, as there is for Aristotle. In the extreme formulation of Socrates at *Republic* VI. 500d4-9, the philosopher will be a good demiurge, i.e. artisan, "of temperance, justice, and all of demotic virtue." Temperance and justice are the key political virtues, but they are part of demotic or vulgar virtue, which is to say that they are fashioned by political demiurges, and in the wished-for case, by the philosophical demiurge.

I believe we can now see why it is a complete mistake to take one's bearings by the expulsion of the mimetic poets in attempting to understand Plato's *Republic*. The mimetic poets are expelled because they are to be replaced by the philosopher-kings, not because poetry is to be replaced by mathematics or dialectic. This is why we cannot take literally or narrowly Socrates' remarks to the effect that he and his interlocutors are not poets but *oikistai poleōs*, "builders of the city" (II. 378e7). They are indeed *builders*; as such, they must abstain from all other demiurgic arts and be *dēmiourgous eleutherias tēs poleōs*, "demiurges of the city's freedom" (III. 395b8-c1). That this art is not to be carried through in metre or verse is irrelevant (see III. 393d8). It is a poetic or productive art; the name "demiurge" is in one sense of course metaphorical, but the point of the metaphor is to make explicit the constructive nature of the art of philosophical politics.

So much for the terminology of the *Republic*, examples of which could easily be multiplied to confirm the points I have made. I turn now to the *Statesman*, a dialogue of Plato's later years, and one in which the Eleatic Stranger replaces Socrates as the main interlocutor. I can only mention here that one meaning of this shift is to indicate the more theoretical or as we would say "abstract" nature of the treatment of politics in the *Statesman* as compared with that in the *Republic*. The very success of the *Republic*, which is due to its resemblance to a kind of comprehensive novel in the style of Tolstoy, has prevented many readers from noticing that it is indeed a philosophical novel, even a fairy tale about the relation between philosophy and politics. We have been deluded on the one hand by Socrates' criticism of mimetic poetry, and on the other by the magic of Plato's mimetic poetry, and so failed to notice that, with all of its discussion of the Ideas, the good, and the role of mathematics in a philosophical education, the *Republic* is a poetical presentation of the problem of philosophical politics.

However this may be, the Eleatic Stranger makes entirely prosaic the Socratic view of politics as demiurgic, rhetorical, or constructive. The central paradigm, or as we would say today, model, of the art of politics is the art of weaving (for details, see my *Plato's Statesman: The Web of Politics*). Weaving makes artificial products like curtains, blankets, and cloaks in order to protect the body from the painful and destructive effects of nature. As the Stranger formulates his final definition of the art of politics, the statesman "rules all of the arts and cares for the laws and everything in the city and weaves tham all together most effectively (305e2-6). The statesman is not a creator *ex nihilo*. Just as the weaver is furnished with wool by nature, so the statesman is furnished by nature with the fundamental materials of the city, namely, two types of human natures (306e9ff), one courageous and sharp, the other moderate and gentle.

These two human types are in a certain manner enemies; a condition of political discord (*stasis*) exists between them. In other words, nature furnishes the philosopher-statesman with discordant materials and leaves it to him to harmonize them or weave them together into a city, which is artificial in the fundamental sense that it is not supplied by nature, and in particular not by our *political* nature. We are by nature enemies, and our cities, unless respun by the philosophical weaver, are in the natural conditions of revolution. As a consideration of the subsequent description of political weaving would show, the philosopher stands to the city as does the divine artificer of the *Timaeus* to the cosmos. It should also be mentioned that from 293b1 almost to the end of the dialogue, the Stranger regularly compares politics with medicine.

The conclusion about weaving together different natures does not deviate substantially from that of the *Republic* (see for example III.410d1ff; the philosophers must harmonise the hard and the soft natures by a judicious mixture of gymnastics and music), but it does bring out certain ambiguities that are left undiscussed and even unmentioned in the earlier Socratic presentation. I will give just one example. According to the Stranger, those who are by nature gentle or moderate try to live in an orderly and quiet way, *ta sphetera autōn prattontes*, "tending to their own affairs" or "minding their own business" (307c9-e4). This is of course almost literally the definition of justice given by Socrates in the *Republic* (III. 433a1-b4). Socrates means that each of us ought to perform the basic type of work for which we are best suited, namely, philosophy, soldiering, or the arts and crafts, farming, and physical labor. But this is a matter of justice, and so of politics, not of personal happiness. Our goal is not to make even the guardians happy, although Socrates assures us that they will be, but rather to make the city as happy as possible (III. 419a1ff).

Apart from the absurdity of a happy city, and indeed one in which happiness is so variously enacted as in the *Republic*, it is clear that in the crucial case, that of the guardians, Socrates intends "one's own" to be interpreted as strictly as possible to mean "the city." In the *Statesman*, on the other hand, "one's own" means whatever makes one personally content or happy. To recur to a previous point, let us assume that it is just in the beautiful city for philosophers to be compelled to rule rather than to devote themselves to the study of philosophy in their private capacities. In order to make this assumption, we must also assume that by shifting from that at which they are best, and which is the best life in itself, namely, philosophy, to something at which they may excel but which is neither best in itself nor does it constitute their personal happiness, the philosopher-kings exemplify justice, or minding their own business, by not minding their own business but that of everyone else, and so by exercising *polupragmosunē* or injustice.

Even on the basis of these assumptions, it remains the case that justice and happiness are not entirely compatible. Perhaps even more important, however, from a political standpoint, is the fact that the correct political interpretation of "minding one's own business" requires violence to be effected upon the sense of "one's own." The problem was clear in the *Republic* as early as Book II, 375a2ff. The guardians must be both sharp or courageous, and so cruel to enemies, but also gentle to friends. In order to support his contention that the mixture of these attributes is not contrary to nature, Socrates cites the example of the noble hound who barks at strangers and is gentle to friends. The choice of a beast as the paradigm of human nature is obviously a joke, not to mention a circular one, since the political attributes of dogs have undoubtedly been inculcated into them by generations of breeding and training. But it is in any case silently rejected by the subsequent reliance upon education of body and soul in order to produce by art what nature herself cannot supply.

It makes no difference that the best interests of the courageous and the moderate alike are served by a harmonious city. The main point is that no such city can arise, or if it arises, last for more than a short interval, unless it has been woven together by the royal art of the statesman. I repeat: the materials are furnished by nature, but not the city. Human beings are by nature enemies, very much as Thomas Hobbes, the ostensible founder of modern political philosophy, who was of course only following Machiavelli, said that they are. To come back to the *Statesman*, the Stranger tells us that the two kinds of human natures "are always marked by the greatest enmity and stasis with respect to each other" and so that "not the least parts of virtue differ from each other by nature, as do the deeds of those who possess them" (308b2-8). This passage is important because it differs from the usual Socratic contention that the virtues are one. The difference can be partly removed by noting that Socrates regards virtue as knowledge; in other words, all virtues other than wisdom are demotic or not genuine virtues. As he says in the *Republic* at VII. 518d9-e3, the so-called virtues, namely, those other than intelligence (*Phronēsis*), are close to those of the body and are produced by habituation and exercise. But that does not alter the political situation. Neither nature nor political prudence in the usual or non-philosophical sense will overcome this natural split between human kinds. That task can be performed by the philosopher alone, the royal weaver, as we may call him.

One has to underline in red the important conclusion that for Plato there is no such thing as political virtue in the Aristotelian sense. There is no practical virtue as distinguished from theoretical and productive virtue. On the contrary, there are only theory and production; this is the principle in accord with which the arts and sciences are regularly classified in the Platonic dialogues. And it remains a deep question in Plato, seldom noticed but nevertheless of crucial importance, whether politics is a theoretical or a practical art. In the division of the arts in the *Statesman*, the interlocutors begin by classifying the statesman under the epistemonic men, i.e. those who are knowers rather than makers (258e4ff). But the dialogue culminates in a comparison of the statesman to the weaver or maker, and it is this paradigm that is followed subsequently. I cannot go into the extraordinary intricacy of the intermediate steps, which I dissect in my book on the *Statesman*, but the result is plain. Just as diaeresis is constructive in the modern sense of the term because it makes new concepts like that of the art of tending horned herd-animals, or whatever else is necessary to the intentions of the diaeretician, so too politics is not pure gnosis but at the same time construction. The very least one could say (in the terminology of the *Statesman*) is that politics is the art of commanding for the sake of a genesis. We cannot simply say that the city requires guardians who exhibit the paradigm of the noble hound; we must construct a new type of huma being by devising a sound set of procedures which are furnished not by political prudence but by the madness of philosophy.

The *Statesman* is a maddeningly obscure as well as a perversely amusing dialogue. It forces us to consider the political analogues of carding and combing wool, as well as of arts like that of the fuller, the mender, and the launderer of clothing. But the central image is not difficult to decipher. The statesman is neither a shepherd of a herd nor a captain of the ship of state, two metaphors for the royal art of politics that are initially suggested and then rejected or radically modified. Herds are not cities, nor are their members human beings; the captain does not participate in the building of his ship nor is he responsible for the deeds performed by his passengers when they reach their destination. Of the many paradigms entertained in the *Statesman*, perhaps that of the architect comes closest to the central figure of the weaver. Both make clear the need to construct artifacts for the protection of the human race against nature. In this sense there is a genuine anticipation of the theme of the modern Enlightenment, although self-defense rather than mastery is the central issue.

I said a while ago that the *Statesman* is more theoretical than the *Republic*. The Stranger makes it clear that the purpose of his conversation about politics is not political but is rather intended to make his pupils "more dialectical" (*dialektikōteros*) about everything. The stated purpose of the *Republic*, on the other hand, is to persuade Glaucon and Adeimantus that the just life is intrinsically more choiceworthy than the unjust life. Everything said by Socrates in this dialogue is accommodated to the understanding and desires of his youthful interlocutors, including the treatment of topics like the Ideas and the good (VI. 504b1-8, 509c3f). Whereas dialectic is the peak of education within the guardian class of the beautiful city (VII. 534e2), it can only be described metaphorically, in images like that of the sun, to Glaucon and his companions. The Stranger, on the other hand, is training in the first instance young mathematicians; he engages in extended exercises of diaeresis if not in Socratic dialect, and he discusses at some length such topics as paradigms and measurement. Throughout this dialogue, the emphasis is upon theoretical construction, another example of how the *Statesman* is surprisingly modern in its themes, despite the archaic form of its rhetoric. I do not want to overstate this point, but it is nevertheless true to a considerable extent that the *Republic* emphasizes the link between mathematics and the eternal, whereas the *Statesman* emphasizes concept-construction or more generally, the constructive nature of the method of diaeresis and conceptual analysis.

One could also say that poetry dominates the presentation of eternity in the *Republic*, whereas in the *Statesman*, eternity is represented predominantly if not always as changing, and in particular in the great myth of the reversed cosmos. It would be very foolish to attempt to draw general inferences about Plato's doctrines from these brief comparisons of the two dialogues, or for that matter from the more extended sample of evidence contained in my lecture as a whole. I have hoped to show, or at least to make plausible the claim that Plato views human being as divided in nature and hence as by nature incomplete, *das noch nicht festgestellte Tier*, as Zarathustra expresses a related thought. This is the teaching of the Greek tragedians, and this is the most important link between Nietzsche's psychology and that of Plato. In this connection, I add that whatever one may think of Nietzsche's criticism of Socrates and his ostensibly anti-dramatic rationalism, Nietzsche does not identify Socrates with Plato as do so many less perceptive classicists.

I want to take up one last point. Since this is in effect my peroration, I ask you to indulge a very general reflection. The fundamental tension in Platonism is that of the possibility of constructing a political artifact in the light of eternal order. I have tacitly suggested that Plato shows us in the two quite different images of the *Republic* and the *Statesman* the same underlying need to create by poetry and philosophical rhetoric a city that serves to protect human beings against themselves as well as the hostile characteristics of the cosmos. Whatever may be the ontological truth about the Platonic Ideas, they serve what most of us would call a "conservative" political function, and the same is finally true of the various technical exercises of the Eleatic Stranger. I hesitate to use the term "conservative" because I do not wish to invoke contemporary political ideologies. I use it to designate the posture of self-defense against nature. Aristotle, who is as sober as Plato is mad, comes much closer to the Stoic notion of living in accord with nature, but only, I think, at the price of suppressing the problems that are so vividly if at the same time ambiguously presented in the Platonic dialogues. I sometimes wonder whether the docrtine that man is by nature the political animal is not Aristotle's noble lie, and whether his soothing rhetoric is not intended to function as the genuine political medicine that replaces and corrects the damage instilled by Plato's excessively daring pharmaceutical recipes.

It is a long story but it would be very rewarding to make a detailed study of why the modern scientific and by extension political revolution begins under the flag of Plato rather than Aristotle. I have tried to tell a part of that story today by bringing out a side to Plato that is seldom given its due, namely, his constructivism and his perception of the uneasy residence of human being within a nature that is hostile as much as it is friendly to our healthy existence. Whereas Aristotle may finally be a better political poet than Plato, I believe that we have more to learn from Plato at this juncture of late modernity (late at least for us). Under no circumstances would I wish to be understood as suggesting that part of what we need to learn is how to construct a healthy city, if this is taken to mean an endorsement of the beautiful city of the *Republic* which is neither beautiful nor a genuine city on its own criterion of unity.

Department of Philosophy, Boston University

STEPHEN SCULLY

THE NATURE OF THE GODS IN EARLY GREEK POETIC THOUGHT

In addressing the question of nature in the ancient Greek world, I thought that I would turn to the gods because they so complicate conventional distinctions between nature and culture.[1] At one end of the spectrum, they embody elements of nature and indeed personify natural forces – Zeus, the Indo-European Sky God, cloud gatherer and hurler of thunderbolts, and at the other end they represent ideals of civilization and moral order: Zeus Polieus, Zeus Euboulos ("the Good Counsellor"), Zeus Xenios ("Protector of Hospitality"); or consider anarchic Aphrodite, the joyous and dangerous goddess of sexual intercourse, often flanked by two winged youths Eros and Himeros, abstract nouns for sexual desire and yearning, as contrasted with Aphrodite, defender of cities and warlike Bringer of Victory (Nikephoros), portrayed in sculpture with a diadem of towers on her head.

What does such dualism tell us about a Greek view of nature? It is not my aim to identify some all encompassing world view, nor shall I attempt to put forth striking new theories, but I shall call to mind certain opposing, but complementary forces within Greek culture which seem familiar but, when contemplated, retain their essential strangeness. As in the examples of Zeus and Aphrodite, the polytheism of Greek religion can be seen as much within the complex identities of each individual god or goddess as it can in the diversity between different deities, old and new, chthonic and airborne, that make up the pantheon: in short, each individual god embodies a kind of polytheism. Even when we think we can identify a deity with one function, we discern opposition, as for example in the case of Artemis who as "Mistress of the Animals" and goddess of untamed nature both protects animals in the wild and hunts them.[2] But Greek religion further complicates our understanding of this goddess, as of all other gods, by attributing to her roles quite different from her "canonical" ones. So at Iasos in Caria, she is protectress of the city (Artemis Astias), and at Olympia, she is Artemis of the Market Place (Artemis Agoraia). Such is the Artemis in Sophocles' *Oedipus Tyrannos* (161-62) who "sits on her round, and far-famed, throne in the agora" at Thebes.

Each of these gods, then, defying our human desire to classify and categorize, embodies contradiction, a *being* both of nature and of culture, absorbing the antithesis between *phusis* and *nomos* that prevails in Greek thought, an

R.S. Cohen and A.I. Tauber (eds.), Philosophies of Nature: The Human Dimension, 163–176.

antithesis that the Greeks repeatedly turned to, to illuminate the uniquely human and the unstable distinctions between human nature and all the rest of nature. *Phusis* I interpret in its broadest sense as the forces and elements of the non-human world which either lie outside of or invade the human environment, and lie beyond the control of humankind; *nomos*, at the opposite pole, may be associated with the human enterprise to order existence by shaping nature to its will and establishing norms and standards, based on the prevalence of rational order over nature's rhythms.[3]

A visual example of the contrast between nature and culture is well represented in the desolate mountains of Arcadia. High up on the slopes of Mt. Lykaion stands a grisly and primitive cult to Zeus Lykaios, Wolf Zeus. Plato identifies this shrine as the place where human sacrifice was commonly practiced (according to tradition, every ninth year), the celebrant being turned into a wolf upon consuming the entrails of the human victim (*Rep*. 8.565d).[4] Lower down on a narrow rocky terrace of Mt. Kotilion at a place known as the "ravines," or Bassai, and within eyesight of the sanctuary to Zeus lies a jewel of a temple, in the Doric order, to the most civilized of gods, Apollo Epikourios, Apollo the Helper. Approached from the city of Phigaleia as it was in ancient times, the temple in its ordered design would suddenly come into view, built of the same cold grey limestone as the mountain behind, but set apart from the broken rock above by its ordered columns and geometric pediment. The temple enlists the aid of the youthful, rational god and the shaping human intelligence as a spiritual boundary, setting ordered space against Zeus' primitive ritual that turns man into beast.

We shall never know how a fifth-century Greek would have interpreted this spectacle. Today the inhabitants of the area simply identify the site as "the columns" (*stous stulos*). In its abstract, measured lines, the temple is, as one art historian had described it, "the god as savior of humanity making his presence felt in the heart of the wild."[5] (I suppose I must confess that this art historian is my very own father, Vincent Scully.) Juxtaposed against the jagged mountain terrain and the ancient cult of Zeus where one of man's first rules, as imagined by the Greeks, that is the taboo against cannibalism, is transgressed, it is hard not to see the site itself as Charles Segal describes it: "an architectural embodiment of the conflict between savage and civilized spaces . . . that pervades Greek thought, art, and religion. It attests to the inexhaustible power with which the Greeks were able to represent this basic antinomy to themselves and to us."[6] And as Scully goes on to note, that same struggle is equally enacted in the dark interior of the temple in the frieze that ran around the *interior* of the temple's cella. There sculpted scenes of Lapiths struggling with Centaurs and of Greeks with Amazons bring, in effect, the surrounding mountains inside and "engage in the contest which is one part of the meaning of the temple as a whole: men against the beast power in nature."[7] In this desolate setting, the savage cannot be defined simply as uncultivated space outside the walled interior of the city, but rather the savage is any form of nature, physical or psychological, within ordered space or not, which the Greeks called *agrios*.

This adjective, meaning "untamed" or "wild," comes from the noun *agros*, for "field" or "country," the land set apart from the town. This semantic shift from the noun *agros*, field, to the adjective *agrios*, savage, untamed, is in one sense puzzling since a vineyard or wheatfield is neither untamed nor unordered, but the transition from field to savage, from noun to adjective, from an economic perspective to a psychological one, reveals an ancient Greek bias. The realm we call "nature," derived from the Latin *natura*, from *nascor*, "to be born," the Greeks were prone to regard as savage, dangerous and formless, outside the safe world of the ordered space of the *polis*.

This view of nature leads me to speculate for a moment about Homeric similes, especially those in the *Iliad*, which typically liken human behavior to that of animals or forces of nature. But it is not all aspects of nature that these similes depict; rather they focus on nature in its violent manifestations, often taking place at the edge of the cultivated world, what Homer calls *agrou ep' eschatiēn*, "at the edge of the field." That is, they are situated at the marginal zone between the human world and nature's realm, and establish by analogy the battlefield at Troy as a similar marginal zone between the city and its destruction.[8] Such similes corroborate my point about the obsessive Greek association of the wilderness with destruction, threatening the sacred enclosure of the city.

It is not surprising, then, that the Greeks would imagine that the followers of Dionysos, when intoxicted with the god, abandon the sheltered world of their homes for the mountains. Both in the geographical wild and in an inner state of Dionysian possession, the Maenads ("Mad-women") tear apart live game with their bare hands and devour their catch raw. This ecstatic ritual of the mountain in all aspects perverts the civilizing norms of sacrifice, overstepping the rituals of slaughter and forsaking the transformation of raw into cultivated food.[9] Like the followers of Zeus Lykaios on another mountain, Dionysos' initiates do not always shrink from savagery, exampled by the daughters of Minyas from Orchomenos in Bioitia who under the influence of the god tore one of their own children to pieces. As in the Zeus cult, the women who take part in this ritual lose human shape, transformed on the mountain into owls and bats.[10] Yet in keeping with the complexities and contradictions of the other Olympians, even this most destabilizing god of ecstasy, the wild, and the chthonic realm, is also identified, however ambiguously, with civic cult and tutelary functions at Thebes.[11]

Far from our contemporary belief, in these ancient examples nature does not make us feel at peace with the world nor does it restore our soul. At best, the Greeks felt great ambivalence towards nature. The gods are part of the natural world. In a fundamental sense they *are* beings in nature and they can awaken the animal or instinctual within us. Rather than finding joy in the ageless rhythm of nature and its promise of release from culture's restraints, however, the Greeks tended to look upon such powers of the gods as *deinos*, both awe-inspiring and horrifying. But the gods are also part of the created human world and they represent humanity's effort to set itself apart from the engulfing surround.[12] The paradox is evident in Greek religious cult, as illustrated in the worship of Zeus Lykaios and Zeus Polieus, one cult primitive in origin, the other a product of a more sophisticated later age, but both preserved in the same religious system, revealing a god who both can lead humankind downward into the animal and make it possible for humans to escape from the bestial and attain the uniquely human. That cultivated space, the Greeks saw clearly, is only an enclave within a larger world which dwarfs our humanity and is indifferent to it. Such a view of the gods, some may object, embodies no moral imperative, but in compensation it offers an intense, if merciless, total order of reality, not easy to embrace in our liberal meliorism, evident in moderns strains of both the Jewish and Christian traditions.

In this paper, I wish to explore that Greek ambivalence by looking at two myths: Hesiod's *Theogony* which traces the history of the divine structure of the universe from the first patriarch Sky (Ouranos) through his son Kronos, head of the Titans, to its culmination in the reign of Zeus, head of the Olympian gods, and the *Homeric Hymn to Demeter*, a hymn unsurpassed in its emotional depth, about Demeter's search for her daughter Persephone, violently seized from her mother by Hades, god of the Underworld. Both are early, the first, contemporary with Homer, dated perhaps to 700 BCE, the second a century or so later. The common motivating force behind each of these two mythoi is restless Eros, on the one hand instinctual and untamed, on the other the energy which creates culture and leads us to higher thought, a paradox found not only in the *Theogony*, as we shall see, but also at the end of the classical tradition in Plato who in the same work can refer to Aphrodite ("things Aphrodisian," *taphrodisia*) as "a raging and savage master" (savage here is *agrios*, 329c), and to "true eros" (*alēthinos eros*, 499c) as the driving force behind philosophy. It is the same paradox made familiar to us in the modern world by Freud, at whose death Auden laments: "sad is eros, builder of cities, / and weeping anarchic Aphrodite."[13]

PART II: THE *Theogony*

As Homer in the *Iliad* and *Odyssey*, so Hesiod in the *Theogony* assumes a stable unchanging universe, the Olympian order firmly established and everlasting. The *Theogony* begins and ends with a celebration of Zeus and his reign of justice (*dikē*) and in that sense the poem is static. But, unlike Homer's poems, this is also a story of *genesis*, of upheaval, instability, and evolution.[14] Paradoxically, then, the *Theogony* imagines both, in philosophical terms, "a thing that is" and a "coming-to-be," framing a world of becoming within an eternal being. Our concern is how Zeus changes the divine order from a world of becoming to a world of being.

Contrary to Genesis in the Hebrew Bible or to the Babylonian creation myth *Enuma Elish*, in Hesiod nature is not a creation, lovingly crafted and endowed with purpose by its Creator. Nature is closer to a cosmic accident, and leaves it to a later creator to make order and harmony out of a world already filled with evil (*kaka*).

The first deities in the Greek cosmos were aspects of the physical world embodied in anthropomorphic form: Earth, broad-bosomed and obviously female, Tartara (neuter pleural), Darkness (male), Night (female), Brightness (male), Day (female), divine equivalents to the primordial divisions of earth from sky and light from darkness in the first two days of Genesis, and similar in kind to the primordial Apsu and Tiamat, fresh and salt water, in the *Enuma Elish.* Alone of all the ancient mediterranean creation myths, however, the *Theogony* also places Eros (male) in this company, Eros, whom Hesiod describes as the limb-loosening god "who dominates the mind and sensible thought in the breasts of all gods and men" (121-22).[15] In the absence of a cosmic architect, Eros may be considered the driving energy of the early universe, arousing male and female gods to unite and through their offspring to bring the universe progressively into existence. Such a desire indeed compels Earth to create out of herself starry Heaven (*Ouranos*), so that she with Sky might produce through sexual intercourse the next generation in the evolutionary chain.

This is not the place to discuss in detail the violent history of that progression, when Sky, in fear of his offspring, pushes his children back into Earth's womb, or Kronos, Sky's son, driven by the same fear, swallows his children as soon as they emerge from their mother Rhea. Zeus, also subject to this fear, swallows not the children but the pregnant mother and then gives birth to that lone child himself, producing Athena from his head, a daughter who is always loyal to her single parent, assuring his invincibility. Since Zeus succeeds in establishing permanency where his forefathers failed, the myth suggests that for the early Greeks conflict between the sexes was seen as even more threatening than the generational struggle for power.

In swallowing Metis (Athena's mother), the language of the *Theogony* would suggest, Zeus hardly differs from Kronos; compare what Zeus does to Metis: "and then deceiving her with a trick (*dolos*), with beguiling words Zeus swallowed her down into his stomach" (889-90; cf. 899) to what Kronos did at Zeus' birth: "then taking him in his hands, Kronos swallowed him down into his stomach" (487; cf. 466). But if Kronos' act may be equalled with Sky's evil when he pushes his children back into Earth's belly, Zeus, swallowing Metis but not holding back the next generation, marks his final achievement before cosmic order and stability can be realized.

Thus, it cannot be said that Zeus' regime is free either from violence or from craftiness. Like Kronos, Zeus employs violence and craft to gain power, even if the nature of his power differs from that of his forefathers as it is based less upon personal might than upon alliance with others, his freeing of the downtrodden, and his fair distribution of power. And contrary to those before him, his final ascent is based upon public proclamation, not force, when the "blessed gods," prompted by Earth's advice, "urged Zeus to be their king and ruler" (881-85).

With Metis lodged inside him, Zeus is free to create Olympus and form the Olympians – the number usually fixed at twelve: three siblings (Demeter, Hera, and Poseidon), five sons and daughters (Artemis, Apollo, Ares, Athena, and Hermes), one "aunt" Aphrodite (born from Sky's genitals; but in many accounts the daughter of Zeus), and Hephaestus (in some myths, Hera's son created parthenogenically in retaliation for Zeus giving birth to Athena; in others, son of Hera and Zeus) (Hestia or Dionysos usually make the twelfth).

Commentators on the *Theogony* and Greek religion, naturally enough, usually equate these Olympians with the Greek family. So, for example, Walter Burkert writes that the "primary differentiations [of the Olympians] are taken from the elementary family groupings: parents and children, male and female, indoors and outdoors."[16] In a similar vein, Marilyn Arthur writing on cultural strategies in the *Theogony* remarks that in this archaic poem "there is no movement, as there is in the *Oresteia* [two-hundred and fifty years later], from family to civic and cosmic realms; in the *Theogony* the family, the cosmos, and the arena for settlement of disputes are one and the same."[17] Such views, however, misrepresent Zeus' achievement in escaping from the fate of his forebears and, by suggesting that Olympus is modelled on the biologically determined family, fail to expose its distance from nature's rhythms.

The stability of Olympus, in contrast to Kronos' and Sky's failures, is predicted on this distance. Central to that stability is Zeus' socialization of eros. In spite of the many sexual unions prior to Zeus' hegemony, two of his first seven sexual partners are called wives, Metis and Hera, the first and the seventh, (*alokhos* and *akoitis*, respectively; cf. 886 and 921), as if with these two women Zeus has introduced to the cosmos a social convention previously unknown to it. These marriages are further framed by Athena whose conception occurs when Zeus sleeps with Metis (894-96), but whose parthenogenic birth (924-26) occurs after Zeus marries Hera. The other five "marriages" are of a different sort, and it is with these partnerships that Zeus introduces the new civic and agricultural order of the cosmos. The narrative framing of these five unions within the marriages of Metis and Hera reveals, as we shall see, a tension between *oikos* and *polis*.[18]

Of these five sexual bonds, those with Themis, Eurynome and Mnemosyne are the most distinctive for our purposes. Each brings forth virginal daughters, in multiples of three, the three Hours (or Seasons) (*Horai*) and the three Fates (*Moirai*) with Themis, the three Graces (*Kharities*) with Eurynome, the nine Muses (*Mousai*) with Mnemosyne. The individual names of these offspring reveal their essential natures, both individually and as collective bodies. The Hours from Themis (That which is laid down by Custom and Law) are Eunomia (Good Law or Orderly Government), Dike (Justice), and flourishing Eirene (Peace). These children offer a counterweight to the parthenogenic children of Strife already born into the world: Dusnomia (Disorderly Government), Pseudea (Lies), Lethe (Forgetfulness), Limos (Famine), Makhai (Wars) (226-32). The children of Themis help define the joyous civic enclave of Olympus that keeps at bay the disorderly strife, wars, and lies that characterized the pre-civic and primitive (might we also say natural?) world before Zeus. Themis herself is an oracular power capable of foretelling a future which is predetermined. Outside the *Theogony*, her children, the Hours, frequently represent the predictable cyclical stability of the seasons,[19] but within this creation myth they represent the regularity and permanence marked by a social, not a natural, order: good law, justice, peace in a social setting.

The three daughers of Eurynome (Widelaw) also express the joy engendered by a harmonious social order: Agalaia (Splendor), Euphrosyne (Gladness), and

Thalia (Festivity) (907-11). The thrice-three Muses from Mnemosyne (Memory) are the signature of that joy in the social order, a chorus who glorify Zeus' *nomoi* (laws) (cf. 65-7) and "to whom festivals (*thaliai*) and the joy of song are pleasing" (917). These multiple children, of "like minds" as Hesiod characterizes the Muses (60), evoke the image of choruses who, in their collective song and dance, embody the public harmony which Zeus as choregos leads.

To illustrate that the three children represent the collective social order and not that of the family, we need only turn to the *Works and Days* where these divinities, in their plural forms, are part of the "just polis" (225-31):

For those who give straight *justices* to strangers and natives alike and do not in any way stray from the path of *justice*, the city flourishes for them; the people within bloom. *Peace*, nurturing the young, prevails; never does Zeus far-seeing, afflict them with anguishing *war*. Never does *famine* or *ruin* accompany men of straight *justice*, but they enjoy the fruits of their labors in *festivals*.

The qualities of a just city (*dikai*, *eirēnē*, *thaliai*), here marked in lower-case letters, are earthly residues of Zeus' children with Themis, Eurynome, and by their inference Mnemosyne whose daughters provide joy at Festivals and keep the wretched children of Strife at a distance from Olympus and from the human just city.

In contrast to this civic order and joyous spirit, Zeus' marriage with Hera in the *Theogony* is marked by extreme enmity and strife (*zamenēse kai ērisen*, 928) unparalleled verbs in the poem. The violence within the house is also evidenced by the quality of their only son, the "city-destroying" (*ptoliporthos*) Ares (936; cf. 922). But the ultimate disgrace of the marriage occurs in Zeus' taking over of Hera's power to give birth. In producing Athena from his head and begetting without eros, Zeus triggers fury in Hera who, imitating her husband more than her primordial grandmother, begets Hephaistos without a mate. The strife that characterized the "marriages" of Earth and Ouranos and of Rhea and Kronos is continued in the third generation, but its force has now been defused, as Zeus has created a new and abstract order characterized by justice, lawfulness, parity, and social harmony.

It also is based on thwarting the biological laws of nature and the primal quality of eros. Zeus' regime depends upon eliminating one of two parents, as Earth explicitly stated. Not only does he do away with the potentially destructive male child, but allowing for the birth of the next generation he also creates a child loyal to one parent alone: the Zeus-like Athena. (This is not the case with all parthenogenic births, as is shown in the case of Dionysos who, born from Zeus' thigh, nevertheless seeks to find his mother's grave, insisting on a memory of the absent mother, a sentiment alien to Athena.) In Zeus' swallowing of Metis, generation is not aborted, but succession is (Apollo will always remain an adolescent; he will never grow a beard) and biology has been circumvented. In the *Theogony*, political stability depends on single parent birth and, in spite of Mnemosyne, on selective forgetting.

In good Freudian fashion, eros is also denatured. Perhaps Zeus' inexhaustible sexuality on Olympus obscures this point, but consider the following. Zeus is the first to have a "wife." His parthenogenic child is obviously relevant in this context. Equally striking is the novel sense with which Zeus uses the term *philotēs* when he asks the 100-arms, whom he has just freed from bondage, to "remember our kindly love," *philotēs* (651), transforming a word, which up to now in the poem always referred to sexual unions, from an erotic to a political meaning. Even eros on Olympus takes on a new nature as Zeus, through his sexual unions with women from the previous generation (Themis and Mnemosyne), from his own generation (Metis, Eurynome, Leto, Demeter, and Hera) and from future generations (the Atlantis Maia, Semele, Alkeme), brings into harmony, as in the political marriage of medieval Europe, generations once at war and former opponents. When Zeus joins erotically with the older Themis, Eurynome, and Mnemosyne, these unions awaken into active agency the latent power of these mighty aunts and cousins. No less striking by his "marriage" to Maia, daughter of his arch enemy Atlas, son of Iapetus, results in the birth of Hermes, a trickster like his uncle Prometheus but one who serves Zeus (937-38).[20]

Zeus, in short, creates a realm apart, called Olympus, defined by a *new* nature, one contrary to the old nature of things, the original constitution of the eros-driven, biologically-reproducing world. Joy, communal dancing and song, harmony, and good order are found in escaping the old nature; there is a politicization of a dynamic, but destabilizing, primal eros; reproduction continues, but not in a form that will overthrow the father; the female is prominent, not as primal Earth, but as loyal daughter and as choruses of daughters who are abstractions and namesakes of the new harmony: Eunomia, Thalia, Euphrosyne, Dike.

PART III: THE HOMERIC HYMN TO DEMETER

This story of great antiquity with many Near Eastern parallels is unlike all others for the intimacy it portrays between mother and child (indeed, it is the only myth of its kind where the dying child is the same gender as the mother). If at its heart this myth, like the more conventional ones of mother and son or woman and lover, is story of fertility, it is also a story of great psychological reach. In the brief account which follows, I shall resist any temptation to mine the *Hymn* for clues it may reveal about the Eleusinian Mysteries and concentrate instead on its own unfolding and tortured narrative.[21]

The myth begins with the daughter of Demeter and Zeus gathering narcissi, and other bulb flowers, in a meadow on the plain of Nysa at the edge of the world near the River Ocean. In the midst of this seemingly innocent activity, the Earth opens up and the daughter, Kore, is "seized" (*hērpaxen*) by Hades who carries her away in his golden chariot to the Underworld. While it may be tempting to see this as a scene of abduction, it is also said in the same breath that Zeus gave (*dōken*) his daughter to Hades, as if this were a marriage (3; cf. 30-2, 77-80). This same paradoxical combination of Persephone's abduction and marriage is told in the *Theogony*: "Then Zeus went to the bed of Demeter who feeds many; she begot white-armed Persephone whom Hades seized from her mother, given to him by Zeus lord of wisdom" (912-14). The *Theogony* leaves the story at that, nothing more said, but the *Homeric Hymn* goes on to narrate Demeter's woe as she desperately searches the world over for her lost daughter, until she comes upon Helios, the god of the Sun, who informs Demeter that Persephone has married the Lord of the Underworld. A "worthy son-in-law" (83-4), Helios says, for the bride Persephone. But if this abduction is also a marriage, there is no indication whatsoever that the mother has approved of the marriage, or indeed that the mother and daughter even know of its plan. The version in the *Theogony* brings this out particularly well, saying that Hades "seized Persephone forcibly from her mother" (*hērpasen hēs para mētros*). The verb *harpazō*, common to both the *Theogony* and the *Hymn*, only refers to abduction in the *Hymn* ("Hades seized Persephone"), but in the *Theogony* when followed by *hēs para mētros*, the verb must mean "snatched from" ("Hades snatched Persephone away from her mother").

The description of this abduction in the *Hymn* is cause for wonder to us, focusing almost completely around the story of the narcissus flower (5-16; cf. 343):

> Playing with the full-breasted daughters of Ocean, (Demeter's daughter) was plucking flowers in the soft meadow: roses and crocuses, beautiful violets, irises and hyacinth, and the narcissus, which (Mother) Earth grew as a trick for the girl who had a face like a budding flower, and this to please lord Hades and by the will of Zeus; the narcissus held a wondrous gleam, object of reverence for all, both immortal gods and mortal men; from its root grew a hundred heads, perfumed. All of broad heaven above and all of earth laughed, as did the salty waves of the sea. The girl in wonder reached with both hands – for the beautiful object of delight. Then the wide earth gaped wide . . .

for Hades to drive out his golden chariot. Persephone is both the girl reaching for the flower and the flower to be seized – the girl who had a face like a budding flower (cf. 8, 79, 108, 420, 427), a likeness that Milton also saw in "Proserpine gathering flow'rs, / Herself a fairer flow'r" (*Paradise Lost* 4.269-70), a hundred years before the *Hymn to Demeter* was re-discovered in the West. The association of the rose with sexuality and Aphrodite on the one hand, and funerary motifs on the other is well known. The other flowers in the company of the rose – the crocus, violet, iris, hyacinth, and narcissus – are all bulb flowers and also associated with the dual realms of sexual allurement and death.[22] Attention needs fall on the allure itself, the narcissus. Some interpret the flower as a "monstrosity of nature," its one-hundred heads analogous to the one-hundred headed Typhoeus whom Earth begot to thwart Zeus' hegemony in a last-ditch effort to overthrow Olympus, but the *Hymn* calls the narcissus an object of reverence for all the world, which causes the heavens, earth, and sea to laugh in what must be construed as a joyous contrivance.[23] If not monstrous then, the many headed flower may still suggest earth's irrepressible fertility and perhaps the dangerous nature of eros itself. For all the Demeter, this joyous allure brings laughter, an object of mediation which, as an *axis mundi*, links earth and sky, upper and lower realms.

It is hard not to see in these details a story of marriage, the flower itself an image of an adolescent girl wakening to her own instinctual sexual desires, her reaching for that flower a source of joy to the whole universe, the narcissus for the girl ambiguously situated between a thing of wonder and a child's toy. In the *Hymn*, when she is seized, Persephone feels nothing but terror as she is dragged, screaming, to the Underworld. I need not recall that screaming and reluctance are inscribed in the ancient Greek epithalamia, or bridal songs. But if this abduction is a marriage in the eyes of Heaven and Earth, Zeus and Hades, for Demeter it is a source of unrelieved grief, the loss of daughter equal to death itself. The goddess wanders over the globe for nine days in despair searching for Persephone. When she finally learns the truth of what has happened, she leaves Olympus in a rage, roaming the earth in the form of an old woman, "long past her ability to bear children or enjoy the gifts of Aphrodite" (101-2) until she comes to Eleusis. From Demeter's perspective, the abduction/marriage tears at the very fabric of her life, not only because she has lost her daughter but, because, one has the sense, she has, at least in the temporary form of her grief, passed symbolically beyond her own child-bearing years. One wonders if there are not multiple causes for Demeter's anguish. But the contrast with the *Theogony* is our main point here. If in Hesiod's vision of Olympus, Zeus can steal Demeter's honor with apparent immunity, negotiating with Hades about his daughter's sexuality without regard for daughter or mother, in the *Homeric Hymn* Demeter's wrath and vengeance redress the assumed privileges of that new order.

In her unrelieved anger, the fertility goddess prevents seeds from sprouting forth from the rich earth for a full year, causing a famine that would have obliterated the human race, had not Zeus intervened (305-13). First he sends Iris requesting that Demeter return to the company of the gods and then, when she fails to appease Demeter, Zeus sends all the other gods, offering Demeter gifts and honors, but to no avail (331-33).

> Never, she said, would she set foot on fragrant Olympus
> Nor would she send forth the fruits of the earth
> Until with her own eyes she had seen her daughter.

Zeus acquiesces to her conditions and sends Hermes to the Underworld to lead Persephone out of Erebos, so that, as Hermes explains to Hades, Demeter "may cease her anger and terrible wrath against the gods" (349-51) (the word for Demeter's wrath is *mēnis*, the same word that opens the *Iliad*, and related to the word describing Hera's rage in the *Theogony*). Persephone is released, but because she has tasted of the pomegranate seed (371-74), a fruit which plays a key role in ancient Greek wedding rituals, she will be obliged to return to her husband for two-thirds of the year. After mother and daughter embrace and reveal to each other what has happened since Persephone's "abduction," Zeus sends Rhea, Demeter's own mother, to bring the goddess back to Olympus, promising "to give her whatever honors she may choose" (443-44). The earth, previously laid barren by Demeter's plans (453), *mēdesi*, a word in the *Theogony* associated exclusively with the ascendency of Zeus, now promises to yield her fruits once again, and bright-crowned Demeter, who has been for a time at odds with her epithet, will once again crowd the rich furrows with long tassels of gold (454, 470-71).

Demeter, a pre-Olympian goddess of the earth and fertility, in this *Hymn* rebels against the new order defined by the *Theogony* in the Olympian creation of a new eros and its socio-political institutions based upon abstract orders of policy, justice, and patriarchy. Her wrath at Zeus' control of their daughter's sexuality causes her to distance herself from the society of the gods and to work her vengeance by bringng nature's rhythms to a halt. A number of readings of this *Hymn* claim that by the poem's end the wrathful goddess has finally been controlled by Zeus. By teaching humanity how to control the earth's fecundity through the techniques of agriculture and by accepting the fact of her daughter's marriage to Hades, Demeter has *de facto* agreed to patriarchal marriage and submitted to Zeus' broad reorganization of nature.[24] But the myth suggests a different form of mediation, I believe. No where does the myth suggest that the end depends on Zeus' plan; to the contrary, it portrays Demeter's power and authority over nature, and over Zeus.[25] Instead of Zeus' will prevailing, Demeter demonstrates that he is powerless against her and her control over nature and that the new laws of endogamy must bow to her blessing. It is Rhea, Demeter's mother, not Zeus, Demeter's sexual partner and lord of Olympus, who shows Demeter the way. In contrast to Zeus' power of reproduction and his extra-ordinary procreation of Athena, the female power of reproduction is true to a different kind of mystery, at different times barren and fertile. The eating of the pomegranate seed initiates Persephone into a new state of her being and equally Rhea recalls for Demeter, Demeter's own place, at once mother and daughter, in the complex tripartite stages of a woman's fertility.[26] The *Homeric Hymn*, I believe, celebrates and honors those rhythms and, in the act of doing so, the *Hymn* offers a counterforce to any confidences one might have about Zeus and Olympus in Hesiod's creation myth.

In the *Theogony* Zeus takes over the power of nature without ambiguity, transforming the primal energies of generative but anarchic eros to patriarchal and political order. The *Hymn*, on the other hand, immensely complicates Hesiod's resolution. In dramatizing Zeus' inability to lure Demeter back into the Olympian order, and in stressing, as well, the dependence of the entire world on Demeter's good will, the *Hymn* demonstrates a far more complex picture of the balance of power between female and male forces. In our own time, some regard marriage as institutionized rape, or as prostitution, and would regard the *Hymn*'s resolution, Demeter's reinstatement on Olympus with new honors, as one more "pay-off" to maintain patriarchy. But perhaps we are all, men and women, slaves to an overwhelming urge that spares only a few, the virginal refusers like Hippolytus, Artemis, Hestia, and Athena who maintain their own kind of liberty. For those less free the two ancient Greek works that we have been considering show the Greek struggle to live within nature by transcending it, but the forms of transcendence differ. We are not likely to find an answer in these myths to the tensions between eros and society that we still wrestle with today, but Greek mythology does afford a chance to view with a stark and unsentimental eye contraditions within us. A great female force erupts in protest against male institutional arrangements; in consequence, those arrangements are forced to adapt, recognizing, symbolically at least, a certain maternal authority linked to the generative powers of the earth.

To conclude by returning to the themes with which we began, Greek mythological thought, a dynamic and evolutionary system especially when various myths are taken in conjunction, is a subtle enough instrument, particularly in its pre-philosophical phase, to register the highly charged and reciprocating dances of *nomos* and *phusis*.

The Greek gods elude systemization into simple categories of nature or civilization. Equally complex is the juxtaposition of myths related to a common story; resolution in one need not be so easily contained in another. The narrative of myth, like the later narrative of Plato, performs its truth, and prevents differences from ossifying into rigid polarities, showing that the double helix of nature and culture can never be disentangled for long.[27]

Department of Classical Studies, Boston University

NOTES

[1] Much of what follows in this paper presents a view of nature that hardly conforms to Prof. Erizam Kohak's eloquent treatise on nature as a source for moral wisdom as defined in *The Embers and the Stars: a philosophical inquiry into the moral sense of nature* (Chicago, 1984). My comments about archaic Greek thought are not intended to refute but to offer contrast and context to his meditations.

[2] Some readers of Greek religion attempt to explain these divergent sides of the Olympians historically, seeing a merging at a very early date in Greek history of pre-Greek and Greek traits; others, looking to the "function" of Greek religion, find such theories of origin inadequate to account for the co-existence of these complementary, or diverse aspects of a single deity at any one time in Greek history. See Louise Bruit Zaidman and Pauline Schmittt Pantel, *Religion in the Ancient Greek City*, tr. by Paul Cartledge (Cambridge, 1992), 5–6.

[3] The studied opposition of *phusis* and *nomos* first occurs in the fifth century, at a time when the word for "law" begins to shift from *themis* and *thesmos* "that which is laid down and established" (by custom or the gods), to *nomos*, "law" or "statue" imposed by human design. At the same time *phusis* comes to describe the "order or law of nature," a word used only once in Homer, describing a thing's essence or in-born quality, the *phusis* or a drug, *Od* 10.303. The standard reference is Felix Heinimann, *Nomos und Physis* (Basel, 1945). See further, W.K.C. Guthrie, *A History of Greek Philosophy* III (Cambridge, 1969), 55–134 and 204–9; Richard Garner, *Law and Society in Classical Athens* (London, 1987), 19–26; Martin Ostwald, *From Popular Sovereignity to the Sovereignity of Law* (Berkeley, 1986). By the end of the fifth century, an initial optimism in the power of *nomos* and human rationality has given way to the view that social norms often oppress and harm the human spirit.

[4] On the question of human sacrifice at this site in Plato's day, see Dennis D. Hughes, *Human Sacrifice in Ancient Greece* (London, 1991), 96–107, and Madeleine Jost, *Sanctuaires et cultes d'Arcadie* (Paris, 1985), 180–85 and 249–69.

[5] Vincent Scully, *The Earth, The Temple, and the Gods* (New York, 1962; rev. 1969), 123.

[6] Segal, *Tragedy and Civilization: An Interpretation of Sophocles* (Cambridge, Mass., 1981), 1. Those who are familiar with the opening essay in this book will know how much I have profited from his account of Bassai and, indeed, from his discussion of nature and culture in Greek thought; see also his 'Nature and the World of Man in Greek Literature', *Arion 2* (1963), 19–53.

[7] Scully (note 5), 126. For a discussion of the temple and its sculpture, see Mary Beard and John Henderson, *The Classics* (Oxford, 1995), *passim*, esp. 75–82.

[8] See Martin Mueller, *The Iliad* (London, 1984), 113–14; cf. James Redfield, *Nature and Culture in the Iliad* (Chicago, 1975; repr. 1995), 189–92. For a foreshortened view of city to *agrios* at Troy in the *Iliad*, see Stephen Scully, *Homer and the Sacred City* (Ithaca, N.Y., 1990), 10–14.

[9] Cf. Zaidman and Schmitt Pantel (note 2), 173–74. Cf. Marcel Detienne, 'Culinary Practices and the Spirit of Sacrifice', in Detienne and Jean-Pierre Vernant, edd., *The Cuisine of Sacrifice among the Greeks*, tr. by Paula Wissing (Chicago, 1989), 1–20.
[10] Cf. Walter Burkert, *Greek Religion*, tr. by John Raffan (Cambridge, Mass., 1985), 164–65.
[11] See Lewis Farnell, *The Cults of the Greek States*, vol. 5 (Oxford, 1909), 135–38; cf. Paul Vicaire, "Place et figure de Dionysos dans la tragédie de Sophocle," *REG* 81 (1968), 60–65 and 69. Richard Seaford, "Dionysus as Destroyer of the Household: Homer, Tragedy, and the Polis," in Thomas Carpenter and Christopher Faraone, edd., *The Masks of Dionysus* (Ithaca, N.Y., 1993), 133–38, argues the point that Dionysus, as a god who destroys the household, "embod[ies] theh communal principle of the polis" (p. 137), even as he threatens the gender divisions of the city. Framing Dionysus within such a dualism softens the fact that this god, as a being primarily of the mountain and of dissolved boundaries, threatens both house and city.
[12] Indicative of the complexity of their thought, the Greeks can also image humankind as most *deinon*, either because of monstrous (female) passion (cf. Aeschylus, *Libation Bearers*, 585–601) or because of a (male) ability to subdue nature (cf. Sophocles, *Antigone*, 332–75); cf. Eilhard Schlesinger, '*Deinotēs*', *Philologus 91* (1936–37), 59–66; Donald Carne-Ross, 'The Beastly House of Atreus', *Kenyon Review* n.s. *3*, (1981), 20–60; Charles Segal, 'Sophocles' Praise of Man and the Conflicts of the *Antigone*', (1964), revised in his *Interpreting Greek Tragedy* (Ithaca, 1986), 137–61.
[13] W.H. Auden, 'In Memory of Sigmund Freud' (1939).
[14] An evolutionary view of the universe is not entirely absent in Homer. For the *Iliad*, see Laura Slatkin, *The Power of Thetis* (Berkeley, 1991) and Jenny Clay, *The Politics of Olympus* (Princeton, 1989), 11–12; for the *Odyssey*, see Charles Segal, *Singers, Heroes, and the Gods in the Odyssey* (Ithaca, N.Y., 1994), 195–227.
[15] See Robert Lamberton, in *Hesiod* (New Haven, 1988), for the primordial nature of Hesiod's Eros (pp. 69–72), but more questionable are his claims that Earth is a destructive force throughout the poem and *dikē* is disassociated from Zeus (pp. 72–3 and 92–95, respectively). For other readings, see Normal O. Brown, *Hesiod, Theogony* (Indianapolis, Ind., 1953), 7–35; Richard Caldwell, *The Origin of the Gods* (Oxford, 1989).
[16] Burkert (note 10), 218.
[17] Arthur, 'Cultural Strategies in Hesiod's *Theogony*: Law, Family, Society', *Arethusa 15* (1982), 63–4; cf. James Redfield, 'The Sexes in Hesiod', *Annals of Scholarship 10* (1993), 31–40.
[18] For discussion of the poem's ending, which I place at line 955, see Richard Hamilton, *The Architecture of Hesiodic Poetry* (Baltimore, 1989), 96–9; cf. Mark Northrup, 'Where did the *Theogony* End?', *Symbolae Olsoensis 58* (1983), 7–13. For *akoitis* also see *Th.*410.
[19] Cf. M. Detienne and J.-P. Vernant, *Cunning Intelligence in Greek Culture and Society*, tr. by Janet Lloyd (Sussex, 1978), 107–9.
[20] Cf. Annie Bonnafé, *Eros et Eris: Marriages divins et mythe de succession chez Hésiode* (Lyon, 1985), 92–7.
[21] Of the extensive bibliography on this *Hymn*, I single out here: Helene Foley, ed., *The Homeric Hymn to Demeter* (Princeton, 1994); Ann Suter, '*Homophrona thumon echousai*: Mothers and Daughters in the *Homeric Hymn to Demeter*', *New England Classical Newsletter and Journal 19* (1991), 13–15; Clay (note 14), 202–65; Cora A. Sowa, *Traditional Themes and the Homeric Hymns* (Chicago, 1984), 95–144; Charles Segal, 'Orality, Repetition and Formulaic Artistry in the Homeric "Hymn to Demeter'," in C. Brillante *et al.*, edd., *I poemi epici non omerici e la tradizione orale* (Padua, 1981), 107–60; Bruce Lincoln, 'The Rape of Persephone: A Greek Scenario of Women's Initiation', *Harvard Theological Review 72* (1979), 223–35; Marilyn Arthur, 'Politics and Pomegranates: An Interpretation of the Homeric Hymn to Demeter' (1977), now in Foley (1994), 214–42, with an important foreword, 212–13; N.J. Richardson, *The Homeric Hymn to Dementer* (Oxford, 1974).
[22] Cf. Foley (note 21), 33–35; Eleanor Irwin, 'The Crocus and the Rose: A Study of the Interrelationship between the Natural and the Divine World in Early Greek Poetry' in Douglas Gerber, ed., *Greek Poetry and Philosophy: Studies in Honour of Leonard Woodbury* (Chico, California, 1984), 147–68; Richardson (note 19), *ad loc.*; Ileana Chirassi, *Elementi di cultura precereale nei miti e riti greci* (Rome, 1968), 91–155; G. Piccaluga, 'Ta Pherephattēs anthologia', *Maia 18* (1966), 241–2.
[23] For the narcissus as monstrosity, see Clay (note 14), 214; Carl Ruck and Danny Staples, *The World of Classical Myth (Durham, N.C., 1994)*, 318, see a pun perhaps between the hundred (*hekaton*) heads of the flower and Hekate, who mediates in the story between Demeter and Persephone.
[24] Cf. Kristina Passman, 'Re(de)fining Women: Language and Power in the Homeric Hymn to Demeter', in Mary Forest, ed., *Woman's Power, Man's Game: Essays on Classical Antiquity in Honor*

of Joy King (Wauconda, Ill., 1993), 54–77; Clay argues that each of the four major Homeric hymns (to Demeter, Apollo, Hermes, Aphrodite) reveals a redistrubtion of honors within the Olympian cosmos according to Zeus' plan as he creates a new and stable order among gods and between mortals and gods. For reasons indicated in this paper, the *Hymn to Demeter* fails to conform to such a systematic and coordinated picture of archaic myth adjusted to a single perspective of cosmic order; see Clay (note 14), 11–15, 210–15, and 247–60; see also, Lincoln (note 21).

[25] Cf. Charles Penglase, *Greek Myths and Mesopotamia* (London, 1994), 154–55; Foley (note 21), 111–12; Segal (note 21), 124–25 and 141–50.

[26] Cf. Foley (note 21), 118–37; as well as both Arthur and Suter (note 21).

[27] I wish to thank Professors Donald Carne-Ross, Carl Ruck, Charles Segal, and James Wiseman, each of whom has helped to bring this paper into its present form.

ABNER SHIMONY

THE RELATIONSHIP BETWEEN PHYSICS AND PHILOSOPHY*

Working physicists, I believe, almost inevitably have strong philosophical interests, regardless of whether they have taken courses labeled "Philosophy" and whether they have liked what they have sampled. Their interest is implicit in the discipline of physics itself. Peter Bergmann, in the introduction to his *Basic Theories of Physics I* attributed to Einstein the view that "a theoretical physicist is . . . a philosopher in workingman's clothes." I would omit the adjective "theoretical" and apply the characterization to experimentalists as well. This claim depends, of course, on a conception of what philosophy is. I propose the following: philosophy is the systematic search for perspective, for connections among aspects of the world, and for depth of explanation. Some physicists are drawn into their profession from the beginning because they have been convinced by a few revealing examples that the procedures of physical investigation help to achieve perspective, connections, and depth of explanation. These are philosophers from the start. Others are drawn in because of their fascination with specific phenomena. In my case, the onset of curiosity about the physical world, so far as I can recall, came at the age of four, when I saw my father siphoning wine out of a barrel, and I was amazed that the wine went up in the siphon before it descended. But even if the route to professional physics is via wonder at specific phenomena, an approach to philosophy, in the sense mentioned, is unavoidable, because the physicist's understanding of phenomena goes beyond the phenomena themselves to underlying causes and to connections.

There are various ways to subdivide the discipline of philosophy, but the following will be convenient for our purposes:

I. Epistemology, which assesses claims to knowledge;

II. Metaphysics, which considers what kinds of things exist and what are the basic principles governing them;

III. Theory of value.

*Reprinted with small modifications from *NASA Conference Publication 3270: 3rd International Workshop of Squeezed States and Uncertainty Relations*, ed. D. Han et al., 1994.

R.S. Cohen and A.I. Tauber (eds.), Philosophies of Nature: The Human Dimension, 177–182.

I'll put greatest stress upon the relations between physics and metaphysics, because they are richer and more surprising than its relation to epistemology and value theory.

I. EPISTEMOLOGY

Among the many problems of epistemology I shall focus on these: (i) what is the proper formulation of scientific method, and to what extent is that method rationally justified? (ii) to what extent can we disentangle subjective from objective contributions in our experience of the world? There were serious and highly intelligent figures in the history of thought, like Descartes, who maintained that these and related questions in epistemology had to be answered before the substantive work of the sciences could begin, because a foundation is necessary before constructing the edifice of science. We can confidently say, I believe, that this architectural metaphor was mistaken, because all attempts to establish the foundations of knowledge without any substantive assertions about the world seem to have failed. Instead, work on the theory of knowledge and work on substantive science had to proceed in tandem, and – judging by the spectacular successes of the natural sciences – we can say that somehow human beings are capable of this remarkable and somewhat paradoxical intertwining of investigations.

As to problem (i), concerning scientific method, the great breakthrough occurred in the sixteeth and seventeenth centuries, and was inseparable from the development of the new physics and astronomy of Copernicus, Kepler, Galileo, Huygens, and Newton. It has been known from the time of the great Greek philosophers, and probably in a vague way many millennia earlier, that somehow human beings can learn by sense experience and also by reasoning. The division between Plato and his followers, on the one hand, and Aristotle and his followers on the other, concerned the primacy of sense experience or reason. In the work of the new physicists, culminating in Newton's *Principia*, the interplay of the contributions of experience and reason was clarified. These physicists recognized that human beings are not so constituted that they can have direct sensory experience of the fundamental structure of matter or of the fundamental forces among material things. However, if tentative propositions about fundamental physics are formulated in mathematical terms, then logical and mathematical reasoning leads to conclusions that can be checked by sense experience, notably the trajectories of terrestrial bodies and the apparent positions of planets in the celestial sphere. Newton's successful prediction that the orbit of Saturn would be distorted when that planet is close to Jupiter, and Halley's successful prediction of the recurrrence of a comet, and other spectacular verifications of predictions, indirectly confirmed the tentative propositions about fundamental physics which were incapable of direct check. Thus, in tandem with the substantive advance in physics and astronomy was the demonstration of the power of the scientific method summarized above, called the "hypothetico-deductive method." This method was not discredited by the significant refinements of scientific methodology in the centuries that followed, notably enrichment with probability theory. Reflection upon scientific practice was largely responsible for these refinements, since it was noticed by Gauss, Bessel, and others that experimental data are subject to random errors, and they realized that in addition to the dominant causes of a natural phenomenon there are always innumerable small perturbations that cannot be completely catalogued and yet can be treated rationally in terms of probability distributions. Further refinements of scientific method have been made in the twentieth century, and there are still controversial questions concerning inductive inference. Nevertheless, I would maintain that the great revolutions of physics of the twentieth century were accomplished essentially with a classical methodology. The propositions of relativity theory, quantum mechanics, elementary particle theory, etc. were established by applications of the hypothetico-deductive method, refined somewhat by probability theory, without a methodological revolution.

As to problem (ii), concerning the disentanglement of subjective from objective factors in experience, there was not – so far as I can see – a single critical breakthrough, but rather a long, intricate, progressive, and still incomplete development. As the human observer was recognized to be a natural system, the experiences of that observer could be studied as the termini of causal chains, some of which are initiated within the observer, while others are initated externally. Of course, all external stimuli eventually impinge upon the neural and psychic machinery of the observer. The naturalistic point of view regarding the human observer has made it possible for psychologists and neurophysiologists to understand such matters as optical illusions, the limited range of the visual and auditory spectra, and the distortion of perceptual judgments by emotional and conceptual bias. In turn, such understanding of the subjective element in experience permits scientists to take counter-measures, by instruments and procedures of observation, in order to enhance the revelation of objective factors in experience. Of course, the ultimate subjective elements of experience, such as sensory qualities, are entirely beyond present science, and will continue to be mysterious until we have a scientific world view that can accommodate consciousness. There has been, however, a capital negative achievement concerning the subjectivity of experience: namely, an accumulation of evidence that the structure of space and time are not imposed upon experience by the operation of the human mind, as Kant maintained in his doctrine of transcendental idealism. Kant's primary argument proceeds from the apparent necessity and universality of geometrical propositions, understood not as mere statements of a formal calculus but as assertions about physical space. The invention of non-Euclidean geometry by Lobatchevsky and Bolyai, the demonstration of its consistency relative to Euclidean geometry by Klein, and the success of its incorporation in general relativity theory are devastating to Kant's argument. Furthermore, when his exemplary geometric instances of synthetic a priori judgments are undercut, then his further claim, that the principles of causality, substance, etc. are necessary because they are imposed by the mind, is seriously weakened; and it is weakened even further by the triumph of a non-deterministic physics. In other words, the epistemological explanation of the basis of science – which was a major part of Kant's program – is undermined, and we are driven to recognize that in so far as science is valid, even as an approximation, it is so because it is a quite good description of the world as it is. The foregoing anti-Kantian argument is regrettably condensed, but it suffices to point to my next thesis: that there are important connections between physics and metaphysics.

II. METAPHYSICS

This thesis may be disconcerting to many people because the word "metaphysics" is commonly used as a pejorative by physicists, even by those who are sympathetic to epistemology. Until the beginning of the eighteenth century there were great scientists who were also metaphysicians, like Aristotle and Leibniz, and a case can be made for adding Newton to the list. The historical consequence of Kant's work at the end of the eighteenth century and Hegel's at the beginning of the nineteenth was a rift between science and philosophy, regardless of their intentions; and Hegel's grandiose and obscurely reasoned attempt at a "System" of thought engendered a negative reaction among many careful and critical scientists. The positivism of Mach and the Vienna Circle was a consequence of the quest for clarity of expression and rigor of demonstration, and to some extent it was a reaction against the excesses of Hegel. The enterprise of systematic philosophy – aiming at perspective, connections, and depth, as I proposed at the beginning of my talk – became generically suspect, because of the weaknesses of exposition, the excessive claims, and the remoteness from scientific practice of the most influential systematic philosopher of the nineteenth century. The Vienna Circle, which originally called itself the "Mach Verein," was nearly unanimous in its condemnation of metaphysics as nonsense and its desire to purge scientific discourse of any residue of metaphysics. At the time of the great scientific revolutions of the early twentieth century, positivism was a pervasive influence among scientifically oriented philosophers and philosophically oriented scientists. It is not surprising, then, that many of the revolutionaries gave an antimetaphysical interpretation to their discoveries. Einstein for a long time expressed great sympathy with Mach's ideas, and some of his expositions of special relativity centered around the operationalist analysis of space-time concepts like "simultaneity." Heisenberg's initial formulation of quantum mechanics dispensed with position and momentum, which he regarded as unobservable on a microscopic scale, and tried to express his theory in terms of observables like frequency and intensity. Both Einstein and Heisenberg eventually deviated from positivism and both espoused some form of physical realism – Einstein in his argument for "elements of physical reality" and Heisenberg in a statement that the wave function of an atom can be regarded as a description of the atom as a "thing-in-itself." But Bohr to the end of his life gave interpretations of quantum mechanics with a positivistic flavor, e.g., saying of the wave function that "we are dealing here with a purely symbolic procedure, the unambiguous physical interpretation of which in the last resort requires a reference to a complete experimental arrangement." Bohr believed that the philosophical implications of quantum mechanics are epistemological, concerning limitations of human knowledge and renunciations of unitary pictures of the kind offered by classical physics. His consistent playing down of metaphysical interpretations of quantum mechanics reflect a pervasive suspicion of metaphysical speculation in the scientific community, and it is therefore not surprising that he maintained a strong grip upon the attitude of physicists towards quantum mechanics until fairly recently. I have to confess that in spite of my thesis about the relation of physics to metaphysics, I have respect and sympathy for the positivists' pursuit of clarity of concepts and rigor of demonstrations. I am skeptical, however, that criteria for the meaningfulness of sentences, such as "verifiability," "confirmability," or "falsifiability," can be formulated a priori without drastically damaging scientific investigation. Mach was right that there was an obscurity in Newton's concept of "absolute motion," but the reason is Galilean invariance, which precludes an absolute distinction between rest and uniform rectilinear motion; however, Galilean invariance is entirely consistent with a clear concept of absolute acceleration. In sum, the clarification of scientific language must proceed in tandem with the progress of scientific knowledge, analogously to the linkage of scientific methodology and substantive science. If physics has progressed to the point where some of the traditional metaphysical problems can be formulated with precision, and even subjected to experimental treatment, then we are presented with a wonderful opportunity, that ought not to be neglected because of a suspicion of the sterility of metaphysics. And I contend that modern physics has reached such a stage of development. We are fortunate enough to be invited to a great feast of ideas, and it would be a self-defeating austerity to decline the invitation.

Here is a partial list of results of modern physics having metaphysical implications. Of course, all of these results are subject to modification as science progresses, but it is most unlikely that any of them will cease to be good approximations (in the sense of the correspondence principle) to their successors. For this reason, the implications that we can draw from them are likely to have a quasi-permanent status.

1. Modern physics (broadly understood to include chemistry) has established the granular character of matter, the present candidates for elementary being quarks, leptons, and various bosons. That the immense variety of kinds in the natural world can be understood in terms of combinations of these elementary units is a great vindication of the atomistic vision of Democritus. But the modern qualifications of Democritean vision are as important metaphysically as the vindication, especially that none of the elementary units is immortal, but all are subject to creation and annihilation in allowable processes; and that certain quantities, notably energy and momentum, are conserved in all of these transformation processes, and others are approximately conserved.

2. The overwhelming evidence that the quantum state of a physical system is a complete description of it, without any need or place for supplementary "hidden variables," implies that various features of the quantum formalism must be attributed to individual physical systems and not just to ensembles. There is objective indefiniteness, because in any quantum state some physical variables have no definite value; there is objective chance, because the different behaviors of various systems with the same quantum state cannot be attributed to differences among hypothetical hidden variables: there is objective probability, because the chance behavior of an ensemble of systems in the same quantum state conforms to definite probability distributions. These three properties conjoined are called "potentiality," in Heisenberg's terminology. That the state of a physical system involves potentialities and cannot be fully understood in terms of actualities is one of the profound metaphysical implications of quantum mechanics. This interpretation of the quantum formalism, it may be noted, is entirely different from Bohr's, which is epistemological in character. For instance, the uncertainty principle is not interpreted in Bohr's fashion as a limitation upon human ability to measure both position and momentum, but rather as an acknowledgment of the objective fact that position and momentum cannot both be actual.

3. Quantum mechanics also has a remarkable implication for the relation between parts and wholes. There are states of composite systems, called "entangled" by Schroedinger, that cannot be expressed as the product of states of the individual components. Entanglement is manifested by the occurrence of correlations that cannot be accounted for by product states. By contrast, classical physics is pervasively analytic and characterizes the state of composite systems in terms of the states of components. Thus quantum mechanics has brought an unprecedented kind of holism into our view of the natural world.

4. When the parts of a composite system are spatially separated, the entanglement of its state implies a kind of non-locality: there is correlated behavior that cannot be explained by any propagation of causal influences that is not superluminal. Nevertheless, the probabilistic nature of quantum mechanics ensures that this kind of non-locality cannot be used to send a signal faster than light. For this reason, quantum non-locality can be characterized facetiously as "passion at a distance" rather than "action at a distance." In ways that are not fully understood, however, we can tentatively say that quantum non-locality requires some modification of classical ideas of causality.

5. The combination of quantum mechanics with the intrinsic indistinguishability of elementary particles of a given type leads to the conclusion that a system of n identical integral spin particles are in an entangled fully

symmetric state (except in the special case that all are in the same one-particle state), whereas n identical half-integral spin particles are in an entangled anti-symmetric state. There are important differences between these two types of entanglement, especially that the Pauli exclusion principle holds in the latter case and not in the former. In either case, however, the status of the individual particle is problematic. The individuality or thisness (Latin "haecceitas") does not seem to be manifested except in interactions with macroscopic systems.

6. Both the special and the general theories of relativity transform our conception of space and time. The special theory asserts an inseparability of space and time in a much stronger sense than in classical kinematics. Classically, space-time is a four-dimensional affine space, with metric properties restricted to space separately and time separately, whereas the space-time of special relativity is a four-dimensional metric space (with a three-one metric). General relativity envisages a dynamical interaction of space-time with matter, instead of regarding the former as a fixed arena in which material dynamics occurs.

7. The interface of physics and biology has a metaphysical implication of great significance: that all biological processes not involving mentality can in principle be understood in terms of physical interactions. Even the program of pre-biotic evolution – of understanding the emergence of mutually catalytic molecules and eventually explaining the evolution of the genetic code in terms of ordinary physical processes – is very promising. I should note that the qualification "not involving mentality" was not inserted for the purpose of hedging but for the positive reason that, in my opinion, a further scientific revolution is needed in order to understand the place of mentality in the natural world.

8. Elementary particle theory and cosmology already have, or are on the verge of having, profound metaphysical implications. But since the assessment of these implications is beyond my expertise. I encourage others to make them explicit.

Finally, I wish to warn against construing this list of metaphysical implications as indicating that all is settled in fundamental physics, and that we need only read off the corollaries in order to do philosophy properly. Rather, there seem to me to be dark clouds showing the need for radical changes in physical theory. One dark cloud is the problem of the reduction of the wave packet, which I prefer to call the problem of the actualization of potentialities. Of course, the advocates of many-worlds and decoherence theories say that there is no such problem, but I am skeptical of their solution (which obliterates the distinction between potentiality and actuality), not for reasons of sophisticated physics, but because of a straightforward philosophical analysis of the relevant concepts. The outcome of this philosophical analysis is to provide a motivation for a modification of quantum dynamics, such as the stochastic modification advocated by Pearle, Ghirardi-Rimini-Weber, Piron, Károlyházy-Frenkel, Gisin, Percival, Diósi, Penrose, and Bell. They all hope to provide a physical, rather than an epistemological, explanation of the actualization of potentialities. Here is an instance in which philosophical considerations may turn out to provide valuable heuristics for physical investigations. Another dark cloud is the problem of mentality. Before one takes too literally the phrase "theory of everything," applied to the hoped-for theory that explains all forces of nature, the spectrum of elementary particles, the values of all parameters, and the structure of space-time, one should inquire whether such a theory could even in principle account for the immediacy of consciousness.

III. THEORY OF VALUE

I shall say little about the relation between physics and theory of value, partly because of my lack of expertise. However, I do have a few strong opinions. There is, I believe, a real danger that our emotional and moral senses will be dulled by the advance of technology. The case is not hopeless, and the humanities and social sciences have serious suggestions for confronting this danger. But I think that the natural sciences can also make a contribution. A sense of wonder is one of the great motivations of the investigations of science, and the discoveries of science should provide much new nourishment of the sense of wonder. If physics is to offer an antidote against emotional flattening, however, it is essential that its discoveries be understood as revelations of the real structure of the world, to a good approximation, and not just as recipes for making laboratory predictions. My personal experience is that penetrating even a little into the secrets of the universe provides an emotional satisfaction that is like the satisfaction traditionally provided by religion. I suspect that others share this experience.

The immense value of a sense of wonder is a point of intersection between Erazim Kohák's vision of the universe and mine.

Department of Philosophy, Boston University

ALFRED I. TAUBER

ECOLOGY AND THE CLAIMS FOR A SCIENCE-BASED ETHICS

I

Erazim Kohák loves the forest. He spends much of his leisure there, and he has written eloquently of it as a paradigm of nature, specifically searching for an ethics based on our relation to nature. He seeks to ascribe meaning to Nature. How is Nature itself, and especially our relation to it, meaningful to us, and perhaps more importantly, what is our responsibility to Nature. Our age has witnessed an increasing insularity from the natural world. Both individually and as a culture, we spend a large part of our national fortune on dissecting natural phenomena in order to control nature for our purported economic and social welfare, and the technological product of that endeavor has had a tremendous price. Erazim has been an important voice in attempting to assess that cost, a powerful witness to how we must still account for a humane philosophy of nature even as we recede from it in the guise of controller.

The fulcrum of my discussion rests on Erazim's passionate articulation of a biocentric[1] ethic. As he recently lectured (1994a),

> Every living being, in its strenuous effort to remain alive . . . testifies that its own life is a value for it . . . The prereflectivity given rule of all life is that life is a value for itself – and as such, a value in itself, internally, quite independently of the existence or of the acts of any other being whatever. Life is good in itself because it is good for itself. Wherever there is life, there is value. That is the point: a biocentric cosmos is not a value-neutral one.

This perspective arises from Erazim's well-known phenomenological approach, where he maintains that we remain within a meaningfully ordered, value-indexed world, and more to the point, this is not simply a result of human reflection, but is constituted by life as such.

I might agree with his overall moral position for different reasons, but my discussion is only peripherally concerned with the ethics, per se. Instead I seek here to situate what I perceive to be a fundamental confusion concerning how that position may be derived inappropriately from biology, how science has been co-opted to serve an ideologic agenda. Erazim is not himself guilty of arguing how an ecological ethics purportedly arises from science,[2] but many do, and my concern here is to dissect how an ideology, even a moral philosophy of nature, would usurp science for its own ends; these final ethical stances

R.S. Cohen and A.I. Tauber (eds.), Philosophies of Nature: The Human Dimension, 185–206.

founded on other principles may be legitimate in themselves, but to use science surreptitiously is dangerous business, and we must be aware of the true basis of the arguments employed to fulfill essentially a political agenda. The issue that raises my ire is simply this: Much of what passes as ecological ethics claims that unlike other ethical ventures based on religious or metaphysical foundations of belief, these ethics are rooted, in fact "proven" or "demonstrated" by the incontestable facts of a new science – ecology. For if laboratories can demonstrate the deleterious effects of aerosols in the atmosphere, or particularly toxic chemicals in our rivers, or automobile fumes in our cities, then is it not senseless, if not immoral to continue along the path of environemental degradation? From this perspective, there is a seamless joint between the findings of ecologists as scientists and the values drawn from their studies.[3]

Why and how is the claim made, and is it legitimate? There are many reasons that might be conjured to support an environmental ethics, and I will only mention two. The first is a utilitarian imperative, which seeks to root ethical decisions in an objective accounting of gains and losses. Here we invoke the power of an objective science to attain that rational ideal; seemingly innocuous interpretations of the objective data offer the obvious utility by which rational choices might be made. But such choices of course have a moral setting, which while framed by our scientific understanding, is hardly neutral. As Luc Ferry puts it, there is a syllogism at work:

1. The biological sciences, including ecology have disclosed that organic nature is systematically integrated:

2a. That mankind is a non-privileged member of the organic continuum, and

2b. Environmental abuse threatens human life, health, and ultimately happiness, therefore

3. We ought not violate the integrity and stability of the environment (1995, p.88)

Like any utilitarian argument, one may contest its relative merits. In this case, the ecological syllogism fails to acknowlege the relativity of the ideal, "health." Moreover, it asserts a nonnormative analysis of what we, de facto, are supposed to love or abhor. "The ethical criteria becomes identified with what empirical anthropology teaches us about human . . . psychology." (ibid. p.89) In assuming that we all share the same ideas and values concerning what is healthy, any deviance then becomes pathological. Ferry's major thesis concerns how such totalitarian thinking has its roots in Nazism and Stalinism, and I refer the interested to his text, *The New Ecological Order*.

It so happens that I am in fact very sympathetic to an environmentally sensitive ethics, and I embrace much of the Green Program from a persuasive utilitarian argument. But this position is vulnerable to other interpretations of the scientific data, as well as other human needs, and thus I understand why a moral imperative might be sought to further the ecological perspective. But in seeking such a standing, the moral argument takes an ominous twist vis a vis seeking a scientific basis for the ethical argument; it is at this juncture that I would defend science against any such intrusion. The rest of this paper is concerned with how certain arguments for environmental ethics would use a scientific rationale to raise the ecological ethical position over other competing ones, because the former presumes to be *scientifically* based. Aside from the question of co-opting the science inappropriately, the irony is that the biology invoked hardly fulfills the standard of the day.

Thus I will follow a different tack from that chosen by Luc Ferry in dissecting a second reason that would establish an "objective" foundation for ethics, one which would hope to take the moral high ground that could not be assailed by the relative merits of any particular utilitarian position. The rationale concerns the idea that life has a telos, that is, goals by which organisms organize their behavior and physiological organization that suports end-seeking functions. This is a complex issue, and is one that I will attempt to flush our more thoroughly, not only because it is most germain to the topic of the symposium, namely a Philosophy of Nature, but because it is perhaps the deeper and more obscure rationale for a science-based ethics. That argument begins with asserting that Nature in itself contains certain objectives, for instance the preservation instinct, independent of our humane or subjective opinions. From this position, teleology blossoms like a prolific bush, sprouting assertions in many directions. Let me sample a few representative opinions:

First, an empathetic element is introduced. Paul Taylor writes how our science documents the life cycle of the species, its ecological interactions, etc., and at the same time we recognize the uniqueness of each individual, which he believes we may easily convert to a moral perspective. As he writes:

> This progressive development from objective, detached knowledge to the recognition of indivuality, and from the recognition of individuality to a full awareness of an organism's standpoint, is a process of heightening our consciousness of what it means to be an individual living thing, We conceive of the organism as a teleological center of life, striving to preserve itself and realize its good in its own unique way. (1986, pp.120–121)

The telos of the organism defines what is good for it, and it is the shared sense that organisms are individuals, like human agents, that confers a moral standing to them. In short, teleological centers of life serve as the foundation of value. And the entire enterprise rests on the science of biology, for as Taylor asserts,

> certainly our acquiring scientific knowledge about certain kinds of animals and plants can help us enormously in the attempt to understand objectively the everyday existence of particular individuals of those kinds. (ibid., p.126)

Thus a biocentric value is inserted into the biology. Note what is occuring: According to various critics, the purportedly objective science is in fact documenting value as constructed within an evolutionary and physiological context. For instance, Holmes Rolston, in *Environmental Ethics*, writes that the organism is an "axiological system," an "evaluative system" as it grows, reproduces, repairs its wounds, and resists death. He then slips in the V word, value, and here I quote:

Value is present in this achievement. Vital seems a better word for it than *biological*. We will want to recognize that we are not dealing simply with another individual defending its solitary life but with an individual having situated fitness in an ecosystem it inhabits. Still, we want to affirm here that the living indiviual . . . is per se an intrinsic value . . . The organism has something it is conserving, something for which it is standing its life. (1988, p.100)

Then the ethics become explicit:

There seems no reason why such own-standing normative organisms are not morally significant. (ibid., p.100) [Thus a] tree has a *telos* before the logger arrives, and the logger destroys it. It is *autotelic*, it has a law (Greek: *nomos*) on its own (=*autonomos*). (ibid., p.105)

Telos, essentially a descriptive mode that biologists employ to describe function, Rolston has extended to a moral category. What he calls the good of the organism, for instance striving to preserve itself, a biologist, even employing a teleological orientation, would say is a survival behavior which follows the rules of biological fitness, where fitness means the ability of the organism to ensure the continued existence of its genes in future individuals. This is a statement of evolutionary survival, but Rolston assigns a moral value to it:

Value is not just an economic, psychological, social, and political word but also a biological one. Value_{or} or what is good for the organism . . . is for the organism a telic end state, an intrinsic value, not always a felt preference. (ibid., p.257)

Of course such value is interpreted, and again teleology offers the rationale for our assuming that judgment. The famous philosopher of biology, Hans Jonas, in *The Imperative of Responsibility* (1984), assumes man to be the most elevated product of nature, who is thus capable of deciphering and taking responsibility for the world we cohabitate. Truly, a divine inspired Adam, for on this view, Man reads the law of nature and adjudicates according to his wisdom. Michael Serres opines similarly when he asserts:

The life of the entire species is in our hands; it is a basis as true and faithful to things as that of the sciences themselves. We are entering a period in which morality is becoming objective. (*Le Monde*, January 21, 1992; quoted by Ferry [1995] p.87)

I am very disturbed by such assertions, not necessarily because I disagree with the ethical conclusions, but rather because I deplore the usurption of science to support the ethical claims and their "objectifying" argument. I will attempt to show that the two realms, biology and ethics, are fundamentally different discourses, and to connect them is to make a category error with wide ramifications and dangers. Let us delve a bit into the so-called "boundary question" concerning the elusive separation of science and ethics to draw a clearer perspective on the problem at hand.

II

In the strictest sense, biology is a science, and a science in its purest state is limited to a description of natural phenomena, whereas ethics derives from metaphysics and its domain is delimited to a sphere of thought that cannot be confused or conflated with the scientific enterprise. Optimally, each, legitimate in their own domain, should remain discrete and separate. Obviously they are often merged to varying degrees, and we have come to a heightened awareness of how science is not only co-opted by politics and ideologies, but also is more subtly influenced by its cultural milieu and moral environment. In fact, one might argue that the social constructivist view of science is a critical supportive pillar of environmental ethics. If science is afterall constructed from cultural elements inseparable from the ostensibly objective procedures of scientific inquiry, then we must both acknowledge and accept the intrusion of additional ethical demands upon what science might regard as its own sacrosanct independent practices. On this view, science never stood alone in an objective cultural chamber insulated from the social pressures surrounding research and theory. Better to openly bring science into its full moral context, and by reciprocity apply science to its full social potential. In this case, let ecology support environmental ethics.

This discussion rests on disentangling two often confused issues: Firstly, what are science's ostensible values and how do they govern scientific practice? And secondly, how does science influence our broader morality? Each of these concerns relate to the boundary question, because the ethics of science arise within broader cultural ideals, and the culture may in turn be influenced by institutions of science, whose moral code may serve as a model of behavior extended to other domains. Despite the intimate historical links between scientific and moral discourses (Shapin, 1994), we must recall how science ostensibly was born in the effort to free itself from the conflations between fact and value, between natural and supernatural, between body and spirit. Even at the dawn of modern science, this issue of mapping the domain of knowledge proper was recognized as crucial, given the rise of political agendas concerned with how science could be used to serve particular social and economic interests. And it was in the recognition of science's power that its early institutionalization was structured to guarantee its independence from meddling politicians, who in turn extracted the promise that scientists would remain disinterested in wordly (i.e., political) concerns (e.g., Shapin and Schaffer, 1985). Afterall, in the search for Truth and Reality, the scientist should be immune from the messy debate of how the fruits of his labor were to be applied (whether in warfare, medicine, or technology – at large) or the possible dire consequences of his discoveries for the environment and for the individual. From Science's lofty laboratories, only unperturbed truth seekers would explore Nature's secrets, oblivious to the political, social, and economic needs of the supporting culture.[4] As patient spectators, the "common" people would reap the material harvest, although some would argue it was a Faustian pact. Objectivity is a value, and we will now look at its various manifestations.

The neutrality of science depends on regarding Nature as holding no value: value is rooted in human needs and desires, whereas nature is stripped of qualities, teleology, and meaning, leaving it devalorized, secularized, and disenchanted. The crucial philosophical distinction is between "what ought to be" and "what is." The attempt to free facts from value was to liberate science from its Medieval theological roots, and remains the linchpin for scientists pleading autonomy under the rubric of "objectivity," as well as for their critics, who decry the violation of neutrality of science, which obviously serves particular social agendas. But as Robert Proctor has congently observed,

> [N]eutrality and objectivity are not the same thing. Neutrality refers to whether science takes a stand; objectivity, to whether science merits claims to reliability. The two need not have anything to do with each other. Certain sciences may be completely "objective" – that is, valid – and yet designed to serve certain political interests. Geologists know more about oil-bearing shales than about many other rocks, but the knowledge is thereby no less reliable. Counterinsurgency theorists know how to manipulate populations in revolt, but the fact that their knowledge is goal-directed does not mean it doesn't work.
>
> The appropriate critique of these sciences is not that they are not "objective" but that they are partial, narrow, or directed towards ends which one opposes. In general, knowledge is no less objective (that is true, or reliable) being in the service of interests. (1991, p.10)

Given these distinctions, science nevertheless can hardly be separated from its political support, and scrutiny of that political context has increasingly raised a critical chorus since World War II. Citizens maintaining a vigilant watch over scientific aspirations and purported successes, no longer accept as gospel the claims and promises of a growing scientific lobby. Critics have successfully halted or modified multi-billion dollar Big Science projects in the 1990s, in what some regard as an anti-scientific conservatism, and others the appropriate constraint of a ravenously imperialistic Science. Controversies surrounding public policy concerning investment in major scientific projects that are touted as the penultimate, if not the ultimate, climax of scientific progress (e.g., Gilbert [1992]; Weinberg [1992]) are largely connected with three domains: the disappointment in past, similar programs, such as the failed War on Cancer and other overly optimistic projects, which have promised to deliver solutions that were unrealistic; the growing concern that resources should be more carefully allocated towards directed application and more modestly achievable goals; and finally the recognition that naive positivist ideals have been abandoned. Further, despite the reiterated disavowals of a value-laden science, critics have exposed this innocent view.

This critical stance is based on the assertion that science as practiced is not a *free*-standing enterprise, but is socially based and subject to the needs and values of its supporting culture. This public domain of science refers not only to the renewal and support that our society gives scientific institutions, but the recognition that science serves in a political culture, supporting various economic and political interests. From a political perspective, the relevant issue, beyond defining the social origins of knowledge, is the requirement for a philosophy that focuses on the forms of power in and around the sciences:

"Why do we know what we know and why don't we know what we don't know? What *should* we know and what shouldn't we know? How might we know differently?" (Proctor, 1991, p. 13) In short, we must be cognizant of the complexities of science as politics and not confuse the social applications of science with the business of scientific inquiry. It is natural that extrapolations be made, afterall, science is obviously a social activity, but my plea is that we must acknowledge those interpretations, as interpretations. Ecological ethics is an important case in point. We can support and even claim that moral position as our own, but at the same time we must recognize how we justify those judgements as interpretive. In the process, we preserve the openness of scientific inquiry and its pluralistic attitude towards knowledge. That is an ethic that dominates any ideology that would use science for its own purposes.

Recognizing the complex moral context in which science operates, I wish to now explore and expose why biological teleology has been connected to ecological ethics, and why they must remain separate. First, an important caveat. I suspect that ecological ethics arise from the same source in which we seek other forms of meaning in our biology. I myself have explored that domain in the connection of science and beauty, the aesthetic dimension by which we perceive order and grandeur in the complexity and simplicity of the living (Tauber, 1996). Whether we appreciate the extraordinary order of a crystalized protein or the hierarchical interconnections of an ecosystem, as sensitive men and women, we find meaning in those experiences. These are human appreciations of nature's beauty, and there is no reason to exclude a scientist's work from the aesthetic realm. Quite to the contrary, many of us, as biologists, seek aesthetic experience in our research of organic phenomena, and find there an important component of satisfaction, as well, perhaps, a less well-defined rationale for a theory's construction and its acceptance. But generally, except at the deepest psychological level, I regard this aesthetic connection as distinct from the usual business of science proper. Not that I necessarily believe they can be entirely separated, I do maintain that *meaning* per se, whether aesthetic, metaphysical, or ethical, is not the appropriate question for science – as an epistemology – to pose or explore.

Now, let us return to the particular issue at hand. In the following section, I will explore the status of teleology in biology today, briefly reviewing its history and current utility. The irony of a purportedly science-based ethics built on the teleological status of living organisms is that such a formulation in fact is hardly part of current scientific discourse. Teleology is a concept conscripted for describing Life, part of the complex humanistic endeavor to find meaning in Nature, wheras biology is concerned not with "life," but with the biophysics and biochemistry of the organic. In other words, teleology is integral to an older natural philosophy that truly sought a philosophy of nature in human terms. Current biology is a science devoted to providing mechanistic explanations, and to the extent that a telos orients its theory, we witness vestiges of this older metaphysics. To put the matter simply, although we detect teleological descriptions in modern biology, they represent the inadequacies of our true scientific aspirations of offering mechanical explanations for organic processes, whether physiological, developmental, or evolutionary. Although I believe that such teleological descriptions cannot be entirely purged from biology and must continue to serve a complementary role to mechanical explanations, we must acknowledge the actual scientific status of telos. When we do, then it is evident how ethicists might so easily co-opt teleology for their own purposes.

III

Many would see biology as caught in a fundamental paradox upon which the ethical project rests. Namely, biology is characterized by two ways of thinking, an objective, positivist program that seeks a description of biological processes, and a teleological mode of addressing those phenomena in terms of function. It is from this latter logic that broader questions of meaning arise. I, in turn, would like to ask whether teleology in fact represents a paradoxical mode of thinking in biology, and in regarding our immediate concern with environmental ethics, whether it legitimately may be used to support this broader moral agenda. The question is asked in the context of a scientific program that has sought to purge the issue of human significance altogether.

Ethics falls under the rubric of the subjective, and the purging of subjectivity from science has a long history. Most would date its proclaimed origins with the positivist movement in France during the 1820s, but there is an older heritage dating back to the very birth of modern science.[5] Positivism of course has been duly challenged in our own time, and a continuous concern with "objectivity" remains a critical topic of debate in science studies today. I will not delve into this matter in any detail, but it does set the stage for my own topic, for the general question of "meaning" arises subtly from biology, and to discern those roots is to expose the intimate linkage that environmental ethics has been able to make with its supporting science, namely the teleology discerned in biology.[6]

Strictly speaking, teleology is the idea of giving an account of something by reference to an end or goal. Teleological explanations span the entire range of biology, from the apparent goal-directedness of embryological development, to the adaptive character of traits and organic systems, to the purposiveness of behavior. The teleology hovering over both molecular descriptions such as enzymatic cascades and complex social behaviors is to some a "spectre" of a tainted mode of thinking. Why? In answering that question, both in its historical and analytical context, we might discern more clearly the source of confusing teleological descriptions with "meaningful" ones that give rise to *value*.

The issue is perhaps best presented by Jacques Monod in *Chance and Necessity*, where he explicitly states the modern biologist's embarassment, or what he calls biology's "epistemological contradiction." According to Monod, the contradiction concerns the two ways of knowing that the scientist must employ to study the organic realm. These Monod called, "objectivity" and "teleonomy," respectively. (See the Appendix for a historical description of teleonomy.) First, Monod admits the *sine qua non* of goal-directedness in biology and the centrality of teleonomy:

One of the fundamental characteristics common to all living beings without exception [is] that of being *objects endowed with a purpose or a project*, which at the same time they exhibit in their structure and carry out through their performance . . .

Rather than reject this idea (as certain biologists have tried to do) it is indispensible to recognize that it is essential to the very definition of living beings. We shall maintain that the latter are distinct from all other structures or systems present in the universe through this characteristic property, which we shall call teleonomy. (1971, p. 9)

But this definition, while necessary is not sufficient, since as Monod further observed, teleonomy does not provide any objective criteria for distinguishing between living and non-living things. Here he introduces the requirement for a self-organizing system, programmed by chance and necessity, but paradoxically committed to purpose. Monod, asked, is this a "miracle?"

No, the real difficulty is not the physics of the phenomenon; it lies elsewhere, and deeper, involving our own understanding, our intuition of it. There is, really, no paradox or miracle; but a flagrant *epistemological contradiction.*

The cornerstone of the scientific method is the postulate that nature is objective. In other words, the *systematic* denial that "true" knowledge can be got at by interpreting phenomena in terms of final causes – that is to say, of "purpose" . . . But science as we understand it today could not have developed upon those foundations alone. It required the unbending stricture implicit in the postulate of objectivity – ironclad, pure, forever undemonstrable. For it is obviously impossible to imagine an experiment which could prove the nonexistence anywhere in nature of a purpose, of a pursued end.

But the postulate of objectivity is cosubstantial with science; it has guided the whole of its prodigious development for three centuries. There is no way to be rid of it, even tentatively or in a limited area, without departing from the domain of science itself.

Objectivity nevertheless obliges us to recognize the teleonomic character of living organisms, to admit that in their structure and performance they act purposively . . . Here therefore, at least in appearance, lies a profound epistemological contradiction. In fact the central problem of biology lies with this very contradiction, which if it is only apparent, must be resolved; or else proven to be utterly insoluble, if that should turn out indeed to be the case. (ibid. pp. 21–22)

It is this issue that I wish now to fully explore. Let us begin with a short historical survey of what Darwin called "One long argument." The philosophical issue arises naturally from the perspective of natural selection, the fundamental theory of biology. We will find there, in Darwin's *On the Origin of Species*, not only a mechanistic explanation that accounts for evolution and the rationale upon which we assign functional characteristics of extant organisms, but also the essential elements for our current materialistic, purportely non-teleological understanding of biology. I will proceed by showing where the science admits supposed paradoxical elements, and will then finally attempt to resolve what Monod calls a contradiction as simply a complementary way to describe biological phenomena.

Biology, until the mid-nineteenth century, was the exploration of an organic world described by a divine intelligence. The myriad species, ordered in their complex interplay, found their position in the great chain of being, because of an ordered universe variously viewed as a sublime garden of God's creation or a vastly complicated machine, so perfectly construed that upon its initial conception, it worked like a perfect clock by a divine plan we might study, but never fully understand. Although the history of mechanistic explanation is highly germaine to this discussion, let us quickly skip to the end of the eighteenth century, and consider the state of biology that was by this time already largely committed to materialistic and mechanistic accounts of function, but still remained explicitly coupled to an older metaphysics. Kant's teleology was the critical argument that attempted to offer a resolution. As he understood the issue, teleology must serve as a regulative principle by which mechanisms might be examined and understood. Teleological explanations orient scientific inquiry that would discern the mechanism's workings. But, and this is the crucial pre-Darwinian position, we might never discern the ultimate basis of design. Kant believed, as did virtually all of his contemporaries, that there never would be a Newton who might explain how a single blade of grass grows or how it appeared in creation. There had to be some final intelligence that conferred organization and design, the function, the ends of particular organic entities, and their behavior as fundamental to their being.

With Darwin, a mechanism was offered that no longer sought to comprehend a divine Design. There was no need to call upon a pre-ordained Plan, and there was no Purpose. Life simply was. With natural selection, a process of variation and selection explained the evolution of species, their particular anatomies and physiologies. Very simply, since Darwin, teleological explanations as construed from Aristotle to Kant have ostensibly been purged from biology. The natural selection theory of evolution, more specifically, its neo-Darwinian formulation, has declared any intentional interpretation as unnecessary for a mechanistic (which would include a stochastic causality) explanation, which is both necessary and sufficient to account for the current complexity and phylogenetic history of our organic world. But the organic could not so easily be purged of what appeared most characteristic of it, namely that 1) the design of organisms are seemingly purposeful, being directed toward particular ends, and 2) they exhibit active behavior to achieve certain goals. This is what was referred to earlier as "teleonomy," the term coined 35 years ago to acknowledge end-seeking function without a superimposed design. (See Appendix.) The intention arises from the evolutionary history of the species upon a genetic blueprint that designates functional capabilities and potentials, and also allows adaptive and learning behavior. So, contemporary biology incorporates simple and probabilistic causalities with ranges of choices to allow that life processes indeed have a telos – an end, purpose, goal, state of completion – for the sake of which an activity takes place or for which change is produced or sought. Teleonomy has thus replaced teleology, and the issue is to explain how such goals (and the capacity to reach them) are structured within the genetic design.

Nevertheless, teleos – perhaps because it has such a sordid history in biology (Mayr 1982), and more particularly due to its "semantic instability" (having so many interpretations and inferences) – continues to "taint" biology. The extraordinary power and effectiveness of a science adept at chemical and physical characterizations is indisputable. The issue is to formulate and maintain strict positivist descriptive criteria for elucidating seemingly intentional life processes, seemingly far removed into the future and thereby determined by causalities quite distinct from those normally encountered in characterizing biochemical and biophysical processes. "Purpose" smacks of a subjective projection or interpretation. Monod, like Francois Jacob and Ernst Mayr, among many others, have embraced teleonomy as arising from a genetic program determined by evolutionary forces to account for such seemingly purposeful behavior. I will not attempt a critique of that position here, but suffice it to note that on this view, objectivist descriptions, of which teleonomy may stake its own claims, should suffice for biology as it does for physics and chemistry.

Nevertheless, "teleology is like a mistress to a biologist: he cannot live without her but he's unwilling to be seen with her in public" (attributed to J.B.S. Haldane by Pittendrigh in Mayr, 1988, p. 63). Despite the scruples of a science divorced from teleological explanation, it must nevertheless rely on telos to order its theory and methodologies. We must acknowledge that telos may function as the projection of our own particular rational understanding, and although potentially distorting, such descriptions orient our inquiry. Let me emphasize, teleological *descriptions* cannot substitute for mechanistic *explanations*, but because of the incomplete status of our life sciences, such descriptions are complementary to physical and chemical explanations of functions. This duality remains a legitimate strategy. There is no contradiction; there are simply two complementary ways of describing biological phenomena. The basis for asserting the central though elided role for teleology in biology at the end of the twentieth century resides in the need both to maintain our positivistic – Monod's "objective," non-teleological – explanations, and to allow some form of telos to guide our theory and methods. There is no contradiction if we recognize that we are employing complementary ways of describing organic function. The difficulty, due to the incompleteness of our biology, is that we cannot truly separate the two, and that our 'understanding' – whatever scientific erudition our biology offers us, intrinsically interwines our biophysical and biochemical descriptions with our teleological orientation.

We might just accept this strange marriage, either because 1) we cannot completely divorce teleological from mechanistic explanations, primarily because the levels of function approached by each are disparate and the science invoked is as yet inadequately developed (Taylor 1970),[7] or 2) teleological explanations are in fact causal despite their future orientation (Wright, 1973; 1976).[8] In both Taylor's and Wright's accounts, what makes behavior teleological is that the process in question can be shown to occur *because* it is required for the achievement of the goal state (Lennox, 1992, p. 332). In each case, a heuristic value is claimed. Taylor acknowledges the paradox and adopts a wait and see attitude, while Wright falls into an adaptation mode of explanation. I regard this matter somewhat differently. I submit that teleology, although often argued as a logical question, more appropriately is situated as an epistemological issue, resting on the very fault line between *fact* and its *interpretation*. Teleological descriptions are still necessary, serving as proto-theories, and thus requisite to placing the "objective" data in order. To expel teleological descriptions, and here I am referring only to descriptions of end-seeking function, would be to strip biology of its logos. The science must explain function mechanistically, i.e. with factual descriptions, but those facts are placed in some functional edifice. There is no contradiction; there is a tensioned complementary between problems stated in functional terms and explanations given in mechanistic one.

The crux of the matter is the "aura" of interpretation that surrounds facts. The motivation for a purely objective account resides in the foundations of positivist science that, in its starkest statement, attempts to give factual accounts devoid of interpretation. But it is a dictum now that fact cannot be divorced from its supporting theory. The problem of theory-laden facts has been extensively explored since Goethe's dictum, "to grasp that everything factual is already theory" (Tauber, 1993). Goethe was also aware of the converse problem of theory distorting observations to conform to the prejudice of theory. The attempt to purge teleology from biology in the nineteenth century was motivated by this concern, for biologists were aware of the ominous distortion that projection of an interpretation might have on the creation of facts, and thereby corrupting theory construction built upon positivist ideals. On this view, teleological explanations might serve as an undeclared theory posing as fact, and in such constructions, facts situated by theory are then used to erect the supposed objective ediface to support that theory in a circular chain of reasoning.[9] But if we allow an objective science to be judged essentially on the success of its ability to predict phenomena and cohere to explanatory principles deemed constant with its own theoretical construct, then we may legitimately regard teleological descriptions as part of the very fabric of biological science. It serves to order our observations and orient our theory, which then is tested anew.

In short, we organize our observations and place our putative facts within a theory or model to create some structure that confers order to phenomena. Although we do so at the peril of distorting – or should I say, potentially falling short of our essential scientific mandate, namely objectivity – this is the irreconcilable tension of science. (It is also the basis of the evolution of scientific knowledge.) Operative in all sciences to a certain extent, this concern is most relevant to biology, because the phenomena cannot simply be described in isolation from end-seeking, purposiveful function. The positivist ideal is shaken, for at any level, such intentionality is interpretive, even projective of our own bias. We are thus shackled to an epistemology that fails the most stringent positivist requirements.

To summarize, I have attemptted to clarify that biology as an objective science rebukes teleology, and the history of the field over the past two hundred years has been marked by that struggle. Quite simply, biology, or should I say biologists, cannot completely purge teleological structures, whether called teleonomy or left unannounced as implicit in the discipline's thematic infra-structure. Rather than lament this duality, I accept the complementarity of positivist accounts in a science committed to teleological orientation. We are left with a biology that builds its world view, or what Wittgenstein designated as a picture, from a complex dance between its epistemological poles. 'Practicing' biologists, usually unaware, see as Janus, whose vision is formed from both vistas. They unconsciously *operate* with both 'objective' and 'teleonomic' explanations to structure their facts and build their theories. 'Theoretical' biologists recognize the limits of a scientific formalism and being wary of escaping into some metaphysical interpretation, while perched on the edge separating them, peer self-consciously first in one direction and then another. I suggest we accept objectivity and its supporting telos for the service each performs, acknowledging their inherent limits and at the same time recognizing their particular contribution to the scientific enterprise. The mistake is to extend telos to support some broader meaning; for that extrapolation, teleological thinking has been justly attacked.

IV

Let us now return to our central concern. As illustrated, there are those who use the so-called "teleological centers of activity" (to quote Taylor [1986]) as a node of value and from there it is a short step to assigning a true moral standing. That step is taken because it is supported by a metaphysical picture that has a strong grip upon our culture, one upon which the epistemological basis of teleology resides. The problem lies in large measure with the "semantic instability" of the term, teleology. As already discussed, teleological descriptions are by their very nature interpretative. In a strict biological context, they are used to define end-seeking function. The category mistake occurs when one proposes that purpose has moral standing. Certainly human intentionality has moral standing: we make choices embedded in an ethical construct. Why do purposful animals have moral standing? The ecological ethicist must believe that beyond sharing purpose, animals share with humans the same biological origins of morality itself, the very matrix by which purpose, whether human or animal, attains its moral meaning.[10] But more specifically, we are concerned with the central role telos plays in the seemingly effortless jump from human to animal morality.

I have already described the scientific role teleological descriptions offer us in ordering biological phenomena. But now we see the semantic instability of teleology broadening its rubric to offer a foundation upon which a more global application of teleology is presented. From those piles, an ethical edifice, purportedly built from science, is erected. From Alfred North Whitehead (1925) to Daniel Dennett (1981, e.g. p. 28), numerous commentators have noted that the intentions we ascribe to natural systems are abstractions of a sort, in which a rationality resembling our own is ascribed to natural systems or behavior. It is this rationality that enables us to understand and to explain them using our own intelligent faculties. It reflects a deep-seated confidence in the rational order of nature, and we thus profoundly both endorse and recapitulate Kant's project. As Herbert Simon (1969) has noted, this amounts to analyzing natural systems as if they were artifacts, attributing to them the same type of rational adaptation of means to ends that we employ in the design of our machines. Beyond what Gregory Bateson (1980, e.g. p. 299) and others have underscored as naivete of believing that nature conforms to our particular scientific rationality (Atlan, 1993, note #83, p. 91), there is another startling philosophical revelation.

As Henri Atlan has observed, this projection of our rationality upon biology "is an even stronger postulate than that of a rational intelligibility of nature, namely, that of an *intentional* rationality in nature" (1993, p. 75); imposing goals on nature results from anthropomorphological reasoning. Intentionality, used to characterize psychological or social action of conscious beings, is according to him misapplied to apparently intelligent behaviors adapted to the achievement of some goals observed *in* individuals, whether human or animal (1994, p. 74).[11] And here we can now return to our original query concerning meaning in nature. This need for meaning, this relic of an ancient metaphysics, resides deeply within our psychologies, for all that has been repressed and circumscribed by a scientific culture that aspires to objectivity (if not some positivist ideal) and technical virtuosity.[12]

To speak of a philosophy of nature, we draw from both the objective scientific accounts as well as the relational aspects derived from this powerful metaphysical imperative. Herein, in this latter faculty, lies our quest for *meaning*. In this latter task, we ultimately speak of ourselves. Rather than shun our metaphysical assumptions, perhaps we should delve more deeply to understand them. For although we resist alluding to metaphysical constructions in this post-metaphysical age, we seemingly cannot escape their grip. In acknowledging their hold on us, we will more clearly comprehend their pervasive, and often disguised influence arising from their diverse origins and complex structures.

APPENDIX – FROM TELEOLOGY TO TELEONOMY

Soon after Watson and Crick had offered the key to the genetic puzzle in 1953, Colin Pittendrigh offered a new term and argument that attempted to eclipse the spectre of teleology from the life sciences. He wrote in 1958,

> Today the concern of adaptation is beginning to enjoy an improved respectability for several reasons: it is seen as less than perfect: natural selection is better understood; and the engineering physicist in building end-seeking automata has sanctified the use of teleological jargon. It seems unfortunate that the term "teleology" should be resurrrected and, as I think, abused in this way. The biologists long-standing confusion would be more fully removed if all end-directed systems were described by some other term, like "teleonomic," in order to emphasize that the recognition and description of end-directedness does not carry a committment to Aristotelian teleology as an efficient causal principle (pp. 393–394).

Pittendrigh's entreaty was to evoke a serious debate. There were those like Bernard Davis and Gaylord Simpson who viewed the term as virtually synonymous with adaptiveness. Simpson wrote that teleology may implicitly negate rather than express the biological conclusion that organization in organisms is with respect to utility to each separate species at the time when it occurs, and not with respect to any other species or any future time (Simpson, 1958, p. 520).

Such a definition uses teleonomy only to explain adaptation in nature as the result of natural selection. Teleonomic in this view becomes a description for a system or structure which is the product of evolution and selective advantage.

The chorus seemed to agree: teleology as intention had to vacate the premises, even while the descriptions of biological processes required an acknowledgement of their goal-seeking qualities. Teleonomy had been coined in the attempt to rid the science of its ancient metaphysical baggage and still retain the characteristic feature of the organic, namely there were ostensible goals to which organisms strived. Yet, as mentioned, biologists initially were employing the new term as a synonym for adaptation. Pittendrigh complained in a letter to Ernst Mayr about his frustration in threading the philosophical needle:

What it was the biologist could not escape was the plain fact – or rather the fundamental fact – which he must (as scientist) explain: that the objects of biological analysis are organizations (he calls them organisms) and, as such, are end-directed. Organization is more than mere order; order lacks end-directedness; organization is end-directed . . .

I wanted a word that would allow me (all of us biologists) to describe, stress or simply to allude to – without any offense – this end-directedness of a perfectly respectable mechanistic system. Teleology would not do, carrying with it that implication that the end is causally effective in the current operation of the machine. Teleonomic, it is hoped, escapes that plain falsity which is anyhow unnecessary. (quoted by Mayr, 1988 pp. 63–64)

Ernst Mayr shared these sentiments, but believed that Pittendrigh's opposition to Aristotelian teleology was unfortunate, since most of Aristotle's references to end-directed processes were still appropriate (ibid., pp. 56–7). This is a most interesting admission, for it reaches back to a classical debate and highlights the critical issues for modern conceptualizations.[13] Essentially, the modern commentators wanted teleology, but without some backward cause. A more elaborate strategy was required.[14] Following Pittendrigh, Mayr attempted to consolidate the notion of teleonomy as a seminal conception of biology by fully embracing the new enthusiasm about cybernetics of that period. He actually built on the earlier formulation suggested by the fathers of cybernetics. In 1943, Rosenbleuth, Wiener and Bigelow posited that negative feedback would obtain information in the environment necessary to correct any perturbations in a system moving toward a goal. Thus purpose was to be explained by a new neutrality.[15] Mayr became a cybernetic enthusiast, and defined a teleonomic process or behavior as one which owes its goal-directedness to the operation of a *program*. "Individual development and . . . seemingly goal-directed behavior . . . depend on some endpoint or goal which is foreseen in the program regulating the behavior (Mayr, 1982, p. 48). Teleonomy was thus regarded as a non-intentional, non-purposiveness, end-seeking function guided by a program, an executive formation residing in the cell's nucleus and determined by natural selection.[16]

Mayr was careful in attempting to avoid the obvious objections (Mayr, 1988). He embraced the idea of "program" as 1) something material, and 2) as existing prior to the teleonomic process and thus consistent with the temporal directionality of a causal explanation. ("It is the endpoints that produce the selection pressures which cause the historical construction of the genetic program" [Mayr, 1982, p. 49].) Heavily endowed with the newly emerging application of information theory to biology, he pursued the metaphor of a code to its full application. More than simply a blueprint, the code also conveyed instructions of how to use the information of the blueprint, remaining segregated from the cell's machinery that enacted the program. Further, Mayr shrewdly allowed for both 'closed' programs, i.e. the strict instructional component of the genetic code laid down in the genotype, and an 'open' program, which guaranteed the incorporation of additional information. The open program would allow for learning, and although such information may not be acquired through selection, it clearly was to be in part responsible for teleonomic behavior. The history of the acquisition of a program was not part of the definition of teleonomic, and Mayr argued that the origin of the program was irrelevant to the definition, since the program might be the product of evolution, or acquired through the open channels of information. This allowed, he thought for both somatic and epigenetic events, that is a 'program' rigidly wired or acquired by experience, respectively.

Whether such an open-ended definition of program was warranted or only served as a fanciful metaphor was the subject of dispute. Early critics like Erwin Schrödinger, writing at the time of cybernetics' birth in 1944, noted the difficulty of accounting for a self-written code.[17] But the power of the code metaphor persisted and was modified in the 1960s to specify 'program' to account for how the gene acquired its "specific value." Whereas Schrödinger had been puzzled by the dual attribution of "law code and executive power," Francois Jacob enthusiastically adopted the paradox with an important twist:

> What are transmitted from generation to generation are the 'instructions' specifying the molecular structures: the architectural plans of the future organism. They are also the means of executing these plans and of coordinating the activities of the system . . . The organism thus becomes the realization of a programme prescribed by its heredity. The intention of the psyche has been replaced by the translation of a message (1974, pp. 1–2).

The metaphor is here working subtly to offer a mechanical account of organic function, but critics now argue that there is a critical short circuit: control and the controlled have been conflated.[18] The salient point is that computer scientists need not deal with meaning as such, but more simply with the technical problem of transmission. Thus for all practical purposes a program may be treated as pure information. But organisms are unable to bracket meaning, which is intrinsic to what we metaphorically refer to as a program. On this view, programs and information must be distinguished in the organism. There is unresolved debate concerning what kind of system is capable of generating its own program – its goals, meaning, tasks. What indeed is a natural machine? Champions of the cybernetic model drew parallels with self-directing machines (like guided missiles or heat-seeking bombs): "At any time, the machine that executes its programme is capable of directing its action, of correcting or even interrupting, in accordance with the message received" (Jacob, p. 253). Although Jacob recognized that the metaphor had limits, it was left to others to add important critical caveats. For example, Henri Atlan wrote regarding the organism determined by the informational structure of its genome:

This widespread idea should be accepted as a working hypothesis or a metaphor and not taken literally, since DNA as a genetic "program" needs the products of its reading and execution – RNA and regulatory proteins – to be read and executed. In living cells, the "program," if any, seems to be identified with the whole cell. The DNA molecule look[s] rather like memories where parts of this program are stored (1983, p. 28).[19]

We need not further review the more interesting rebuttals and counter theoretical offers currently in discussion, but close with Jacques Monod, who extended the concept of teleonomy by formally introducing the problematics of self-organization and stochastic properties to the original formulation, but retained the essential argument: organisms achieve their goals mechanically, like a computer, with the genetic program offering the necessary direction. Again, only evolutionary necessity provided the organization for such function, for natural selection was ultimately responsible for writing this program. This would represent the "strong" anti-teleological position. Bereft of the design connotations of teleology, teleonomy was relegated to simply a vestige of a descriptive biology, eventually (and optimally) to be replaced by a purely mechanical account. Given the power of this formulation for the *science* of life and its philosophical implications, environmental ethicists might well ponder the wisdom of grounding their own arguments on the foundation offered by *telos* and its various expressions.

ACKNOWLEDGEMENTS

This paper is dedicated to Erazim Kohák, whose clarification of diverse philosophical issues and encouragement to pursue others is gratefully acknowledged. Although not responsible for the argument and conclusions discussed here, I wish to thank Henri Atlan, Robert Cohen, Eileen Crist, Burton Dreben, David Kazhdan, and Scott Podolsky for their critical comments.

Boston University

NOTES

1 "Biocentric" appeared approximately a century ago: R. Meldola in *Nature* (Jan. 5, 1899) wrote, "In brief, there has arisen a set of ideas which are even broader than 'anthropocentric', and which might fairly be designated *biocentric*." In 1913, L.J. Henderson charted those limits in *Fitness and Environment*, "The biologist might now rightly regard the universe in its very essence as biocentric."

(p. 312) (Quoted from *Oxford English Dictionary, Supplement A–G*. Oxford: Oxford University Press, 1972, p. 265).

2 Kohák makes clear in many places in his oeuvre that our ethical relationship with nature is determined by a direct correspondence of our phenomenological encounter, not by any other mediating function such as scientific knowledge. For instance, he recently wrote regarding "nature as experience":

> Here let us stress sharply: this is *not* a matter of dealing with nature "as we interpret it" in contrast with nature "as it really is." Nature as experience *is* how nature really is . . . That is why, as the starting point of this inquiry, we take neither scientific theory nor the art of argumentation but rather a descriptive phenomenology of the ways humans experience nature . . . and how they experience nature's distress of which we speak as *the ecological crisis* (Kohák, 1994b, p. 2).

3 As Luc Ferry has observed

> We are witnessing the development of the idea that knowledge of the secrets of the universe or of biological organisms endows those who possess it with a new form of wisdom, superior to that of mere mortals. But it is probably in the area of ecology that the feeling that the natural sciences will deliver *ready-made* teachings applicable to ethics and politics seems to be most confidently asserted (1995, p. 84).

4 Some critics, such as Theodore Roszak writing in *Where the Wasteland Ends* (1972), have deplored the moral consequences of this posture:

> Bacon went in search of a philosophy of alienation. They [scientists] broke faith with their environment by establishing between it and themselves the alientative dichotomy called 'objectivity'. By that means they sought to increase their power, with nothing – no sensitivity to others or the environment – to bar their access to "the delicate mysteries of man and nature." The cult of objectivity has led scientists and the general public, to think of everything around us – people and biosphere – as "mere things on which we exercise power." Objectivity is in practice a cloak for callousness (p. 169).

5 I prefer Goethe to Comte: modern positivists could hardly state this concern with bias more eloquently than did Goethe at the end of the eighteenth century:

> Every piece of empirical evidence, every experiment, must be viewed as isolated, yet the human faculty of thought forcibly strives to unite all external objects known to it. It is easy to see the risk we run when we try to connect a single bit of evidence with an idea already formed . . . Such efforts generally give rise to theories and systems which are a tribute to their author's intelligence. But with undue applause or protracted support they soon begin to hinder and harm the very progress of the human mind they had earlier assisted. (Goethe, 1792).

Goethe here espouses a purging of the subjective, although he himself was certainly guilty of projecting his own subjective bias (see Tauber, 1993).

6 Most would argue that the very success of biology and medicine has been in orienting the organic to a physical and genetic reductionism (Tauber and Sarkar, 1993), with the espousal of a strong positivism. This orientation began in the German physiology laboratories of the 1840s, which endeavored to rid biology of both vitalism and teleology as determining factors in the life processes. That strategy to provide a comprehensive physico-chemical, and later genetic explanation of the organic had profound epistemological and metaphysical consequences. With the ascendency of neo-Darwinism, offering a global mechanical-causal explanation, the scientific community, and the lay public, has by and large recognized the power of such explanations and the descriptions of nature provided by them. I am deeply dissatisified with this description, not because I believe it is wrong within its own province, but because I strongly suspect it is inadequate as a full science of biology. This is a problem I have already addressed in other places (e.g. Tauber, 1994).

7 Schaffner, although adopting a more sympathetic view towards reductive analyses than Taylor, acknowledges the current heuristic value of teleological explanations as appropriate to this stage of biology:

> A natural consequence has to be *chosen* as a goal or purpose on analogies with intentions or ends-in-view in human design situations. Until we have a powerful and general reductionist theory of human learning and action which would license very complex identities, I think that a

> functional analysis will be irreducible to causal analysis. This, however, is *not* an argument for *in principle irreducibility.* (1993, p. 379)

Atlan assumes a similar position (1994).

8 To capture both teleology's causality and its future orientation, Larry Wright's "consequence-etiology" attempts to integrate goal-seeking behavior as occurring (and thus legitimate for our descriptions), because it has been causally efficacious in the past (1973; 1976). In other words, goal-directed behavior becomes structured by, or better, has become adaptive as a result of the history of the species or organism as it engages its various goals. Intentionality is thus incorporated into the cause by its past efficacy. Note, on this view, teleology and completely mechanistic accounts of goal-directed behavior may co-exist, and more strongly, any biological function must contain, as intrinsic to such descriptions, a teleological character. But as already noted, teleology is being used here as a description of adaptive behavioir, and like other evolutionary accounts, does not adequately expose the nature of biological causality independent of purpose. (Criticisms reviewed by Schaffner [1993, pp. 396–399].)

9 Again, Goethe said it best:

> Thus we can never be too careful in our efforts to avoid drawing hasty conclusions from experiments or using them directly as proof to bear out some theory. For here at this pass, this transition from empirical evidence to judgement, cognition to application, all the inner enemies of man lie in wait: imagination, which sweeps him away on its wings before he knows his feet have left the ground; impatience; haste; self-satisfaction; rigidity; formalistic thought; prejudice; ease; frivolity; fickleness – this whole throng and its retinue. Here they lie in ambush and surprise not only the active observer but also the contemplative one who appears safe from all passion. (1972).

On the other hand, teleological thinking has also been productive. A striking historical case in point is that of Elie Metchnikoff, whose revolutionary view of the organism spawned a bitter dispute with German reductionists who accused him of teleological sin, yet his heuristic framework provided the foundation of the new discipline of immunology. (See Tauber and Chernyak [1991] for an historical account of this most illustrative case.)

10 Ecological ethics builds, usually implicitly, on the foundations of evolutionary ethics which argues how human morality arises from our biological heritage. The argument is well-stated by Michael Ruse:

> The new scientific claims are as simple as this. We now know that despite an evolutionary process, centering on a struggle for existence, organisms are not necessarily perpetually at conflict . . . in particular, cooperation can be a good biological strategy . . . Now let us unpack the science. We begin with the general claims about cooperation, or as today's evolutionists . . . like to call it, altruism . . . Both the theory and the empirical evidence that biological 'altruism' is widespread and promoted by natural selection is very secure and well documented. The simple fact of the matter is that . . . one is frequently better off if one decides to accept a cake shared rather than gambling on the possibility of a whole cake but one which might be lost entirely. (Ruse, 1993, p. 502).

Thus morality on this view is founded on an evolutionary-derived experience common to myriad species and hardly unique to humans. (See Nitecki and Nitecki [1993] for various presentations.) On this view, altruism, like telos, is a shared biological characteristic with other animals, and just as we are moral creatures because of those biological endowments, so are our animal brethren. Thus there is an important linkage between evoutionary and environmental ethics, but we need not further delve into that relationship. I only wished to show their common philosophical roots.

11 According to Atlan this metaphysical posture results from the deep-seated desire that science provide a comprehensive explanation of nature. Atlan refers to this as a mystical aspiration, and writes,

> *The need for an explanation of reality is, fundamentally, antiscientific.* The satisfactory explanation is a bonus, the esthetic pinnacle that accompanies and sometimes completes . . . the result truly sought: technical performance . . . For the practitioners of contemporary science . . . the need for explanation is merely a relic of metaphysical, indeed religious, wonder. (Atlan, 1993, p. 193).

12 Some, like Gaston Bachelard (1934), rather than lamenting the contamination of such a metaphysical remnant, celebrates its role. He saw in the pursuit of meaning the motive force of

research, animating our scientific query. It is in this psychological conviction that nature not only has a rationality that we might discover, but more intimately relate to as meaningful that imbues our endeavor as significant. Here then we witness the ever-present uneasy, overlap between objectivity and subjectivity (both acknowledged and justified) that characterizes science, a most human endeavor.

[13] According to Lennox (1992), Plato, in the *Timaeus*, depicted the natural world as the product of a divine craftsman, who designed nature according to an approximation of the eternal good. In this scheme, "external teleology" 1) relies on an *agent* whose goals are achieved external to the object that is being explained teleologically, and 2) the *value* aimed at is the agent's, not the object's. On the other hand, Aristotle's teleology (which Lennox terms "internal teleology") is embedded in a very different theory of causality and explanation (ibid.). In contrast to Platonic doctrine, there is no external agent (the dominant notion of natural theology until Darwin), for goals are "internally" driven and reflect only the nature of the organism; value is suspended and remains "immanent." The Aristotelian formulation is remarkably akin to modern adaptational explanations, for they both share the basic view that if a part comes to be because of its contribution to the organism, "that sanctions its identification as there for the sake of that contribution" (ibid. p. 327).

[14] I make no attempt to even sketch here the major contours of the debate concerning teleology. My focus is on the biologists' comments, but perhaps an even more interesting dialogue may be found in the philosophical literature. Kenneth Schaffner (1993, pp. 362 ff.) develops insightful, and to my reading, generally appropriate criticisms of post-World War II philosophical discussions of teleology, and the reader is directed there for a good summary. I would only comment here that much of the effort to purge teleological descriptions of teleology have been curiously misplaced. For instance, regarding Pittendrigh's sentiments, Ernst Mayr believed that the opposition to Aristotelian teleology was unfortunate, since most of Aristotle's references to end-directed processes were still appropriate (Mayr, 1988, p. 27). Here is a striking admission from an advocate of "teleonomy," and we realize the issue is not goal directedness per se, but rather some outside "design." The pervasiveness of teleological language in biochemical descriptions is an important case in point. Rosenberg (1985, p. 255) cogently illustrates the impossibility of ridding functional or goal-directed descriptions even in those biological disciplines devoted to physical constructions.

[15] Ernest Nagel embraced such neutrality, but unlike Mayr and other biologists, pursued a logical analysis. Nagel coined the term "directively organized," whose analysis would putatively circumvent the pitfalls of teleological projections. Regarding Nagel's analysis, Schaffner makes the salient point that there are two distinguishable components, one *causal* and the other a *goalstate* component (p. 370). In introducing the goal concept, there is no justification except by postulate, and without the introduction of the goal, the account generates an infinite regress, "since any system in which a goal state is not defined needs to be embedded in an inclusive system in which the same consideration, defining a goal concept, holds" (p. 368). Schaffner goes on to argue that evolutionary explications of functions are, like cybernetic analyses, incomplete, since 1) evolutionary theory cannot be used to infer a purpose, and 2) evolutionary explications of functional analyses covertly introduce (by postulate) a goal state (p. 373 ff). A truly mechanistic theory of evolution is a purely efficient causal theory and cannot warrent goal ascriptions required for functional analyses. "In sum, evolutionary theory accounts for the development of entities for which we extrinsically provide goals, but per se evolutionary theory does not provide any goals" (p. 379). Schaffner thus sees no exit for teleological accounts, but which he still values as heuristic.

[16] Evolution is not necessarily adaptive: There are clearly neutral evolutionary mechanisms, and convergent evolution may be the result of non-adaptive similar physical constaints acting in two different pathways. I note these points only to emphasize Mayr's own commitment to a particular neo-Darwinian formulation, of which he was a key architect.

[17] As early as 1944, Erwin Schrödinger, while embracing the genetic code qua code, recognized the analogy's problem as applied to the organism: He wrote in *What is Life*?

> It is these chromosomes . . . that contain in some kind of code-script the entire pattern of the individual's future development and of its functioning in the mature state . . . But the term code-script is, of course, too narrow. The chromosome structures are at the same time instrumental in bringing about the development they foreshadow. They are law-code and executive power – or, to use another simile, they are architect's plan and builder's craft – in one. (1944, p. 22).

[18] Susan Oyama (1985) notes that the collapse of information and program worked to introduce again the "argument by design," and Evelyn Fox-Keller (1995) observes that the homunculus has been effectively reinscribed into the gene. The sleight of hand is effected by the collapsing of a computer and an organism. As Henri Atlan and others have observed, the function and goals of

computers are externally prescribed, whereas organisms generate their own behavior, what he calls, the "self-creation of meaning" (1994).

[19] Others have reiterated this circumspect view of the program metaphor. For example, Francisco Varela and Jean-Pierre Dupuy recently wrote,

> if one takes the notion of a genetic program literally one falls into a strange loop: one has a program that needs its own product in order to be executed. In fact, every step of DNA maintenance and transcription is mediated by proteins, which are precisely what is encoded. To carry on the program it must already have been executed! (1992, p. 4).

REFERENCES

Atlan, H. (1983) 'Information Theory' in *Cybernetics. Theory and Applications*, edited by R. Trappl. New York: Hemisphere Publishing Corp. and Berlin: Springer Verlag.

Atlan, H. ([1986]; 1993). *Enlightenment to Enlightenment. Intercritique of Science and Myth.* Translated by L.J. Schramm. Albany: State University of New York Press.

Atlan, H. (1994) 'Intentionality in Nature. Against an All-Encompassing Evolutionary Paradigm: Evolutionary and Cognitive Processes are not Instances of the Same Process'. *J. Theory Soc. Behav. 24*, 67–87.

Bachelard, G. ([1934], 1984) *The New Scientific Spirit.* Translated by A. Goldhammer. Boston: Beacon Press.

Bateson G. (1980) *Men are Grass: Metaphor and the World of Mental Process.* West Stockbridge: Lindisfarne Press.

Dennett, D. (1981) *Brainstorms. Philosophical Essays on Mind and Psychology.* Cambridge: MIT Press.

Dupuy, J.P. and Varela, F.J. (1992) 'Understanding Origins: An Introduction', in *Understanding Origins*, F.J. Varela and J-P Dupuy (eds.) Dordrecht: Kluwer Academic Publishers. pp. 1–25.

Ferry, L. (1995) *The New Ecological Order.* Chicago: The University of Chicago Press.

Fox-Keller, E. (1995) 'A Vision of the Grail' in *The Code of Codes. Scientific and Social Issues in the Human Genome Project*, edited by D.J. Kevles and L. Hood. Cambridge: Harvard University Press. pp. 83–97.

Gilbert ([1972]; 1988).

Goethe, J.W. ([1792]; 1988) 'The Experiment as Mediator between Object and Subject', in *Scientific Studies*, edited and translated by D. Miller. New York: Suhrkamp Publishers.

Jacob, F. (1974) *The Logic of Life.* New York: Pantheon Press.

Jonas, H. (1984) *The Imperative of Responsibility.* Chicago: The University of Chicago Press.

Kohák, E. (1994a) 'The Ecological Dilemma: Ethical Categories in a Biocentric World', delivered at the Northern New England Philosophy Association, Dartmouth College, October 21, 1994.

Kohák, E. (1994b) 'Varieties of Ecological Experience' (delivered at the American Philosophical Association, December 28, 1994). Environ. Ethics, In press.

Lennox, J.G. (1992) 'Teleology', in *Keywords in Evolutionary Biology*, E. Fox-Keller and E.A. Lloyd (eds.). Cambridge: Harvard University Press. pp. 324–333.

Mayr, E. (1982) *The Growth of Biological Thought. Diversity, Evolution, and Inheritance.* Cambridge: Harvard University Press.

Mayr, E. (1988) 'The Multiple Meanings of Teleology', in *Towards a New Philosophy of Biology.* Cambridge: Harvard University Press. pp. 38–66.

Monod, J. (1971) *Chance and Necessity.* New York: Vintage.

Nitecki, M.H. and Nitecki, D.V. (1993) *Evolutionary Ethics.* Albany: State University of New York Press.

Oyama, S. (1985) *The Ontogeny of Information.* Cambridge: Cambridge University Press.

Pittendrigh, C.S. (1958) 'Adaptation, Natural Selection, and Behavior' in *Behavior and Evolution.* A Roe and G.G. Simpson (eds.). New Haven: Yale University Press. pp. 390–416.

Proctor, R.N. (1991) *Value-Free Science? Purity and Power in Modern Knowledge.* Cambridge: Harvard University Press.

Rosenberg, A. (1985) *The Structure of Biological Science.* Cambridge: Cambridge University Press.

Rosenbleuth, A., Wiener, N. and Bigelow, J. (1943) 'Behavior, Purpose and Teleology'. *Phil. Sci. 10*: 18–24.

Rolston, H. (1988) *Environmental Ethics. Duties to and Values in the Natural World.* Philadelphia: Temple University Press.

Roszak, T. (1972) *Where the Wasteland Ends.* Garden City, N.Y.: Doubleday.

Ruse, M. (1993) 'The Significance of Evolution', in *A Companion to Ethics*, edited by P. Singer. Oxford: Blackwell. pp. 500–510.

Schaffner, K.F. (1993) *Discovery and Explanation in Biology and Medicine.* Chicago: The University of Chicago Press.

Shapin, S. (1994) *A Social History of Truth. Civility and Science in Seventeenth-Century England.* Chicago: The University of Chicago Press.

Shapin, S. and Schaffer, S. (1985). *Leviathan and the Air Pump: Hobbes, Boyle and the Experimental Life.* Princeton: Princeton University Press.

Schrödinger, E. (1944; 1967) *What is Life?* Cambridge: Cambridge University Press.

Simon, H. (1969) *The Science of the Artificial.* Cambridge: MIT Press.

Simpson, G.G. (1958) 'Behavior and Evolution', in *Behavior and Evolution*, A. Roe and G.G. Simpson (eds.) New Haven: Yale University Press. pp. 507–535.

Tauber, A.I. (1993) 'Goethe's Philosophy of Science: Modern Resonances'. *Persp. Biol. Med. 36*: 244–257.

Tauber, A.I. (1994) *The Immune Self. Theory or Metaphor?* Cambridge and New York: Cambridge University Press.

Tauber, A.I. (1996) 'From Descartes' Dream to Husserl's Nightmare', in *The Elusive Synthesis: Aesthetics and Science*, edited by A.I. Tauber. Dordrecht: Kluwer Academic Publishers.

Tauber, A.I. and Chernyak, L. (1991) *Metchnikoff and the Origins of Immunology.* Oxford and New York: Oxford University Press.

Tauber, A.I. and Sarkar, S. (1993) 'The Ideological Basis of the Human Genome Project'. *J. Roy. Soc. Med. 86*: 537–540.

Taylor, C. (1970) 'The Explanation of Purposive Behavior', in *Explanation in the Behavioral Sciences. Confrontations.* R. Borger and F. Cioffi (eds.) Cambridge: Cambridge University Press.

Taylor, P.W. (1986) *Respect for Nature. A Theory of Environmental Ethics.* Princeton: Princeton University Press.

Varela, F. and Dupuy, J-P. (1992) *Understanding Origins: Contemporary Views on the Origin of Life, Mind, and Society.* [Boston Studies in the Philosophy of Science, Vol. 130] Dordrecht: Kluwer Academic Publishers.

Weinberg, S. (1992) *Dreams of a Final Theory.* New York: Pantheon.

Whitehead, A.N. (1925) *Science and the Modern World.* New York: Macmillan. Chapter 1.

Wright, L. (1973) 'Functions'. *Philosophical Review 82*: 139–168.

Wright, L. (1976) *Teleological Explanations.* Berkeley: University of California Press.

II.

SELECTED ESSAYS OF ERAZIM KOHÁK

ERAZIM KOHÁK

PHENOMENOLOGY AND ECOLOGY: DEPENDENCE AND CO-DEPENDENCY

In spite of its frivolous title, this paper has a serious purpose. I would like to suggest that, in their original intent, phenomenology and ecology are mutually supportive, but that whenever they yield to a nostalgic longing to return to a purity of Beyng or Nature, their interrelation becomes destructive. An ecological philosophy appears to me viable only within reality perceived phenomenologically, as life's world, while the moral implications of such a conception inevitably lead to ecological conclusions. However, in their romantic mode, phenomenology and ecology can become locked in a pathological co-dependency which ultimately leads to a metacrisis.

That metacrisis is the ironic situation that today we are more aware than ever that we face an ecological crisis but less than ever inclined to modify our behavior to avoid it. At conference after conference, the data presented by scientists are drastic, the response of politicians a polite disinterest at best. So we are destroying the Earth – so what? We have jobs and votes to worry about. The ecological crisis is facing a crisis of credibility.

Lest we become entangled in a web of associated commonplaces, let me propose some working definitions. By *ecology* I shall mean the study of the functioning of all there is as an interactive system, with a special emphasis on the impact of humans thereon – and with a tacit commitment to optimizing the harmonious functioning of the whole. Thus *ecological concern* shall mean concern for that functioning and *ecological crisis* the situation in which that functioning is critically impaired. By *philosophical ecology* I shall mean that aspect of that study which sees the problem not as one of the tools of human interaction with the world – say, of excessive or defective technology – but rather in its civilizational strategy: in our conceptions of our place and task in the world. Fınally, by *phenomenology* I shall mean . . . what *shall* I mean by phenomenology?

For me, the most distinctive trait of all phenomenology is its insistance on dealing with the meaning structures of lived experience at its most tangible, expressed by the husserlian slogan, *Zu den Sachen selbst!*, dealing with reality as we actually live it. Philosophy has an implicit tendency to turn away from that lived reality to theories and memories, to metaphysical speculation and to historicism. Phenomenology's passionate concern was to bring it back to

R.S. Cohen and A.I. Tauber (eds.), Philosophies of Nature: The Human Dimension, 209–219.

actual lived experience. That is not to deny the reality of the historical or the psychophysical framework of our purposive activity, but rather to recognize it as the *context* of our action, not some "real basis" to which the reality of lived experience could be reduced. It is the overt lived experience, purposive activity, that is definitive of the human mode of being. Philosophy needs to understand the way humans live and experience their lives – that is, reality as experience, *Sein als Bewußt-sein* – to understand their historical and psychophysical grounding, not the other way around. Humans are the agents of their being, not patients of fate and passion. Philosophy, then, is the act of assuming responsibilty for the direction of our actions rather than leaving it up to instinct or tradition.

As for reality, whatever it may be, in our experience it is initially *phenomenon*, being constituted as in intelligible whole by the presence of purposive agents. Phenomenology is most of all a revolt against both historicism and psychologism/physicalism, a reassertion of the primary of lived experience. Though we act in a historical and a psychophysical context, it is *we*, not History or the Unconscious, who act.

In that sense, *phenomenology represents most fundamentally a reassertion of the decision for life in the responsibilty of our freedom*. Life in responsibilty: it is a formulation much misused and oft maligned since Husserl used it in the Prague and Vienna lectures that later became the *Crisis of European Science*.[1] It is, though, as basic as philosophy. Humans may not be a higher species, a master race, but they *are* different. *H. sapiens* is of all animals least able to rely on a pregiven instinctual framework.[2] Most human communities seek to find a substitute for such comforting framework in custom and tradition, the precipitate of successful action. Where tradition breaks down, they turn to the passions of faith or nation to preserve the illusion of life in instinct.

An illusion it is. Yet humans have another possibility – to form, consciously and critically, a conception of how what there is fits together and what role humans shall assume therein. That is what philosophy is all about – and phenomenology is first and foremost a reaffirmation of that task. It is in that sense that it is a *decision for a life in the responsibilty for freedom*.

Phenomenology's radical turn to lived experience has one other consequence – the recognition of the primacy of subject-related categories. If it is the presence of purposive agents, be they humans, marmosets or earthworms, that constitutes what-is as a meaningful whole, then it is the categories of value and meaning that define the initial intelligibility of what there is. Where purposive agency is present, there the world is not neutral. It is a help or a hindrance, significant or irrelevant, good or bad, anything but meaningless. A mechanistic explanation of its functioning may be legitimate for certain purposes, but it is always a construct, not a datum. To purposive agents, experienced reality initially presents itself as meaningfully, not simply causally ordered. Value is not something added on to a value-free nature. Where there is life, there is value, simply because life wants to live – and experiences whatever aids it as good, whatever hinders it as bad, even when it does not express that experience in reflective categories.

This provides a key to the age-old puzzle of so-called "intrinsic value." *Intrinsic* means self-contained, independent of any relation to any other entity. Value, though, is by its very nature a relational reality: to be good means to be *good for*, relative to some purposive agency. *Intrinsic value* would have to mean an unrelated relation, making about as much sense as a headless cabbage. Or so it would, except for the phenomenological insight that *life* – or perhaps simply being – *is a value for itself.* Whatever lives/is, wants to keep on living/being, striving to remain alive. Certainly, there are empirical cases of individuals who choose to end their lives, either to escape suffering or to serve a higher good. The systemic rule, though, is that life wants to live: it is a *value for itself.* Because it is a value *for itself* – that is, because the value-constituting relation is internal – it is a value independently of aught else, and so a *value in itself.* Wherever there is life – or perhaps simply *being* – there is intrinsic value.

From this recognition, two consequences follow in turn. One is that the non-human world as autonomous of humans, as a *Fremdwelt*, is nonetheless not devoid of value. In destroying it, we are not simply "using up" a value-free raw material. We are destroying something good in itself. Moral considerations apply not only to human transactions with humans but to human dealings with the rest of the world because what is, is intrinsically valuable.

The second consequence is that the world as our environment, as our *Heimwelt*, is not something that simply happens to us, but rather a way we constitute what-is as a meaningful, value laden whole by our purposive presence. Such value indexing may not be the result of a reflective decision, much as fresh air is good not because we decide so but because we are oxygen inhaling beings. That valuation is prereflective, pre-conscious, a result simply of our *functioning*, not of our personal preference, though it is still contingent on our being and doing. As *life's world*, the correlate of purposive activity, the world is prereflectively meaningfully ordered and value laden and so our transactions with it are subject to ethical constraints. We are, in a word, *responsible* for the impact of our acts on the system as a whole.

If phenomenology is a radical acceptance of responsibilty, is not ecology its kin?[3] The attempt to preserve the conflicted harmony of nature, to prevent and repair the damage we do and to devise long range sustainable modes of human coexistence with the rest of nature – is not that precisely what life in the responsibilty of our freedom means? The ecological problem, after all, is not just that humans prey on other species and make use of the nature around them. All species do that. It is that in claiming our freedom we have rejected natural restraints without accepting the responsibilty for self-restraint.

Fundamentally, because of human freedom, human use of nature is not subject to the automatic restraints which protect other species from destroying their environment. A coyote need not worry about overhunting its territory. If the prairie dog population drops to a critical survival level, so will the number of coyotes who prey on them – until the prairie dogs have a chance to replenish their numbers. By contrast, when humans reduce certain fish populations to the critical survival point, human numbers do not drop. Humans can invent more efficient factory ships capable of keeping yields high, in spite of the danger signals of overfishing, until the last fish is taken. Metaphorically speaking, *humans are the species that has to impose its own fishing quotas.* We are the species that needs to regulate its own numbers and its consumption itself. Ethics, as Aldo Leopold[4] would have it, means voluntary self-restraint for the good of others; ecological ethics is a matter of extending ethics to human relationship with non-human nature – or, in the terminology we have been using, of accepting the responsibilty for our freedom. In principle at least, ecology, like phenomenology, is a call to a life in responsibilty.

The idea of *responsibility before the fact*, so to speak, has never figured prominently in ecological thought. Humans have typically recognized nature as precious only *after* devastating it. It was the decadence of ancient Rome that evoked a longing for a simpler bucolic time. It was the urbanization of the renaissance period which produced Shakespeare's *As you like it.* It was the dehumanization of the industrial revolution that evoked the intense nostalgia for the pre-industrial era while the anonymity of mass production is provoking a nostalgia for the age of crafts and trades today. The persistent theme of ecophilosophy, some of my own contributions included, has tended to be not responsibilty but nostalgia, a longing for a return to nature and with it to a simpler, less problematic time.

That theme appears to be definitive of romanticism: *nostalgia as a posture, return as a strategy.* Romanticism, as I shall be using that term, is fundamentally a perspective which sees humans as alienated from some primaeval unity, their condition as one of pining for it and their salvation in restoring it. Schematically speaking, romanticism is any philosophy that imagines an undisturbed harmony, a unity and a purity of being, as the primordial reality. It then sees the flawed present state as merely a temporary deviation from that original and essential condition and, most importantly, the future hope as one of returning to that original innocence.

This, basically, is the theme taken up by certain strands of hermeneutic phenomenology. It is the theme of truth as *anamnesis*: a recovery of what had been known once, before the fall, and has been forgotten, but can be recollected. It is also the theme of *a-letheia*, of uncovering what once was known but has been hidden by Lethe's murky waters, a conception deeply rooted in our Greek heritage. Phenomenology – and philosophy at large – become deeply retrospective.

Is it, incidentally, also a part of our Biblical heritage? Not necessarily. Early Christians, borrowing their reflective categories from Hellenistic thought, did at times so read the story of Genesis. Yet the Garden of Eden of the Biblical story is not yet a state of perfection to which we would want to return. It is only the beginning. Nor is salvation a return to a dreamlike innocence but a *new achievement.* The promise is one of a new heaven and a new earth. We can read the Biblical story as one of a new achievement. Romanticism looks backwards, hoping for a return to the alleged original innocence of pre-metaphysical philosophy, of prereflective emotions or of pre-Columbian Americans.

Philosophical ecology has, in great part, tended to be romantic in this sense: intensely grieved – with every good reason – at the present state of the world around us and longing for a return to an earlier, more innocent time. Nor was this necessarily a flaw. Humans have difficulties imagining counterfactual possibilities. They do so most easily in the mode of quasi-memory, as if *once upon a time* they had been. Not idly did philosophical ecology raise the slogan "Back to Nature!" and set off to retrieve a lost innocence.

Where, though, is that innocence to be found? To the survivors of urban shipwrecks in the early 'seventies, as I can attest,[5] the answer seemed prereflectively obvious: innocence is as close as the level of crafts, of hand-hewn beams, hand drawn water from a dug well, not to mention Tolkien and *The Little House on the Prairie*. We took to the woods at the end of the Vietnam era in a quest of the lost roots we seldom actually had, yet we did find a personal renewal therein – or some of us did, in any case. We loved our backwoods cabins, the warmth of the wood stove and the light of the oil lamp – and prefered not to think of our dependence on chainsaws or hospitals. It was a great gift and an immense privilege to live through those years in backwoods America and not, say, in rural Cambodia.

However, while what we found may have been a personal renewal, it was neither a social strategy nor the putative lost innocence of yesteryear. By no means would I wish to question the appeal of rural life or of its distinctive virtues. They are real, and often the ones we most sadly lack. The vices, though, are as real as the virtues, the greed, the petty malice, the circumscribed horizon and intolerance of the rural in-group as stifling as their urban counterpart. A return to the land may be a personal renewal, but is not a solution for our environmental problems. Rather, as the *Khmer rouge* found out, it is a nightmare.

To some writers, the rise of agriculture actually appears as the fall from grace while life-style of hunters and gatherers seems the symbol of antelapsarian humanity, free of ecological problems.[6] The image is familiar and powerful – native Americans in all their nobility, deviously corrupted by Europeans bearing fire water and fire arms. Nor is that simply the stuff of romance and of *Dancing with Wolves*. The native American peoples – and, for that matter, the hunter/gatherer cultures only now perishing in the remote corners of the globe – did display a genuine nobility of spirit and a sense of harmony with nature. Their life-style demanded and their technology permitted far less destructive intervention in the cycles of nature. The ruthlessness with which Europeans exploited and devastated their comfortably traditional style is truly chilling.

Still, to attribute to hunter/gatherer cultures an antelapsarian innocence is again rather problematic. Already the perennial tribal warfare of pre-Columban America or the Mongol raiders in central Europe, marked by as much ruthlessness, treachery and cruelty as anything Bosnia has to offer, give pause. Intertribal warfare is rare in other mammal species and common enough among humans to be considered a specific human trait long before the emergence of farming and private ownership of land. Certainly, the hunter/gatherer communities are rigorously traditional and custom bound and so can exhibit an absence of internal conflict approximating that of purely instinctual insect communities. However, the readiness with which that seemingly solid order breaks down testifies that we are indeed dealing with custom, not instinct, with tradition, not perfection.

Return to the ways of the hunter/gatherers? I doubt it. Even if it were feasible for six millions of humans to revert to a mode possible when the entire globe held fewer than six million, it would not be a return to innocence. Its sole advantage might be in decreasing the human ability to destroy, not human willingness to do so. The ambiguity of being human, both intimately a part of the natural world and stranger within it, seems a part of being human from the very beginning, already at the hunter-gatherer stage of human development. Return to nature, however personally restorative and commendable, is not a solution for our environmental problems.

Undeniably, the contribution of romantic ecology is crucial in making humans aware that the devastation that seemed the matter-of-fact price of progress was neither matter of fact nor a sign of progress, that being human is not about being a *Herrenvolk*, dominating and exploiting nature unhindered by any *Sklavenmoral.* It helps restore a sensitivity to the autonomous worth of non-human nature and of humans as but one among the species equally sharing this world. Ultimately, as its quest for innocence took it from the industrial age to the ages of the craftsmen, the farmer, and the hunter, ecological romanticism would run up against the recognition that the solution to the ecological crisis is not a return to innocence but a step forward to the responsibilty for our freedom. In our earlier metaphor, it is the imposition of fishing limits: the forging of a culture compatible with nature.

This, though, is where the co-dependence of (romantic) ecology with (romantic) phenomenology enters in. We can call phenomenology romantic when, like ecology, it turns away from the *Sachen selbst* to a quest for a putative "uncovering" of alleged supra-human (historical, fatal) or pre-human (lididinal, biological) hidden ground, true reality or authentic Beyng.[7] Once phenomenology ceases to be a revolt against historicism and speculation in the name of lived reality and becomes an esoteric uncovering of alleged hidden realities, it reinforces the fatal illusion of romantic ecology – that the task and the solution of the ecological crisis is not accepting responsibilty but regaining innocence. There is a paradise lost to be regained, only it is not in history but now in the alleged *depth* of our collective unconscious, of our ancestral memories, of our prereflective myths or perhaps of our primitive emotions. In the name of hermeneutics, possibly taken in vain, phenomenology here encourages a new version of the romantic quest for lost innocence in a return to nature not without but within us.

It seems vastly beside the point to attempt to pin down the turning point of phenomenology from a decision for life in responsibilty to a quest for suprahuman or prehuman factors that could be held responsible instead. The crucial point is that when phenomenology claims not to see clearly and describe accurately the overt being of humans in the life-world but rather to uncover hidden truths inaccessible to ordinary mortals, it reinforces ecology's tendency to long nostalgically for a lost innocence by providing a new mythological innocence just as we realised that there is none to be found in the empirical past.

It is a vicious co-dependency. Nostalgia is a destructive posture and a fatal strategy because it condemns itself to failure. There is no innocence to be recovered, there is no paradise to be regained. The alienation of humans from instinctual nature is not a transient deviation from our authentic being human. It is what we are – in Neil Evernden's words, *the natural aliens*.[8] Only once we recognize not only our radical kinship with the rest of nature – and that is the great and absolutely crucial contribution of romantic ecology – but also our distinctness as free and responsible can the ecological fate of our world cease to seem hopeless. We are not doomed by historical necessity or by Beyng's withdrawal. We are challenged to accept the responsibility for our freedom.

This is by no means to deny or deride the very real and very major contribution of the various "deep" dimensions of our self-understanding to philosophy or to ecology. Sigmund Freud, C.G. Jung, Martin Heidegger or contemporary "deep" ecologists, however tenuous their cognitive claims, play an important role in what is also a legitimate philosophical task, that of the transition from prearticulate experience to critical reflection, from *mythos* to *logos*. Putting pre-verbal experience into words – what Heidegger somewhat fancifully calls "naming the gods" – is a fundamental philosophical task. So is the sensitizing of humans to other dimensions of their experience than that most readily quantified and so most prominent in an age of quantification. We need all that, and need it badly.

Still, there is a difference, and the fortunes of the term "deep ecology" exemplify it quite clearly. As Arne Naess[9] first presented it, it means simply what we have been calling philosophical ecology, that part of ecology which seeks the clue to human disharmony and the hope of harmony in the way humans think of themselves and their world. As Bill Devall and Paul Sessions[10] elaborate that theme, there is a notable shift in emphasis to the emotive aspect of human relation to nature, though still within Naess's intent. In Pat Seed *et al., Thinking like a Mountain*,[11] the emphasis has become emphatically romantic, stressing a return to a pre-reflective collective consciousness of all life – deep ecology here becomes a depth ecology on the model of C.G. Jung's psychology.

Actually, two books appeared, using the same title, *Thinking like a Mountain*. Aldo Leopold first used that phrase in his "Land Ethic"[12] in the sense of taking into account the balance of the ecosystem rather than the interest of a particular species, be they humans, deer or wolves. Susan L. Fladers used it in the title of her excellent study, *Thinking like a Mountain: Aldo Leopold and the Evolution of an Ecological Attitude toward Deer, Wolves and Forests*.[13] The subtitle tells the story: the author traces the development of Leopold's attitude to nature from that of a practical forester to that of a thinker accepting a sense of responsibilty for the nature. Pat Seed *et al.* used the same phrase three years later in the title of their handbook for sensitizing humans to nature's plight, *Thinking like a Mountain: Toward a Council of All Beings*. Here "thinking like a mountain" has come to mean delving beneath the dominating rational consciousness of our daily being and immersing ourselves in the pre-reflective, precritical collective unconscious of all beings.

The contrast could not be more startling. According to the latter version of "deep" ecology, the problem is not a mode of being but a mode of thought. The fall from grace occurs not with a particular mode of physical sustenance but with the emergence of reflective consciousness, here somewhat misleadingly labeled "reason." The fundamental alienation is not that of humans from the rest of living nature by their productive activities of whatever sort, but of humans from the depth of their own being by the act of reflection. Reason is said to be individual, feelings collective, shared by all of humankind and all living beings. Humans – or, according to some authors, specifically males – become masters and exploiters when their reason cuts them off from the collective unconscious of all being, from all that is dark and soft and receptive, all that is caring and sharing in us.[14] The fall, metaphorically speaking, takes place when humans place the cognitive distance of reflection between themselves and the rest of the creation – when they become distinctly human. Salvation would then indeed need to mean going back, though not to any historical stage of human development but rather to the prereflective emotional depth common to all beings.

Again, the criticism is often painfully accurate and we would ignore it at our peril, just as only at our peril can be remain insensitive to the echoes of the hunter, the herder and the craftsperson in our experience. Still, speaking of a co-dependency of romantic ecology and phenomenology is not an overfanciful metaphor. The bonding of the two here presupposes not simply the innocence of prereflective consciousness (or perhals Ucs.) but also a primordial goodness of emotions so that we could wholly entrust our decisions to them. "Don't think, *feel!*." Or again, "*Trust your feelings . . .*" Knowing how prone to error reflective thought is makes such advice tempting. Yet hate, envy, greed are also feelings, and powerful ones at that. Freud identified aggression and sexuality as the archaic contents of the unconscious, not as something reason first invents. Fighters swept by an orgy of destruction report feeling exhillirated rather than alienated.[15] They for the most part report sensing themselves most spontaneously, anthentically themselves, free of all constraints, nothing held back. The fanatic movements of our century repeatedly appeal to the *sound feeling of the common people*, to our *proletarian class consciousness*, to religious and national passions.

Still, the most powerful argument against the romantic strategy of a quest for innocence lost may be systemic rather than empirical. It is a strategy that condemns humans to helplessness and ultimately to necessary failure. As Heraclitus knew already, there are no returns, eternal or otherwise. You may look homeward, angel, but you cannot go home again. There can be new beginnings but no returns, especially to the prereflective innocence of a collective unconscious, free of all alienation. Humans are the *natural aliens*: the cognitive distance which distinguishes humans from the rest of nature is not an accidental temporary deviation but the inmost way of being human.

That need not entail vicious conflict or destructive exploitation of non-human nature. We can live in harmony with nature, but that harmony will not just happen. Nor can we simply *return* to it because it is something that never was. Harmony with the earth is something we need to achieve. Ecophilosophy cannot be econostalgia. It needs to be a conscious effort to forge sustainable modes of cohabitation of humans with the rest of nature.

In one sense, the task is really quite simple. We need first to accept responsibilty for reducing human populations drastically to preserve habitats for other species while preserving the genuine achievements of culture like personal safety, social justice, respect for diversity. We need to accept responsibilty for reducing the per capita consumption among the overconsuming segments of the globe to perhaps 25% of the present level. That, by the way, is far short of poverty: it is approximately the level of affluence of the United States in 1940 or Czechoslovakia in 1965. We need to accept responsibilty for bringing up the consumption of the rest of humanity to the same standard. There can be neither peace of arms nor peace of mind while some feast and others starve. Far fewer people, consuming far less and distributing it far more evenly – that is really all it would take for the humans to live at peace with the rest of nature.

To anyone at all familiar with human psychology and human politics such a proposal may seem wholly unrealistic, and with good reason. As a species, we have never *experienced a reason* for responsibilty. For millenia of our presence on earth, there were so few of us and our power to exploit and destroy so puny that nature could easily restore the damage we had wrought. We could lull ourselves into believing that we could have the privilege of freedom without the concommitant responsibilty. There had always seemed to be so much more nature than we could destroy! Moral exhortation to responsibilty, though at least as old as biblical prophets, remained something theoretical, ethereal.

In our time, though, we have had sterner warnings. The first warning came with the cloud over Hiroshima and the development of nuclear power. For the first time, humankind acquired the means for releasing a practically unlimited quantity of energy. The safeguard of relative weakness ceased to function. Humans became as powerful as, in their freedom, they had always been dangerous. The Earth in effect ceased to be human-proof.

The demographic explosion that followed the war provided the second warning. The point is not a theoretical consideration of how many milliards of humans the Earth could "support," whatever we take that to mean. Rather, it became obvious that we are expanding into habitats unsuited for human habitation without prodigious expenditures of energy, that we are crowding other species out of their habitats and that we are reproducing so rapidly that we are no longer able to pass on our social survival skills to succeeding generations. Anyone who has seen the child soldiers of the third world, the shantytowns ringing great cities – or American inner cities themselves – cannot but hear that warning.

The consumer revolution is the third warning. In a world of flagrant inequalities, we have managed to keep social peace, however imperfectly, only by the promise of *ever more.* Future has come to play the role that heaven played in the middle ages, the promised reward allowing humans to put up with present deprivation. The wars of the third world, fought with the weapons of the first, show the strategy is failing. However, it fails no less when it succeeds – because infinite expansion on a finite earth cannot but produce a catastrophe. We not only can destroy the Earth as habitat for humans and our kin, we are actually doing it at a frightening rate.

That is a global reality which, however, is still experienced rather selectively and locally. It is difficult to speak of overpopulation in Oregon, of hunger in Belgium or social injustice in Denmark. With the growth of global communications, however, that is changing. Our experience, too, is becoming global. Whether it is becoming so sufficiently to save us from our own greed and irresponsibility, only time will tell. It is, however, becoming possible to speak not only of the reality, but also of the *experience* of an ecological crisis – and so of its philosophical significance.

That significance, I would submit, is not that we need to "go back to nature," either the living nature around us or the nature we bear within. Our separation from the whole, the individuation we experience as alienation, is not yet the original sin. It is the moment of creation, when, metaphorically speaking, the fingertips touch on the ceiling of the Sistine Chapel and humans are called forth out of the continum of nature into freedom. No, it does not make them a "higher" species, only one that bears a greater responsibilty. The moment of the fall comes later, when humans accept the freedom and *fail to accept the responsibilty* – as in claiming "property rights" to land in total disregard of its aboriginal and non-human inhabitants or of the land itself. *That* is the original sin. The ecological crisis is an accumulation of such responsibilties deferred. The survival of the human race and its mammal and vertebrate kin upon this earth depends on our willingness to accept the responsibilty which goes with our freedom.

Are we capable of it? I do not know, though I do not find our inability to deal with problems as simple as carbon monoxide emissions or population planning encouraging. Perhaps not: perhaps nature will have to replace us with a species less free and less dangerous. That, though, is no excuse for philosophy to tell comforting tales. The task of the gadfly with which Socrates charged it seems to me more relevant than ever. What matters now is for philosophy to challenge humanity to accept the responsibilty of its freedom. In their moments of greatness, both phenomenology and ecological philosophy do that.

Philosophical Faculty, Charles University, Prague/CZ (Boston University, Boston, MA, USA, emeritus) (1995)

NOTES

[1] My conception of phenomenology is evidently deeply indebted to Edmund Husserl, especially to the "Wiener Vortrag" which appears as a supplement to the Husserliana edition of *Die Krisis der europäischen Wissenschaften* (Hua VI, Walter Biemel, ed.; The Hague: Martinus Nijhoff, 1972). However, my understanding of it was powerfully molded by the recent collection of texts from the *Nachlaß, Die Krisis der europäischen Wissenschaften: Ergänzungsband* (Reinhold N. Smid, ed; The Hague: Kluwer, 1933). In this rich volume, I would call the reader's attention to Nr. 6,

"Fundierende Subjektivität, und objektive," pp. 60–76, to the text of the 1935 Prague lectures which became the basis of *Krisis*, Nr. 10, "Prague Vorträge," pp. 103–39, and to Nr. 26, pp. 302–16, and Nr. 32, 362–420 – though the entire volume is a powerful experience.

[2] I have argued this in "Varieties of Ecological Experience," presented at APA Eastern Division Conference, 28 December 1994; *Environmental Ethics* 19.2 (Summer 1997): 151–73.

[3] While my reading of Husserl's phenomenology is derived from Husserl's post-1935 writings, my application of it to ecology is deeply indebted to Hans Jonas, *Das Prinzip Verantwortung*, (Frankfurt a.M.: Insel Verlag, 1979), English by author and David Herr, *The Imperative of Responsibility*, (Chicago: University of Chicago Press, 1984), esp. 46–50 and 79–89.

[4] "A Land Ethic" in *Sand County Almanach*, (New York: Oxford University Press, 1987). Note also Susan L. Flader's exposition of it noted in note 13 below.

[5] Erazim Kohák, *The Embers and the Stars*, (Chicago, University of Chicago Press, 1984), note esp pp. 27ff. As evidence I was not alone, see Max Oelschläger, ed., *The Wilderness Condition* (San Francisco: Sierra Club, 1992).

[6] Locus classicus is Paul Shepard, *The Tender Carnivore and the Sacred Game* (New York: Scribners, 1973) which condemns agriculture as the source of the environmental crisis and holds out a hunter/gatherer model as the once and future option.

[7] Martin Heidegger appears to me as the towering example of romantic phenomenology. While it is his work as a whole I have in mind, the most relevant texts appear in *The Question Concerning Technology and Other Essays* (trans. William Lovitt; New York: Harper & Row, 1977).

[8] Neil Evernden, *The Natural Alien: Humankind and the Environment* (Toronto: University of Toronto Press, 1985).

[9] Arne Naess, *Ecology, Community and Life Style* (Cambridge, MA: Cambridge University Press, 1988).

[10] Bill Devall and George Sessions, *Deep Ecology: Living as if Nature Mattered* (Salt Lake City: Peregrine Smith Books, 1985).

[11] John Seed, Joanna Macy, Pat Fleming, Arne Naess, *Thinking like a Mountain: Towards a Council of All Beings* (Philadelphia: New Society Publishers, 1988), note esp. 45–52. Not to be confused with study by same title noted in note 13 below.

[12] Above, note 4.

[13] Susan L. Flader, *Thinking like a Mountain: Aldo Leopold and the Evolution of an Ecological Attitude toward Deer, Wolves and Forests* (Colombia, MO: The University of Missouri Press, 1974). No relation to work by same title mentioned in note 11 above.

[14] The classic statement of the position, capturing its sensitivity without sliding into irrationality, is Carolyn Merchant's *The Death of Nature: Women, Ecology and the Scientific Revolution* (San Francisco: Harper & Row, 1980).

[15] Cf. the analysis of the front line experience in Jan Patočka, *Heretical Essays in the History of Philosophy* (Chicago: Open Court Press, 1997): 119–37.

ERAZIM KOHÁK

HUMAN RIGHTS AND NATURE'S RIGHTNESS

There is something rather incongruous about the idea of inalienable human rights coming to the fore in the late 20th century. Sociologically, it is quite understandable: this is the age of the me-generation, and the weak, as Nietzsche saw, ever claim in the name of "justice" what they cannot seize by force. Philosophically, though, it is rather embarassing. Not that the idea of human rights is not notable, inspirational and generally righteous altogether: it is all of that. However, it is also badly out of step with Western philosophical development over the past two centuries.

The claim that humans, simply as humans, can be said to have certain "rights," intrinsically and absolutely, whether a given society chooses to respect or even to acknowledge such "rights," presupposes an orderly universe, value laden and morally ordered, from whose order, more basic than conventional societal rules, such rights could be derived. It presupposes a universe whose order is more basic than human order, whose rules are more basic than human rules and preferences, individual or collective.

Western thought, for at least two centuries and longer, has strenuously resisted the suggestion that the universe is that kind of place. Contemporary philosophy views any claim that there may be values independent of human preference with suspicion, and for the most part denies the very possibility of unambiguously true statements of any kind. A recent French grafitto summed up the sentiment with admirable clarity: "Truth is fascism – all power to the imagination"!

Against that kind of a background, against the conviction that the universe in itself is value neutral and that values are arbitrary, a function of human preference and consensus, any claim about absolute, inalienable "human rights" sounds like no more than a particularly shrill individual demand. The me-generation not only wants it all and wants it now, but wants it as its "inalienable right" . . .

I am going to suggest that there is an irresoluable contradiction between our conception of reality and our conception of human rights. We can, if we wish, hold that the universe is a place that has no moral rules so that, assuming we are strong enough to take it, we can have it all and have it now, a place where our arbitrary will is sovereign and subject to no restraint: *homo mensura*. In that case, however, we cannot consistently claim that we have "rights." In such

R.S. Cohen and A.I. Tauber (eds.), Philosophies of Nature: The Human Dimension, 221–230.

a universe, claims of "human rights" can represent at best an arbitrary demand that we be treated in conformity with certain rules. The validity of that demand, however, depends entirely on our ability to convince or to coerce others to treat us accordingly. If they do not, we cannot claim that our "rights" have been violated, only that we have not succeeded in establishing our claim.

Conversely, we can consistently claim that our being as human, simply as human, carries with it certain rights, certain rules in accordance with which human beings ought to be treated. To make such a claim, however, we have to claim that the universe is a place that has rules, a place that is value laden and morally ordered, and that our presence in it is not arbitrary but defined by such rules, entailing both rights and obligations. If that is the case, however, we cannot claim to be ourselves exempt from those rules. The rules which guarantee our "rights" impose obligations on us as well – or, in other words, no, we cannot have it all and have it now. We have to play by the rules.

Or, in philosophical shorthand, we can be consistently Hobbesian or we can be consistently Lockean, but we cannot consistently combine a Hobbesian conception of the universe with a Lockean conception of natural rights, claiming a relativity of value and the sacredness of human rights in the same breath. It is not the case, as Ivan Karamazov supposed, that if there were no God, everything would be permitted. If there are no rules, *nothing* can be said to be "permitted" – or, for that matter, prohibited. We can have it all and have it now, but we cannot claim anything as a right. If we wish to claim to have "rights," we have to recognize that we can have only that to which we have a right, and only when we have it.

That, I would suggest, is the appropriate setting for any discussion of human rights – the recognition that the very idea of a right is inseparable from that of a rule-governed context within which rights and obligations arise, and that the idea of a *human* right is inseparable from the idea of reality as itself such a context, value laden and rule governed.

Following common usage, I take a "right" to refer to a legitimate expectation of pursuing a course of action without having to justify it to anyone, and of expecting others not to hinder such a pursuit – or, perhaps, even of expecting them to aid us therein. If freedom of expression is a right, I need not justify speaking my mind or apologise for it, I can legitimately expect others not to interfere with my doing so and, perhaps, can expect others so to structure our life together as to facilitate my doing it. Or if we say that economic security is a right, we are saying that we can legitimately expect to be adequately rewarded for our labors, can expect others not to interfere and, perhaps, can expect them so to structure society as to shield us from destitution.

In common practice, such legitimate expectations arise within a community in accordance with explicit or, more often, unspoken rules representing that community's consensus. Most human communities agree, explicitly or tacidly, that certain prior performance establishes such a right. So digging, maintaining and using a well may establish a right to that well's water. Having dug a well, I need not justify watering my flocks from it to anyone, and I can expect others to refrain from interfering with that activity. If they do, I can claim not only that I have been injured, exacting revenge, but also that my *right* has been violated. The right here is contingent on my prior performance – having dug the well – and on the tacit assumption that the act establishes a legitimate claim: I have earned the right.

Alternately, again within tacit community consensus, a right may be based on an explicit bestowal. Thus a civil right – say, that of voting – may be linked to certain prior performance – in New Hampshire upon the payment of ten dollars poll tax on the first of April. What establishes that right, however, is not the performance but a bestowal. That right has been bestowed by the constitution of the State of New Hampshire, together with certain obligations, down to displaying a rather overstated state motto on my license plate. As a citizen of New Hampshire, abiding by its constitution – and proclaiming "Life Free or Die" on my number plate – I can claim a right to vote. I need not justify wishing to do so, I can legitimately expect others not to keep me from the polls – and perhaps can even expect the State to make the polling place wheel-chair accessible.

Our ordinary as well as our legal discourse on the topic of rights presupposes that basic framework of explicit or tacit rules, and within it justifies the expectations we call "rights" either by prior performance or by prior bestowal. Do the handicapped have a right of access to public facilities? Definitely: they are tax payers whose prior performance – the payment of taxes – has earned them that right. Do the indigent have a right to vote? Definitely: though they may pay no taxes, they are citizens and the right to vote has been bestowed on them. We might, to be sure, invent clever puzzles so dear to some of us – for instance, does a handicapped indigent alien have a right to access to public facilities? But those are games, not affecting the matter of substance. The truth of the matter is that in ordinary discourse as well as ordinary practice we get along quite adequately, for the most part, with the conception of rights as established within a prior explicit or tacit framework of rules by prior performance and/or prior bestowal.

That facile and familiar justification, however, becomes rather problematic when we try to extend it to "human" rights, said to pertain to all humans simply in virtue of their humanity, even if they live in a community which does not recognize any a priori rules governing human behavior to their fellow humans. We need not look to Nazi Germany or Sadam Hussein's Iraq for examples. The prevalent conviction of our own society that all questions of right and wrong are relative to an individual speaker's preference – some people are nature lovers, others prefer pumping oil: it all depends on what you like – makes us an example of such a society. How can we speak of *human* rights? Civil rights, certainly: constitution bestows them and custom establishes them. But *human* rights? Don't they depend on how you look at it?

Certainly human rights cannot be said to be based on a prior bestowal by any authority. The very idea of a human right challenges the claim of any authority to bestow – and so be entitled to deny – such rights. A right bestowed would cease to be an intrinsic *human* right. So, though, would a right earned by prior performance. When we speak of a right as *human*, we claim that they are rights we have simply in virtue of being human, without having to do anything to earn it.

The right to life we claim in the Declaration of Independence as the inalienable right to "life, liberty and pursuit of happiness" – paraphrasing Locke's earlier "life, liberty and pursuit of property" – provides a clear instance. No body politic needs bestow that right, we do not need to earn it. We assume that simply because we *are* we also have a *right* to be. We need not justify being, we can legitimately expect others not to interfere with our being and, perhaps, can even expect others to support us in the task of being. We are, and so we assume we also have a *right* to be, simply because we are.

Our being as humans, however, is no simple matter. It takes a great deal to sustain us in being, though we seldom consider that. Ordinarily, we assume that because we are and so have a "human" right to be, we also have a right to whatever is necessary for our sustenance. That is no small claim. Humans are notoriously expensive to maintain, incomparably more so than other species which also are and, it would seem, can, by the same logic, claim a right to be. Yet we expect them to bear the cost of our sustenance, and not of sustenance alone. We assume that we have a right not only to meet our needs but also to gratify our whims, down to VCRs, 2 mpg speedboats and every manner of conspicuous waste. Simply because we *are* we assume we have a *right* to anything we want – an inalienable human right to life on whatever terms we envision it, subject to no constraints, no obligation: a right not only to be but to be anything we please, whatever the cost to all else that also is.

Even upon the least reflection, that is a rather problematic claim. Humans are, but does that also mean they ought to be? That they have a right to be? It is entirely arguable that, for instance, the American southwest would be far better off if the fifty or so million *belagaana* settlers *were not*, leaving the place to buffalo and a few thousand Navajo shepherds with a handful of Zuni thrown in for good measure. Those white settlers have proved immensely destructive, their life-style immensely expensive. Certainly, they *are* and so, presumably, have a right to be – and need a place to live. But should that need be met when the cost is so incredibly exorbitant, some seventy times that of the average human inhabitants of this globe and many more times the cost of a buffalo? On what basis can they claim a "right" to sustenance?

As America's founding fathers set about raping the virgin continent, they were not unaware of the problem. The right they claimed as human – or in the terminology of the time, as "natural" – was bestowed upon them, they declared, by their Almightly God, the law giver of the universe . . . presumably without consultation with the gods of the Navajo. Yet that God bestowed not only rights but also obligations. Given the benefits God provided, including a "right" to some highly desirable real estate, we, their heirs, might well have been willing to put up with God's idiosyncratic preferences, as for instanace that prudish commandment about not committing adultery. When, however, God sought to command us, "Thou shalt not covet," God struck at the very heart of our free enterprise system. We believe, after all, in an expanding economy, and coveting is the driving force of that expansion. Almighty God simply had to go: as Laplace said it for us, we have no need for that hypothesis.

With the departure of the Almighty God, we have found ourselves free of all the awkward constraints, but also bereft of anyone to sustain our claim to rights. We are, that is all. There is no one to bestow "rights" upon us, not even the right to be, and we certainly cannot claim to have earned that right by our performance. If we wish to claim human rights as rights, not simply as one more arrogant claim of the me-generation, we have to claim that, even without a God to bestow them, those rights are somehow built into the very nature of reality, into the structure of the universe. We have to claim that our world is the kind of place within which, independently of us, there are values and rules, more basic than human desires, from which we can derive legitimate expectations or "rights."

Therein, however, lies the rub. For some four centuries, we of the West have devoted much of our philosophic efforts to denying that the universe is that sort of place. Starting at least with the Spanish colonization of the Carribean, philosophy has been called upon to provide us with a conception of reality that would be consistent with wanting it all and wanting it now – and was willing to oblige. The 16th century Spanish physician, Gomez Pereira, first suggested that living beings are mere mechanisms unless endowed with an immortal soul by the Almighty God and the Church of Rome, enabling the *conquisadores* to exterminate the Taino Indians with the same easy conscience with which they clear-cut forests. Two generations later, a French soldier of fortune in the Imperial army that laid waste to the Czech lands, Renée Descartes or Cartesius, extended that formula to all reality. The universe became a place without rules more than ever.

In justice to Descartes, we ought to note that he was not himself a "Cartesian" in the sense which the term was to acquire in later usage. In his own thought, mental reality, the *res cogitans*, in which inhere all values and all moral order, was neither less than real nor private or arbitrary. As with Kant's noumenal realm, it was as universal as reason itself and as real as extensional reality. However, it was distinct. The reality accessible to our senses – and so accessible to the rapidly rising empirical sciences – was said to be solely an extensional reality, devoid equally of value and meaning, the *res extensa* which we would come to call "objective" reality. That reality, the world around us and even our bodies within that world were said to be an aggregate of objects in space, with no properties other than extension and no relation other than their mathematically describably interaction: things bumping into each other in predictable ways in space time. So it would be an error to attribute to that reality any value, meaning or moral order, much less rules that might constain our arbitrary will – or ground any human rights.

Value, meaning, moral order there may be, to be sure – humans certainly speak as if there were. However, on the cartesian scheme, they are not a part of extensional or "objective" reality. They belong to a wholly different order of being, that of the *res cogitans*, which much later we would come to describe as "merely subjective." For Descartes, that realm of being was still no less real than that of extensional reality. After all, it includes the one absolute certainty, the self-certifying I, the *Cogito*, as well as God, who for Descartes was still very real indeed. It is, however, an entirely separate order for which there is no room in the world of extension. The sciences which study that extended world – in later usage, the "objective" world – need have no recourse to the world of thought and value. Nor, conversely, does the study of the realm of the mental, of the *res cogitans*, have any relevance for the realm of extension. Plessy vs. Ferguson applies: the two are separate but equal.

Yet, as a later ruling would hold, separate is inherently unequal. For Descartes' heirs, very much including ourselves, reality came to be equated with the realm of the value-free, merely causally ordered realm of extension, the "objective" reality. The realm of the mental shrank to a "merely subjective" reflexion of that "real" world, and rather less than real itself. While the object world of causal relations in space-time came to be conceived as reality, the realm of meaning came to be conceived of as only "real for me," and the rhetoric of bifurcation persists to our day. "Subjectively," we may recoil from the devastation of Arctic ecosystems, but "objectively" there is oil there and we need oil: subjective sentiment may be a private privilege, but must not constrain objective reality. But if that bifurcation holds, can human rights be anything more than subjective sentiment?

Descartes' bifurcated vision of reality, to be sure, did not go unchallenged even in his own time. Already his great contemporary, Jan Amos Komenský or Comenius, rejected it, insisting that values, meanings and the moral order of the universe are no less a part of reality than extension or causality. Comenius had reason to appeal to rights: he was an exile, driven from his devastated homeland by the armies in which Cartesius served. But others have taken up the theme – Goethe, Schelling, Masaryk, Husserl. Like Comenius, they remained voices crying in the wilderness. The consensus of modernity emerged as strongly Cartesian: reality is objective, value-free, simply there, while values are strictly private, a personal matter rendered public at most by consensus and convention.

The reason why the Cartesian model prevailed may not be altogether philosophical. That model has an obvious advantage: a value-free reality is one we can plunder and devastate without a second thought: it is, after all, only raw material. Anyone shedding tears over the devastation of a forest or the agony of non-human sensient beings is simply expressing a subjective preference, real only for him/her – must be a "nature lover." But the advantage has an obverse: the ways we treat each other, the rules of social interaction and the rights we can claim as human can, on this model, also be at best "subjective," a matter of private preference or of community consensus. Some people like picnics, others like pogroms: no accounting for taste. For, on this model, humans, too, are "objectively" only raw materials, whatever their "merely subjective" feelings.

Could it be that the recent interest in "human rights" is not just another expression of the arrogant greed of the me-generation but a desperate attempt to exempt humans from the universal rule of availability as raw material? If it is, it may be noble, but it is also futile. There is no basis for such a claim in the reality construct we accept as normative. A normative ethics – and the attempt to establish human rights as binding regardless of anyone's wishes – is possible only on the basis of a very different conception of reality, as value laden and so as morally ordered. We cannot do normative ethics unless we first do a very different metaphysics, learning to think of reality and of the universe as a place of values and rules from which we can derive human rights.

With that recognition, our inquiry becomes philosophical not only in an analytic or a descriptive, but also in a speculative sense. Can we generate a conception of reality which would do justice to all the evidence, including the experience of transpersonal value and of a moral order of which human rights are a part, and do it without resorting to a *deus ex machina*? That is not simply a matter of tacking on a "subjective" addendum onto our conventional conception of "objective" reality the way school systems tack on a course in "values" onto a curriculum of physics or business management. Rather, it means setting aside our habitual assumptions about "objective" reality and "subjective" values and starting afresh, as perennial beginners, with what actually presents itself to us, seeing clearly and faithfully articulating its fundamental structure as it presents itself to us and only as it presents itself to us – setting aside our preconceptions about what is "really real" and opening our eyes to the reality amid which we in truth live.

A useful starting point for such an endeavor may be one evident yet largely ignored recognition – that the reality in and amid which we live is not "dead matter" but rather, initially and intrinsically, the interwoven, meaningfully ordered context of life. Life, purposive, active being, is an intrinsic and inseparable part to it, not a fortuitous addition which we could omit to avoid observer bias. The Cartesian value-free "objective" reality is a highly sophisticated construct, an abstraction. The world of our actual experience is very much laden and purposefully ordered. It is neither "subject free" nor "value free." It is permeated by subject beings – by self-directing, purposive life, by woodchucks, paramecea, trees, porcupines and, yes, human beings. We can never say of it that it simply is. From the start, it is also and always for subjects who have purposes, who value the world as serving them and as hindering them. We might well borrow the convenient term *Lebenswelt* to describe it, but for the unfortunate cluster of associated commonplaces it has acquired in English. The usual English translation of *Lebenswelt* is "life-world," a term no less usually taken to mean "the world of our experience," that is, the "world for us," world as we humans experience it. That, though, is still a Cartesian reading which treats the "life world" as our "mental" reflexion of extensional reality. The point here is that reality is *not*, as Descartes would have it, first extensional and only subsequently reflected in consciousness. Rather, the reality of our lived experience is the meaningful context constituted by the presence of intended subjects, constituted intrinsically as life's world, the world of life, a whole defined and structured by the presence of life.

When we set aside our Cartesian preconceptions and set about seeing clearly and articulating faithfully, that is the first recognition: reality is a *Lebenswelt*, llife's world, the world of life. That recognition in turn leads to another: as life's world, reality is not in the least value neutral or meaningless, merely extensionally or causally ordered. Simply in virtue of the subject's presence as purposefully acting being, long before any act of reflective cognition or preferential choice, reality not only *is* but *means*. Quite simply, because we humans are present in the world as organisms inhaling oxygen and exhaling carbon dioxide, airy places that include green plants and dank dungeons do not present themselves as value neutral – depends on what you like – but as definitely ordered. That order is not a matter of individual reflective interpretation or preferential choice: it is given, simply because we are what we are. Or again, because we are phototrophic organisms, orienting to light, sunny places and dark caves are ordered in a non-arbitrary way, their value a relational reality, related to us, but in no sense "relative" in the sense of being an arbitrary "subjective" preference in a value free world. As life's world, the world is intrinsically value structured, simply in virtue of what we are as acting human subjects.

Nor, we should note, are we humans the only subjects. So are all living, purposive beings, endowing reality with a value structure in virtue of their presence. Grass is never simply a value-free *res extensa*: it is valuable as the woodchuck's fodder. Nor is the sunshine that warms the sunflower ever just a wave or a stream of particles or whatever dr. DeBroglie concluded. It is also good as it warms the sunflower and fosters its life, photosynthesis in turn sustaining all life. Where there is life, there is goodness and badness. A value structure – and, more broadly, a meaning structure – is an irreducible aspect of life's world. We may first become aware of it as human – what the world means for me or for us. Yet it is not a function of human consciouenss or even simply of human presence, but of the presence of all subject beings. Husserl described it as the *absolut fungierende Subjektivität* [Hua VI.265].The scholastics used a different language. *Esse qua esse*, they held, *bonum est*, *esse est unum*, *bonum et verum*: being, the infinitely rich activity of all that is, the life of woodchucks and porcupines, of sunflowers and humans, is good simply in being, in the activity of being. Life is good because it gives rise to goodness.

When we come so to see the world we see it no longer as "objective" and value free, but as value laden and meaningfully ordered, as having a *rightness* of its own – or more abstractly, having an order of being. There is a rightness in the oxygen-rich mountain stream in relation to the rainbow trout: poluted water is not simply a "fact" – it is *wrong*. There is a rightness to the activities of the beaver as he sets purposefully about felling birches, damming a stream and storing the young slash in the bottom against the winter. There is a rightness, too, to the freshet that washes out the dam, and to the parasites that control beaver populations. It is an ordered, value-laden world.

In the context of reality perceived and conceived of as a life's world, value laden and meaningfully ordered, as having its rightness, we can speak of being as not only a fact but also a *right* – or a legitimate expectation that since we are, we need not justify our being and should not be hindered therein. The beaver can be said to have a right to a habitat in which he can assume his rightful place in the economy of life's world. So do the birches. For that is the basis of all "rights" which are neither earned by prior performance nor bestowed by a higher authority. They are an expression of the legitimate expectation of acting out a being's *rightness* of place within the order of life's world.

If we learn so to think of reality – and we do have the choice, just as we can think of it as a storehouse of raw material – then the idea of human rights ceases to be anything incongruous. For humans, too, no less than all other beings, have their rightness of place within the order of life's world. Like all other beings, they have their task. They can know truth, they can do the good, they can cherish goodness and grieve its passing, they can preserve life and heal its wounds. It is not pointless to be human, any more than it is pointless to be a porcupine or a beaver, and it entails no less a right to be. And no more.

More formally, we can now supply the suppressed middle premiss in what at the start of our inquiry appeared as a short-circuited argument. We had noted the common assumption that humans have a right to be simply because they are. Supplying the middle premise we would need to say, Humans are, there is a *rightness* to their being within the order of the cosmos, and therefore they have a right to be: the legitimate expectation, which need not be justified and ought not be hindered, of doing what is distinctive to their way of being, knowing the true, doing the good, cherishing beauty. If we see humans not as alien conquerors of a meaningless world but as dwellers in a meaningful one, the idea of human rights no longer appears as arrogance but as a recognition of our place in the cosmos.

Yet in such a cosmos we are not the only beings who have a rightness and a right to be. The same meaning-structure which ensures our right to be at the same time circumscribes it with the right of other beings who are and have a right to be. In a cosmos so conceived, human rights, while solidly grounded, simply are not unlimited but contingent on a respect for the value structure of life's world. Specifically, the right we claim so confidently, to "life, liberty and pursuit of happiness," depends on the rightness of that life, on its harmony with the web of all of life, and cannot be used as a justification for the heedless destruction of all other life. As we establish the legitimacy of the idea of human rights, we raise, at the same time, the question of how those rights may legitimately be justified.

This is a difficult area, left largely unexplored throughout our Cartesian centuries. Even much contemporary discussion remains restricted to the question of establishing human rights – once established, its satisfaction is assumed to be unproblematic. Yet that is precisely where the problems arise. So as an instance, humans may claim a right to kill a buffalo for their sustenance. That is a part of their rightness, and, as the Zuni hunting song recognizes, it is a part of the buffalo's rightness as well. That right, however, becomes a travesty when we use it to justify William "Buffalo Bill" Cody's indiscriminate extermination of the entire species. Or again, in the rightness of their being, in knowing the truth, doing the good, cherishing the beautiful, humans may have a right to draw upon the earth for sustenance, but not without taking account the rightness and so the "rights" of other beings. We can justify use, but, in a cosmos that has a rightness – and in which we alone we can say to have rights – can we ever justify conspicuous consumption? A right to mobility, perhaps, but can that right conceivably justify private automobiles that get 18 miles to a gallon when both public transportation and 50 mile per gallon technology are available? Or, more basically still, can the human right to life be extended to justify overpopulating the globe and driving other species to extinction? A right holds as long as it expresses a rightness: what are the limits of human rightness?

That, I would like to suggest, needs be the agenda for future debates about human rights. What are the limits of human rightness? Human rights, we have sought to argue, are no more than a smoke-screen for human avarice unless they are grounded in the recognition of the value of all being and in the rightness of being human in a value laden, morally ordered cosmos. Within such a cosmos humans can claim the right of their special role. But that role is not simply one of perpetuating the species and subduing the earth – that is a claim to conquest denying the very idea of rights. It is, rather, the role of knowing the truth, doing the good, cherishing the beauty – and, most of all, of knowing and respecting the rightness of the cosmos.

And so, in concluding, I would suggest that while there may be a great many people for whom the question of human rights is one of securing the minimal conditions of *human* survival, of the rightness of being human, not simply surviving as an organism from day to day, for us the question needs be different. We are members of an incredibly privileged and very, very narrow segment of humanity and of life's world, greedily consuming wholly disproportionate share of the earth's resources. For us the question needs be that of the limits of our rights, of obligations to the rest of the cosmos. For, if our analysis has any merit, if we are not willing to recognize the limits that the rights of other beings impose upon our rights, we shall have no grounds on which to claim any rights in our own behalf.

Philosophical Faculty, Charles University, Prague/CZ (Boston University, Boston, MA, USA, emeritus) (1992)

ERAZIM KOHÁK

TRANSCENDENTAL EXPERIENCE, EVERYDAY PHILOSOPHY

In § 72 of his *Krisis der europäischen Wissenschaften*, Edmund Husserl introduces the term, *die absolut fungierende Subjektivität*, telling us that it refers not to human subjectivity but to subjectivity as such which only manifests itself as human subjectivity. The task and achievement of transcendental phenomenology, he tell us, was

"... die absolut fungierende Subjektivität zu entdecken, nicht als die menschliche sondern als die in der menschlichen, oder zunächst in der menschlichen, sich selbst objektivierende." [Hua VI:265]

Or, in David Carr's noble English translation,

". . . to discover the absolutely functioning subjectivity, not as human subjectivity but as the subjectivity which objectifies itself, (at least) at first, in human subjectivity." [Carr 1970:262]

Whether in German or in English, the passage has since become something of a *locus classicus* for all who wish to claim that transcendental philosophy is an esoteric obscurantism, hopelessly removed from everyday experience.

In the present paper, I would like to suggest that it may be our understanding of transcendental philosophy and of everyday experience rather than Husserl's text that is at fault. If we come to understand those two terms clearly, in at least approximately the way Husserl appears to have intended them, I believe that transcendental philosophy will prove rather an everyday affair, not at all obscure or esoteric, while our everyday experience will turn out to have a distinctly transcendental dimension overlooked by traditional empiricism.

Unfortunately, the terms transcendental philosophy and everyday experience do not glitter with a crystal-like clarity even in our philosophical usage, much less in our ordinary discourse. Both evoke such a rich cluster of associated commonplaces that it is possible to use them in an appalling variety of ways – or even without any clearly specified meaning at all. Thus our first task must be one of conceptual and experiential clarification. All our associated commonplaces aside, how do we wish to have the two terms function in our philosophic discourse? And, beyond that, what in our presentive experience do we wish them to evoke?

R.S. Cohen and A.I. Tauber (eds.), Philosophies of Nature: The Human Dimension, 231–247.

For all its lack of colloquial familiarity, the term transcendental philosophy may be the one more easily clarified. We commonly mean by it – and in our inquiry we shall mean by it – any philosophy which seeks to derive from the structure of our experiencing normative propositions which can be said to be true or false independently of any particular factual (and so contingent) content of that experience.

Were we to resort to time-honored Kantian idiom, we could call any philosophy transcendental if it considers synthetic a priori propositions possible. For the purposes of our inquiry, however, we would like to interpret the term synthetic a priori more broadly than is usual in Kantian discourse, to include any proposition derived from the structure – or perhaps from the *ideal possibilities* – of experiencing, whereof more anon, and applied to its factual contents or *Sachverhalt*. Bracketing prior knowledge for the sake of a convenient if inaccurate example, we would like to use as a metaphor the rather non-Kantian assertion that, "Whatever human explorers may see on Mars will be of a color between infrared and ultraviolet." That is a synthetic statement, since it asserts something about the *fact content* of experience. However, it does so not on empirical basis but rather on the basis of the *structure* or *ideal possibilities* of human experiencing. The human eye can only register as color wave frequencies between the two limits. Admittedly, on a higher level of generalization that is itself a factual rather than an ideal possibility. Still, we can use it on its own level as an illustration of the principle of a priori synthetic judgements – asserting something about the content of experience on the basis of its ideal or structural possibility.

However, while our interpretation of *synthetic a priori* is purposefully broad, our interest is far more specific. We are concerned with the possibility of normative value judgements of the type "Causing unprovoked injury is morally wrong." Within the limits of traditional empiricism, such judgements would be clearly illegitimate. Those limits would permit an empirical generalization, for instance that humans in most or perhaps in all cases do in fact tend to regard causing unprovoked injury as wrong. A command such as "Refrain from causing unprovoked injury" might then be legitimate as an expression of personal revulsion, an elaborate version of "ugh," or perhaps as an expression of a wish that others refrain from such acts. It would make no sense as a description of a putative "truth" that acts of unprovoked injury in some sense *are* wrong, of themselves, intrinsically, wholly independently of what this or that historical community of speakers might or might not say of them.

Normative value judgements of the order, "Causing unprovoked injury is wrong" could become possible, if at all, only on grounds other than those of empirical observation, that is, either on *transcendent* or on *transcendental* grounds. We might invoke transcendent grounds, claiming, for instance, that our normative judgements are justified wholly independently of observed empirical conditions by conforming to some putative higher or deeper reality said to transcend our (ordinary) experience. Such transcendent reality may then be said to be the Will of God, a Platonic realm of forms, Spinoza's *Deus sive natura* or perhaps Hegel's or Marx's History. In all cases, however, a second, transcendent (or perhaps "authentic") level of reality is said to be the basis for statements which we can apply normatively to a first, empirical or experiential level of reality.

If, however, with Kant, we were to consider any such putative transcendent reality in principle inaccessible to human cognition or if, like Comte, we were to deny it any reality whatever, then non-conventional normative judgements could be justified only on *transcendental* grounds, as being required by nothing external to the phenomenon but rather by the intrinsic "structure" or *ideal possibility* of being in the world, independently of and prior to any contingent factual fulfilment of that possibility.

Using our earlier metaphor, whatever colors we shall encounter on Mars, we can, in virtue of the structural possibilities of our perceiving, validly assert that they *will necessarily* fall within the range between infrared and ultraviolet. Or, in what may be a less misleading illustration, the ideal possibility of *something to eat* does not derive from stuffed avocadoes or steamed asparagus. It is real as soon as there is a hungry agent in the world. Only thanks to it can the particular entities subsumed subsequently under it *be something*, not simply be. Yet that empty ideal possibility, *something to eat*, imposes definite conditions which anything that would be subsumed under it must meet.

Or perhaps a still less fanciful example is in order. Given the ideal possibilities of our being in the world as both social and vulnerable, we can claim that the very "structure" of our being in the world demands that, whatever our factual preferences may be, we condemn acts of wanton injury as in principle, intrinsically, necessarily wrong. No alleged transcendent reality, but rather the transcendental condition of our being, its *ideal possibility*, justifies and demands certain normative statements, prior to all empirical content. It is, in another idiom, an empty *intentio* waiting to be filled.

We shall yet see what such structure and such justification may be. For the moment, though, we can say that, in principle, a *transcendental philosophy* as we shall use that term shall be any philosophy that holds that some structural or ideal possibility of our being in the world makes it possible to make a priori normative judgements which will be applicable synthetically to the fact-content of our being.

For all its easy familiarity, our other term, *everyday experience*, may prove rather more problematic. Its reference is experiential, not conceptual merely. Thus the question of its meaning is not simply one of how we wish to have it function in our discourse, even though that, too, requires attention. Rather, we need also ask what experienced reality we wish to evoke by it.

Everyday experience is, presumably, something that humans have always had. Whatever else it may be, human life is something humans live day by day, even in the most dramatic periods of their lives. Interestingly, though, throughout most of philosophy's history humans appear to have experienced little need to designate their everyday experience by any special technical term. Over the centuries, philosophers have generated a whole range of terms to designate segments and types of experience of the holy, of revelation, of inspiration or of madness. When it comes to ordinary experience, however, such terms as we do have to designate it tend not so much to describe it in itself as to distinguish it from special experiences, as in the case of the term "secular," the anticlimactic counterpart of the sacred. Terms like everyday, *alltäglich* or *všední* tend to be lackluster neologisms which never acquired the status of *termini technici*. Ordinary everyday experience was apparently just too everyday, too ordinary to warrant special attention or special designation.

The need for a special term designating what appeared not at all special seems to have arisen only with the positivist revolt against the implicit metaphysics of natural science and against metaphysics in general. Only when a special meta-reality came to appear inaccessible or downright non-existent did ordinary reality acquire sufficient importance and dignity to warrant a special designation – and, as history goes, that happened rather more recently than we commonly assume. As conventional wisdom would have it, the great turning point in Western thought was the rise of modern natural sciences in the 17th century, in the work of Descartes, Galileo or Newton. However, as Jan Patočka points out (Kohák 1989: 239–44), while those thinkers did transform the metaphysical commonplaces of Western thought, they did not similarly transform Western methodological commonplaces. While Western assumptions about the nature of reality changed dramatically, those concerning the nature of explanation remained remarkably constant. The representatives of the new science had a boundless scorn for the earlier "prescientific" *scientia*, whose metaphysics they regarded as idle speculation fit only for the Platonic or the Baconian cave, yet the science they themselves put forward remained essentially *metaphysical*.

Here, though, a terminological clarification may be in order. We are using the terms *metaphysics* and *metaphysical* in an etymologically strict sense, to designate any conceptual system which seeks to explain the putatively "physical" reality of everyday experience by relating or reducing it to an allegedly deeper, usually hidden level of reality, literally a meta-reality which allegedly accounts for the superficial, overt or "phenomenal" reality. Thus claiming to explain, say, speech inhibition as the effect of archaic and unconscious mind contents is a classic instance of a metaphysical explanation in our sense. Generally, the rhetoric of *discovery*, for which natural science remained notorious well into the late nineteenth century, betrays its metaphysical cast. Scientific explanation here evidently is not regarded as a matter of formulating descriptive formulae summing up observed regularities. When we speak of a discovery, we imply the uncovering of a hitherto hidden, deeper reality, perhaps of natural forces or natural laws, said to underlie and account for the merely apparent reality of our everyday experience. While the uninitated layman is said to live on the level of mere experience, in a world of tables, loves, chairs, resentments and apple trees, the scientist is said to uncover a hidden, deeper world of atoms, forces, molecules and physical laws – or perhaps of psychic mechanisms or laws of history – which govern and explain the appearances of everyday experience. For all the rhetoric of empiricism, everyday experience remains for it merely apparent, uninteresting in itself, just as it appeared from the perspective of other meta-physical realities, the holy or the mystical. Only a new kind of experience, that of the scientist, has been added, as the one true and reliable means of penetrating at last beyond everyday experience to true reality.

That is worth noting. For all the ostentatious empiricism of modern science and philosophy alike, everyday experience still remained too ordinary to warrant a special designation. Even for the empiricists, everyday experience proved interesting not for its own sake, as ordinary, but again as special, as the privileged means of access to a deeper reality. That special access might be assured by the tools of science, statistics and mathematics. Alternatively, it might be assured by the tools of hermeneutics, reaching past the contemptible *Alltäglichkeit* to the meaning of History or, later, to Beyng. Either way, though, experience remains interesting only as special, as revelatory. In its ordinariness, everyday experience remained inherently uninteresting to the scientist and the philosophy alike.

That everyday experience came to prominence only with the positivist revolt against the meta-physical or "transcendent" mode of explanation. Whatever its later fortunes, positivism did not initially propose a new conception of (meta)reality but rather a new conception of explanation. When Kant denied reason's claim to penetrate beyond the phenomenal to a noumenal realm, he might still have been thinking of Leibniz and of Christian Wolff. When Auguste Comte rejected the "metaphysical stage" of human knowledge as one that should be and has been surpassed, he was referring to the implicit metaphysics of the science of his day which assumed it was discovering a second level reality, one of laws and forces, which would explain the merely phenomenol reality of everyday experience.

In this context, the crucial point of the "positive" conception of science is that sciences do not "discover" any hidden reality but rather record and articulate the observed regularities of everyday experience – and that it is the very same reality present to and observed by the mystic, the metaphysician and the scientist. The scientist was now said to differ from the mystic and the metaphysician not in what he observed but in the way he observed it, "scientifically." Positive science is said to observe everyday experience with utmost objectivity, clarity and rigor, scrupulously avoiding all observer bias and faithfully recording observed regularities of occurrence in the unambiguous language of mathematics. It does not claim to discover hidden laws of nature, only to note observed regularities to make prediction and control possible. For classical positivism, rigorous description *is* explanation, the only explantion possible, and the everyday experience it describes *is* reality. It is not a veil of appearance which true knowledge, be it *scientia* or science, would need to penetrate. Everyday experience has come into its own at last.

It is in this sense that Edmund Husserl was entirely justified in claiming that the phenomenology he put forward in *Die Idee der Phänomenologie* and in *Ideen I* is the only authentic positivism. In the familiar passage in *Ideen I*, he tells us that

"*Sagt* „Positivismus" *soviel wie absolut vorurteilsfreie Gründung aller Wissenschaften auf das „Positive," d.i. originär zu Erfassende, dann sind* wir *die echte Positivisten.*" [Hua III.46] – "If "positivism" is tantamount to an absolutely unprejudiced grounding of all sciences on the "positive," that is to say, on what can be seized upon originaliter, then *we* are the genuine positivists." [Kersten 1983:39].

The original program of phenomenology was simply one of clearly seeing and faithfully expressing the eidetic structures of lived experience. The attempt to treat alleged "phenomenological" experience as a privileged instance which would lead us beyond average everydayness to a deeper authentic reality misses the radical nature of Husserl's project which was genuinely positivistic in description, but fundamentally transcendental in explanation. Husserl sought to see clearly and to express faithfully the experienced reality of everyday experience, not to "discover" a putative transcendent one. However, he sought to see and express it not in its contingent particularity or in a generalization derived from it, but rather in terms of the (empty) ideal possibilities which such particularity contingently fills.

Ironically, Husserl's idiosyncratic use of the term "positivism" – Eugen Fink, at the time innocent of any sensitivity to British terminological conventions, even spoke of "radical empiricism" – led to misunderstanding because the scientific positivism of the time was not nearly positivistic enough. The everyday experience to which it turned was not in the experience of human beings in their everydayness but rather experience reinterpreted and reconstructed to conform to the metaphysics of Cartesianism as Descartes' British heirs came to understand it.

The point at issue here is the place of the subject in the economy of reality. In everyday experience, as we shall note later, the subject and the subject's intentional acts are an intrinsic part of reality itself. That is what Descartes' great contemporary, Comenius, understood so well and bequeathed to his pansophist heirs and, ultimately, to Husserl. Descartes offered a different reading. For him, reality is a self-contained region of *res extendes*, entities whose sole characteristic is spatio-temporal extension and whose only relations are mathematical and causal. The subject is a reality of a wholly different and discontinuous order, the *res cogitans* for which there is no room in extensional reality. Thus any subject-related categories – the intelligible order of experience which we shall later describe as that of the categories of *the near and the dear* – represent a distortion of and an imposition upon reality itself. In cruder, later terminology, they came to be described aas the *observer bias* which science must eliminate to reach reality itself. Reality then comes to be conceived as whatever there would be were there no subjects while experience comes to be reduced to accidental passive recording of such reality on the ephemeral screen of subject mind. On such a reading, the everyday experience to which positivism proposed to turn becomes the "subjective" and somehow less than real reflection of reality. In another idion, our ideas could be said to be faded impressions. Experienced reality is once more determined not by actual experiencing but by a set of metatheoretical presuppositions.

That conception of reality may have awakened Kant from his dogmatic slumbers but it did so at the cost of generating enough philosophical problems to become a nightmare itself. For if experienced reality cannot be explained in terms of a transcendent reality – something on which Kant and Hume fully agreed – and if its immanent structure is solely extensional or mathematical, how can it be qualitatively intelligible at all? Or, in the much later idiom of the Baden school, how can it be *understood*, not merely *explained*?

In his conception of phenomenal reality, Kant makes a valiant attempt to revive a more adequate understanding of experienced reality and use it to resolve the problems generated by Hume's Cartesian bifurcation. Experienced – or "phenomenal," appearing – reality is structured by cognising reason which introduces into it structural elements not subject to the contingency of fact. Later we shall speak of them as ideal possibilities. On the level of sensibility, it is space and time as the necessary forms of sensibility. On the level of understanding, it is the categories of judgement. On the level of reason, it is reason's necessary ideas. The crucial point is that these are ideal possibilities, structural elements and not simply characteristics of an alleged reflection of reality on the screen of *res cogitans*. They are necessarily the characteristics of any experienced reality as such. Experienced reality thus is no longer an epiphenomenal reflection. It is reality, the reality amid which we live, breathe and have our being, a reality of which the subject is an intrinsic part, not merely its epiphenomenal observer. However, because it is structured by the subject, experienced reality can be understood in terms of the necessary structure of any experience, that is, *transcendentally*.

The brilliance of Kant's transcendental solution is impressive, but its effectiveness is contingent on a conception of our everyday experienced reality to which Kant paid little thematic attention. The problem is the conception of the structuring subject. As Kant, in this respect very much an heir of the Enlightenment, saw it, only those aspects of subjectivity can quality as transcendental which are truly universal, wholly free of any admixture of factual contingency. This stands out most clearly in his *Prolegomena to the Metaphysics of Morals* and in the *Critique of Practical Reason*. Only pure reason – or here, pure practical reason – in its absolute universality can be a source of valid a priori synthetic judgements. Or, in another terminology, only universalizable rational subjectivity can be truly transcendental subjectivity, an intrinsic aspect of experienced reality rather than an aspect of its contigent and merely empirical fact content.

A number of commentators, Max Scheler in the *Formalismus in der Ethik* among them, have pointed out the critical weakness of this solution. Our prereflective experiential reality is not irrational. It has a "reason" of its own, as Pascal and Comenius knew. It is intelligibly structured. However, it is prerational in Kant's narrow sense of rationality as a matter of universalizable cognitive judgements. Reason in that latter sense is in fact a late comer in our lived experience. In actual experience, that kind of rationality is a subsequent imposition which can appear as arbitrary as John Rawl's *Gedankenexperiment* or Kohlberg's moral dilemmas superimposed by well meaning teachers on a schoolboy's attempt to discern between good and evil. A viable transcendental philosophy would need a more adequate conception of the intelligible structuring – "rationality" in our sense – of our everyday experience. Transcendental philosophy can be viable if its conception of the meaningful ordering of lived experience - its "transcendental subjectivity" – reaches to something more basic than Kant's conception of pure reason.

Husserl's conception of the meaningful ordering of everyday experience seeks to do just that. His proposal to see clearly and articulate faithfully the meaningful ordering of everyday experience – literally the *logos* of *phenomena* – is not only positivist in the original sense of that word but also, as Iso Kern points out, profoundly Kantian. It is positivist: unlike some of his more romantically inclined successors, Husserl does not seek a privileged access to some putative deeper reality but a clear view of the experienced reality of everyday experience. And it is Kantian: Husserl seeks to see and atriculate not simply the contingent factual content of that experience but its intrinsic modes of being or *Wesen.*

Translating that term as "essence," though not without some justification in Husserl's early writings, is problematic. German philosophical terminology does include the term *Essenz*, yet Husserl never uses it as an equivalent of his *Wesen.* Nor do the philosophical dictionaries of Husserl's time so define it. Eisler's *Handwörterbuch der Philosophie* describes it as the set of rules governing the conduct of any entity falling under them. Husserl himself uses only one German synonym for *Wesen – ideale Möglichkeit.* In English, that would be ideal possibility or, more accurately still, *possibility in principle.* His "rational intuition," *Wesenserschauung*, focuses on the structural aspect of lived experience – on the transcendental ideal possibilities which it acts out and which render it intelligible. Even at his most Kantian, as in the first volume of his *Ideen*, written at the time when he was most intensely coping with Kant, Husserl operated with a conception of reason significantly broader than Kant's. The wonder of wonders, he tells us, is *Bewußtsein*, the selfawareness of lived experience or, less awkwardly, reality as experience, *sein* as *bewußt*, not simply reason. In his late works, in *Krisis* and in the third volume of his *Intersubjektivität*, that becomes clearer still. The focus now is explicitly the intrinsic intelligible structure of *Lebenswelt*, of reality as meaningfully constituted as a world by the presence of life.

The term *Lebenswelt* or "life world" has acquired a cluster of associated commonplaces of its own and has come to cover a multitude of sins. Its genesis, though, is reasonably clear. Husserl is building on a concept he introduced in his pivotal *Ideen II*, that of experienced reality as constituting an intelligible whole or a "world" with reference to the active intentional presence of a subject. In *Ideen II* he speaks of *die Welt als Korrelat der Naturwissenschaften* (Hua IV:4). A dictionary translation would render that as "the world as a correlate of the natural sciences." A less literal but more faithful rendering might speak of "experienced reality ordered as a coherent whole with reference to the activity of the natural sciences." Analogously, we could speak of *Lebenswelt as die Welt als Korrelat des alltäglichen Lebens* – literally, the world as a correlate of everyday living or, more faithfully, experienced reality ordered as a coherent whole by the presence of ordinary life. The long overlooked everyday experience here comes to thematic focus at last.

At this point, the clear seeing and faithful articulation Husserl called for is much to the point. What is the overlooked reality we seek to designate by the innocuous term, "everyday experience"? The point of the adjective "everyday" is, I presume, to exclude any of the special qualifiers introduced by reflection, such as "religious," "objective" or even "phenomenological." We are concerned with experience as we live it, not as we categorize it in what in *Ideen II* was called *the theoretical attitude*, from the standpoint of one of our special concerns. But what is that?

At this most elementary level, let us presume further, the term experience refers to any and all of our aware transactions with the context of our being, from merely observing it to acting within it. It does not, let us stress, refer simply to the awareness of our transactions, but to the transactions themselves. An element of awareness must be present, though in much of our everyday experiencing we would be unintentionally overstating our case were we to claim that the subject is "conscious" of the act. Much of the time, we are preoccupied with something else and not conscious of what we are doing in the usual colloquial sense of the word. Yet, in Paul Ricoeur's felicitous phrase, we are *not unaware* of what we are doing. An event of which we would be wholly unaware, though it might involve us, would not be a part of our experience. A dimension of awareness is necessarily a part of everyday experience.

Yet experience, as we ordinarily use that term, really is not simply a matter of awareness. It really is a transaction with the context of our being. It is not simply an interaction, with the usual mechanistic connotations of that term: John Dewey's term, transaction, captures the sense of our organic presence which Heidegger describes as the existentiale, "being in," rather more accurately. Nor would I wish to speak of it as a transaction with a "world," since it is first that transaction which constitutes what-is in the unity of "the world." Nor does the transaction need to be "external." An awareness of pain, especially if it is a pain occasioned by a memory, is not an interaction with the world. Yet it is a transaction with the context of our being. The pain presents itself as autonomous, as a reality with which I must come to terms in one way of another, integrating it within the whole of my experience even though it may present itself solely as an unwelcome intruder. Everyday experience is thus neither merely an awareness nor merely an interaction with an external context of which we are aware. Rather, everyday experience is an activity constituting a whole within which, on reflection, we distinguish the dimensions of awareness itself and of the object of awareness. Everyday experience is the unity of aware transactions with the context of our being which, in colloquial usage, we designate by the words rendered technical by Heidegger, "being in the world."

Here a minor terminological excursus may be in order. In *Ideen I*, Husserl introduced the term *die natürliche Welt*, somewhat misleadingly translated as "the natural world," and subsequently confused the situation further by introducing a second term, *Lebenswelt* or life-world. In his usage, "the natural world" has nothing to do with nature. "World" here means not a set of entities but rather the way in which our experienced reality fits together in a coherent whole. "Natural" here has the colloquial sense of easy, familiar, spontaneous, much as we speak of a "natural" thing to do and mean by it simply the accustomed, familiar mode of acting which may in fact be entirely conventional, but has become so familiar that we are not even aware of it. Thus "the natural world" of *Ideen I* refers to the way we habitually, unreflectingly make sense of the manifold of our experience. It is "natural" in the sense of being effortless rather than being a self-conscious, reflective theoretical construct. Nonetheless, Husserl pointed out, it is already a low-level theory, not simply the way we experience the world but the way in which we habitually, unreflectingly interpret it. This low level theoretical construct must be set aside or "bracketed" in order to reach experience prior to interpretation, simply as experienced, the "phenomenological residuum" of *Ideen I*.

The *Lebenswelt* of *Krisis III* really is that residuum. However, by this time Husserl has become aware that the world as experience is not a static given but rather something that is constituted – literally, given a constitution which makes it into a coherent whole – by the presence of life, or purposive activity, entirely prior to any conscious reflection. When we become aware of our "being in the world" in a phenomenological attitude, we are becoming aware of what-is as it had been constituted as a meaningful whole by the prereflective purposive presence of life. We are becoming aware of "life's world" or *Lebenswelt*. When we become aware of it from our habitual perspective (the *natürliche Einstellung* of *Ideen I*), we are interpreting that meaningful whole in terms of a series of habitual filters, in the West usually Cartesian ones. The term everyday experience is thus ambiguous. It can mean either being in the world in the sense of aware transaction within the life-world or the habitual interpretation thereof. The question for this inquiry presupposed the latter, our answer is presenting the former meaning.

In any case, it will be wise to pay closer attention to the unity of our experience. The point is that, as we actually live it, our everyday experience is not a buzzing, booming confusion but rather a coherent whole. It may at times lack the lofty transcendent unity we assume when we speak about The Meaning of Life but, as William James noted clearly and persuasively, it is not discontinuous (James 1890: 237–64). Its components fit together. Even its radical transitions are continuities. We can speak about it coherently because, prior to all reflection, it is already coherent – or in Husserl's terminology, it is a life-world. The typewriter is on the table, the table is next to the window, the sun-struck New Mexico landscape continuous beyond it. This morning followed last night as a page follows a page, anticipations turn to memories. Some events stand out while others recede into the background. There is an order.

For the purposes of analysis, however, it may be useful to distinguish two kinds of order or, better, two types of relations which order our experienced reality and render it intelligible. One kind of structuring relation – or essential possibility – is that in which the subject functions simply as an observer. These are the relations which, at least in principle, could be recorded no less effectively by a measuring instrument set up by a visitor from Mars. Thus in the context of everyday experience the bear on the shelf presents itself as standing in a definite spatial relation to the silhouette calendar on the wall. They are 2.76 meters apart. That distance, though we are aware of it, is constant and independent of our awareness. Our measuring is incidental to it. There are other such constant relations, as the regular correlation between a hammer stroke and the penetration of a nail into the wood. As long as we remain most scrupulously aware that we are using a conceptual shorthand, not "discovering" any hidden force, we can speak of such constant correlation of two events as "causality" and describe that order of our experience into which the subject enters only as an accidental observer as *spatio-temporal*, the order of interactions among extended entities in spacetime.

Let us then reserve the term "objectivity" for this mode of ordering experienced reality. So used, *objectivity* is not the code name for an object or a characteristic of an object. Rather, it refers to a characteristic of our everyday experienced reality, namely, that it is spatio-temporally ordered. It is this characteristic that renders experienced reality explicable. We can account for the occurrence of any event within it in terms of the spatio-temporal-causal relations of that event to all other events, all subsumed within an observed general regularity or a "law." On this level, accounting for the occurrence of a particular event is all we mean by explanation. Thus when we ask about a particular event, as when asking "why did my child have to die" about a child crushed by a truck, this ordering of our everyday experience – its objectivity – enables us to provide a full explanation. The compression of vital organs by a pressure exceeding X atmospheres is regularly correlated with the cessation of vital functions. So?

That example, admittedly, is emotively loaded, and for a reason – to make the point that the spatio-temporal ordering of experienced reality, its "objectivity," typically plays a rather secondary role in the ordering of our everyday experience. It is an essential aspect of it, to be sure, but the questions we normally ask and the answers we normally seek are of a different order, having to do with relations in which the subject is not merely an observer but a participant. The question about the death of the child asks about the place of that event within a no less real but a rather different ordering of our everyday experience, a subject-relative one.

For the moment, let us designate such subject-relative relations as those of *the near and the dear*. The relation of *the near* is not a spatio-temporal one, a part of the objectivity of our experience. It is subject relative: the bear on the shelf is nearer *to me* than the calendar. However, neither is that relationship "merely subjective" in the colloquial sense of being merely private and arbitrary. The bear really is nearer, and no amount of wishing otherwise will change that. Though I may get up and change it, from where I now sit, the bear is nearer to me than the calendar. Yet it is a subject-related relation, structuring lived experience with reference to a subject.

Similarly the relationship of *dearer than* is not a "merely subjective" one. The bear and the silhouette are far dearer to me than the table. Though subject relative, that is the way it is. I might, for good and sufficient reason, wish it were otherwise. In time, it might even become otherwise, just as a few steps can change the "nearer than" relation. But, at this moment of experiencing, the table is nearer, the bear is dearer. That is how my experienced reality is ordered. It is not a buzzing, booming confusion, capable only of an explanation, not of understanding, because it presents itself as ordered also with reference to a valuing subject, by relationships of *the near and the dear*.

Let us call this type of relationship a *meaning relation*, using the term meaning to designate the way an entity or an event present themselves as integrated within the unity of experience with reference to a subject. Let us then reserve the term"subjectivity" for the sum total of the meaning relations ordering our everyday experience. As with objectivity, subjectivity so used is not a code name for a subject or a characteristic of a subject. Rather, it refers to a characteristic of experienced reality, that of being ordered not only in spatio-temporal terms, but also in terms of meanings, in relation to a subject. The meaning order is not an order of the mind, imposed by mind – or by "reason" – upon a merely spatio-temporally ordered reality. It is the order of experienced reality itself. As we speak of the objectivity of experienced reality and mean by it its spatio-temporal ordering, so we can speak of the subjectivity of experienced reality and mean by it its meaningful ordering with reference to a subject.

Here another marginal aside might not be amiss. When we subject our everyday experience to more careful scrutiny, as Husserl does in *Ideen II*, we discover that its objectivity – its spatio-temporal ordering – presents itself as one of the dimensions of its subjectivity, that is, as one mode of meaningful ordering with reference to the subject. Husserl speaks of it as the constitution of objectivity within subjectivity. Objectivity is the way our experience is structured when we assume the posture of an observer, in the *natürliche Einstellung*, our habitual posture. It is a valid and useful way of ordering experienced reality, though not a privileged one, somehow "more true." It is justified by a specific purpose, and its usefulness is contingent on that purpose, that of prediction and control. Metaphorically speaking, it is a useful posture when we wish to know how to construct nuclear bombs, but a wholly inappropriate one for asking whether to construct them at all. It is equally inappropriate in the case of our earlier question, "Why did my child have to die"? Thus strictly speaking it is inaccurate to speak of two ways in which experienced reality is ordered, of objectivity and subjectivity. The basic order of experienced reality in lived experience is one, its subjectivity, its meaningful ordering with reference to a subject. One of the possibilities within that basic order, one subcategory, is the order which experienced reality assumes as a correlate of disinterested observation. Reality – what Nicolai Hartman called *die Härte des Realen* – is not created by a subject. We encounter it as an other, in its hardness. Objectivity, however, the particular ordering of that reality, is constituted within subjectivity and is reducible to it. It is subjectivity that is the basic order of experienced reality.

Were it not for the unfortunate cluster of associated commonplaces which tend to link the term *reason* with the ability to establish relations among discrete individuals rather than with the ability to grasp the meaning of a whole, we might well speak of the subjectivity of experienced reality as its *rationality* rather than its subjectivity. That is certainly a time-honored usage, reflecting the recognition that experienced reality has its *ratio*, its rhyme and reason, its rightness. There is not only the flux of changing fact contents but also the *logos* of their relations. It is a usage which people like Comenius or Leibniz would have readily understood. In our time, though, the connotations of economic "rationalization" and of syllogistic reasoning render the term *rationality* problematic. Thus there is good reason for the Husserlian usage which designates the meaningful ordering of reality as its *subjectivity*.

To be sure, that usage presents problems of its own. For a post-Cartesian age, it calls up connotations of privacy and arbitrariness with which we need to deal. At this stage, however, the crucial point is the recognition of the intrinsic subject-relative intelligibility of experienced reality as constituted by the presence of conscious purposive activity – that is, of its *subjectivity* – and the second recognition that this term refers to nothing obscure or esoteric, but rather to something we blithely take for granted in our everyday experience: that it is meaningfully ordered by relations of the near and the dear as well as by those of space and time. Subjectivity is simply ordinary, not esoteric.

Nor does subjectivity become esoteric when we focus on its prereflective aspect and speak of it as *transcendental* subjectivity, even though the choice of that adjective to designate it is admittedly less than fortunate. Still, all we are doing is recognizing that the meaningful ordering of experienced reality is not arbitrary or private to individual subjects but rather that at least some of it is predetermined by the fact that experienced reality is necessarily *a* subject's, *any* subject's experience.

Perhaps the most obvious and most universal transcendental dimension of subjectivity – of the meaningful ordering – of experience is that it is ordered in terms of good and bad. What constitutes experience as a meaningful whole is the presence of purposive activity. What-there-is is therefore *ab initio* ordered as that which aids and that which hinders or is irrelevant to that activity. We can, then, recognize less general categories, for instance those of the threatening and the reassuring or, in Husserl's familiar example, of at home and alien, which are applicable only to conscious purposive agents. Still, they represent the mold into which experience is poured, the *ideal possibilities* which it acts out. The fact content may differ – you are afraid of poisonous spiders, I am afraid of lawyers – but the possibility of fear as a mode of relating precedes any particular content. It is an ideal possibility derived from – or perhaps generated by – the fact of our being as finite, vulnerable beings who wish to live but are aware of their finitude and vulnerability. Given such a being, the possibility of fear as a mode of relating follows – and as such it is not an empirical but a *transcendental* possibility.

Claiming all the possible benefits of every conceivable doubt, we might here still be said to be following a Kantian lead, though with an Husserlian difference, especially as it appears in the *Krisis* and in his posthumously published writings. The ideal possibilities which render the manifold of fact contents intelligible are transcendental in the sense of being a function of the presence of subjects.

However, it is the very presence of subjects – of purposive aware agents – rather than their putative pure reason in cognitive judgements that constitutes what there is as a meaningful whole. The constitutive element is not reflection but rather purposive activity, function. That is how I would interpret Husserl's term that puzzled us at the start, the *fungierende Subjektivität*, or Alfred Schutz's *die wirkende Welt*. In both cases, I take it, we are dealing with experienced reality as constituted as a meaningful whole (or "world") by the presence of functioning subjects – a meaning, incidentally, hopelessly obscured by the usual translation, "the world of work." Whatever term we use, though, the point is that experienced reality is intelligible in virtue of the non-arbitrary qualitative ordering it acquires as the context of subject acts.

Those acts, incidentally, need not be something highly elaborate. It can be something as elementary as breathing. Because we are organisms which inhale oxygen and exhale carbon dioxide, our experienced reality is value-ordered prior to all reflection. Fresh, airy places present themselves as pleasant and desirable, dank dark dungeons as unpleasant and undesirable. That is a valuation which we do not constitute, but literally discover in reflection. And, to be sure, in reflection we might choose to alter or reverse that valuation. When there is an air raid in progress, dank dark dungeons can be distinctly comforting. The point, though, is that reflection operates in a value-ordered world and that such value ordering is not arbitrary. It is derived from the presence of the subject and while its level of generality may vary – something that would have been anathema to Kant – it is, as derived from the subject, not empirical but transcendental.

And that, after all, was the point all along. Though the terminology might be unfortunate, the point is that our everyday experience at its most everyday necessarily contains a transcendental dimension. It is not a buzzing, booming confusion, it is ordered, and its ordering is not only quantitative, but also qualitative. Nor is the qualitative ordering of everyday experience – the order of *the near and the dear* – simply arbitrary, derived from individual preference. Some of it at least is derived from our being as subjects and so is not empirical but transcendental. Rather than of everyday experience, we could speak of *transcendental experience* – and mean something very everyday, wholly non-esoteric by it.

With that we can return to our initial example, that of the judgement that *causing unprovoked injury is morally wrong*, presented as synthetic a priori, valid "transcendentally," independently of contingent fact contents. That, we want to suggest, is not an empirical generalization but a transcendental deduction from the ideal possibilities of our being in the world, from its transcendental subjectivity. We humans are in the world as vulnerable, and aware of our vulnerability. We are also social beings, not as a matter of cultural accident but as a matter of structural necessity. Our young do not become self-sufficient for some dozen years after birth, requiring a relatively stable social support structure. Our linguistic mode of communication similarly requires a community of speakers. Language is a social product. Given the reality of our being as in principle vulnerable and social, we are, again in principle, a priori, directly dependent on our ability to trust the others, who are awake while we sleep, to act in predictably non-injurious ways. An unprovoked and so unpredictable injurious act disrupts the very texture of trust upon which our being depends. Thus the judgement, "causing unprovoked injury is morally wrong" is not a function of individual or cultural preference, but of the *transcendental subjectivity* – the necessary meaning structure, the ideal possibilities – of our everyday experience, of the way it is structured in principle, as an eidetic possibility, in virtue of our active presence in the world as functioning subjects.

Husserl, as we noted, goes beyond Kant in his recognition that the transcendental structure of everyday experience is a function not simply of cognition but of active incarnate presence of subjects. We in turn are pushing Husserl to the limit in speaking of any subject being whatever rather than of human subjects. In most of his work, Husserl speaks as if transcendental subjectivity constituted by our presence were a function of human subjects only. In the third part of *Crisis*, he explicitly links it to the community of all actual and possible human subjects, past, present and future, in the tradition of the Enlightenment dictum that nothing human is alien to us. Still, in the passage we cited at the start of this inquiry, Hua VI, § 72: 265, Husserl suggests a way of going yet a step further when he speaks of the *absolut fungierende Subjektivität* as not human but as only initially manifesting itself to us as human. The monadological texts in the third volume of his posthumously published writings, *Zur Phänomenologie der Intersubjektivität* (Hua XV: 608–10 *et passim*), hint at a way we might interpret it. Not only human subjects, but all subject beings, all beings intentionally active in the world, constitute what-is as the meaningful whole, "world," subsuming it with a transcendental subjectivity. Their presence necessarily entails an a priori value ordering. Even apart from human presence, the world is a correlate of purposeful life, it is a *Lebenswelt* not simply in the weak sense of the world as a correlate of our everyday life but as *die Welt des Lebens*, life's world. As such, it is always already value laden and meaningfully ordered, never simply a world of extended entities in merely spatio-temporal causal relations. Grass is not simply there, a *res extensa* devoid of all attributes other than extension. It is already deeply good as the woodchuck's fodder. The warmth of the sun is never simply a matter of waves and particles. It is also ordered as near and dear as it warms the sunflower and the chickadees. Our human presence does not bestow meaning upon a value-free Cartesian extended reality. We are present in life's world, meaningfully ordered by the presence of subjects. Our subjectivity is a part of the network of subjectivity as such, of the *absolut fungierende Subjectivität*.

We are going beyond Husserl in thinking of the ideal possibilities not as eternally subsisting geometrical relations but as possibilities, variously constituted at different levels of generality, in the course of subject functioning. Even that, though, is not a direction alien to Husserl. He himself suggests it in

Ideen II and subsequently Merleau-Ponty develops it in *La phénomenologie de la perception*. Future research might well discover a warrant for such a generative reading of Husserl's conception of ideal possibility and so of transcendental philosophy in his posthumous texts. In any case, transcendental – a priori synthetic – ordering characterizes specifically our everyday experience at its most everyday.

And so we can return to our starting point and to the topic of this conference, *Transcendental Philosophy and Everyday Experience*. Nearly four centuries of Cartesian commonplaces about the "objective" world and its "merely subjective" reflection predispose us to see in that topic a matter of the relation of obscure, esoteric philosophical constructs on the one hand and of the obvious, lucid givens of our daily life on the other. Our reflections, though, lead to a different conclusion. On closer inspection it turns out that our experience is not at all what the empiricists would have it be, a passive recording of spatio-temporal relations of extended entities, arbitrarily decorated with meaning relations by a "merely subjective" *res cogitans*. It has its intrinsic meaningful, value-laden ordering, there to be seen – its "transcendental subjectivity" – simply in virtue of the purposive presence of agent beings. That is not at all esoteric or obscure. Nor is transcendental philosophy when it formulates normative judgements, both synthetic and a priori, on the basis of that structure of ideal possibilities. It is very everyday philosophy – because the meaning structure of everyday experience is really very transcendental.

So, I would conclude, our time-honored way of speaking of transcendental philosophy and everyday experience is a bit misleading. It would be far more accurate to speak of *transcendental experience* and *everyday philosophy*.

Philosophical Faculty, Charles University, Prague/CZ (Boston University, Boston, MA, USA, emeritus) (1990)

WORKS CITED

[Carr, 1970] Edmund Husserl, *Crisis of the European Sciences and Transcendental Phenomenology*. An English translation of Hua VI by David Carr. Evanston, Northwestern University Press; 1980.

[Dewey, 1929] John Dewey, *Experience and Nature*, London, George Allen and Unwin; 1929.

[Eisler, 1903] Rudolf Eisler, *Handwörterbuch der Philosophie*, Berlin, Mittler & Sohn; 1903.

[Hgg, 1927] Martin Heidegger, *Sein und Zeit*, Halle, Niemeyer; 1927.

[Hua II] Edmund Husserl, *Die Idee der Phänomenologie*. Husserliana, vol. II. Walter Beimel, ed., den Hagg, Martinus Nijhoff; 1947, 1973.

[Hua III] Edmund Husserl, *Ideen zu einer reinen Phänomenologie und phänomenologischen Philosophie*, Erstes Buch, *Allgemeine Einführung in die reine Phänomenologie*. Husserliana, vol. III. Walter Biemel, ed., den Hagg, Martinus Nijhoff; 1950.

[Hua IV] Edmund Husserl, *Ibid*, Zweites Buch, *Phänomenologische Untersuchungen zur Konstitution*. Marly Biemel, ed., den Hagg, Martinus Nijhoff; 1952.

[Hua VI] Edmund Husserl, *Die Krisis der europäischen Wissenschaften und die transcendentale Phänomenologie*. Husserliana, vol. VI. Walter Biemel, ed., den Hagg, Martinus Nijhoff; 1962.

[Hua XV] Edmund Husserl, *Zur Phänomenologie der Intersubjektivität: Texte aus dem Nachlass*. Dritter Teil: 1929–35. Iso Kern, ed., den Hagg, Martinus Nijhoff; 1973.

[James, 1890] William James, *The Principles of Psychology*, vol. I. New York, Henry Holt; 1890, reprint, New York, Dover; 1950.

[Kant, 1947] Immanuel Kant, *Grundlagen zur Metaphysik der Sitten*, Leipzig, Meiner; 1947.

[Kant, 1951] Immanuel Kant, *Kritik der praktischen Vernunft*, Leipzig, Meiner; 1951.

[Kern, 1964] Iso Kern, *Husserl und Kant*, den Hagg, Martinus Nijhoff; 1964.

[Kersten, 1983] Edmund Husserl, *Ideas Pertaining to a Pure Phenomenology and to a Phenomenological Philosophy: The First Book*. An English translation of Hua III by Fred Kersten. The Hague, Martinus Nijhoff; 1983.

[Kohák, 1989] Erazim Kohák, *Jan Patočka, his Thought and Writings*. Chicago, The University of Chicago Press; 1989.

[MM-P, 1945] Maurice Merleau-Ponty, *La Phénomenologie de la perception*. Paris, Gallimard; 1945.

[Ricoeur, 1960] Paul Ricoeur, *Le volontaire et l'involontaire*. Paris, Éditions du Soleil; 1960.

[Scheler, 1912] Max Scheler, *Formalismus in der Ethik und die materiale Wertethik*. Halle, Niemeyer; 1912, 1921.

[Schutz, 1932] Alfred Schutz, *Der sinnhafte Aufbau der sozialen Welt*. Wien, Springer; 1932, 1970.

ERAZIM KOHÁK

SHOULD AULD ACQUAINTANCE BE FORGOT . . .?[1]

Should auld acquaintance be forgot and never brought to mind? With all due respect to Robert Burns, I am beginning to think it might be a rather good idea. While cherishing old friends and long ago places, I grow ever more doubtful about remembrance as a philosophical strategy. Do we do well to think of knowledge as *anamnesis*, the recollection of what we had known long ago, before the world began to spin backwards? Or as *a-letheia*, uncovering the archaic truths concealed by Lethe's swirling waters? Should we seek to recapture a putative purity in the depth of our psyche or in the darkness of our archaic past? Or is truth rather before us, an achievement rather than a recollection? If so, *should auld acquaintance be forgot and never brought to mind*?

I have lived three score years remembering, longing for a past that, for the most part, had never been, my gaze firmly fixed on the fading star with Dvořák's *New World Symphony* as background music. Keeping the faith with something or other became a habit, living between a remembrance of paradise lost and a hope of paradise regained, with nostalgic longing as the most becoming posture in between. Such is an exile's lot and over the years I have lived it all, including, rather improbably, the concluding act of *going home*, of paradise regained. It is, I can assure you, heady stuff. "I must go down to the sea again, to the lonely sea and the sky . . ." Come now, beats there a heart so craven as to remain unmoved by that call? Like Edith Piaf, *Je ne regrette rien*. And yet, in retrospect, I wonder whether such romantic nostalgia is really the optimal posture for being human. Is life really about returning – or is it about going forward, about self-transcending?

All that would be no more than an aging man's melancholy were not the image of paradise (or "authenticity") lost so damnably pervasive in contemporary thought, including the area which concerns me most, that of ecological thought. Understandably so: we all bear within memories of what once we were and loved. Metaphors of paradise lost represent a powerful articulation of what Paul Ricoeur once called *la tristesse du fini*, the sorrow of finitude.[2] Yet for all their appeal as fond recollection, the metaphors of return, be it to nature, innocence, or authenticity, are proving rather dangerous. However humanly understandable, in their call for a return to innocence lost they set an impossible goal and so engender the sense of impotence embodied by another old man's sigh, *Nur noch ein Gott kann uns retten*! – and we are unbelievers.

R.S. Cohen and A.I. Tauber (eds.), Philosophies of Nature: The Human Dimension, 249–256.

Actually, a quest for return would be problematic even for a firmly believing fundamentalist. Innocence, once lost, is notoriously impossible to restore, even by itinerant deities. Philosophies which strive to regain innocence doom themselves to failure, simply because there are no returns, eternal or otherwise. You may look homeward, angel, but you can't go home again. Heraclitus knew that long before Thomas Wolfe. And yet, like Lot's wife, we cannot resist that backward glance and would rather meditate upon our deep and archaic authenticity than act effectively in something as mundane as the actual world of our lives. Curious indeed.

Contemporary ecological thought provides a model instance. That its problems are real appears to me beyond question. Yes, there are those who would deny it, much as others would deny there ever was an Auschwitz or that cigarette smoking has any ill effects on human health. However, both empirical and experiential evidence is mounting around us, presented by the scientific community and solemnly acknowledged by governments at one conference after another. Infinite expansion on a finite planet cannot but lead to a collision. Our prevalent conception of this world and of our role therein is likely to prove dangerous to children and other living things. We love and need this earth, we know we are destroying it, yet we feel as powerless to desist as to heed the warning that cigarette smoking causes premature death. Whence that sense of importance in an age that thinks itself omnipotent? Curiouser and curiouser.

Perhaps ecological literature can provide a clue. From Virgil's *Bucolica* to contemporary writings – including some of my own contributions[3] – its dominant metaphor has been precisely a romantic one, that of paradise lost, with its concomitants, a posture of nostalgia, a strategy of return – and an inescapable sense of doom. For a return is always a failure – or, when successful, it is so as a new venture, only thinly veiled, for decency's sake, by fond recollection. You really can't go home again. The home is no longer there, and you are no longer the person who once dwelt there. And yet ecological thought finds the metaphor of *return* – to nature, to the collective unconscious, to prereflective innocence – ever tempting.

Let me stress, I fully respect the reason why ecological activists fall back on longing for lost innocence as a powerful impetus to action. However ingenious, however well wrought the arguments in learned ecological journals may be, they are extremely unlikely to induce anyone actually to *do* something. Yes, it is absolutely critical, especially in matters ecological, that human acts be guided by a cool light of reason and thoughtful reflection rather than by blind and righteous passion. Still, their force is more likely to be an empathy with a devastated world, perceived as kin. The plight of one orphaned whale will evoke greater popular response than all the statistics on marine fisheries. The *polis*, a stately steed, seldom moves without the sting of the gadfly of such a response. As an activist's tactic, an appeal to nostalgia may well be most desirable.

As a quest for solutions, however, it is rather less so. Much though I admire and cherish Gary Snyder, I would submit that *the ecological crisis is not a temporary deviation from a "normal" state to which we could return. Rather, it is something radically new, calling for new responses, not just for a recovery of the old ways.*[4] Even if we could go home, we would no longer be at home there. We need to go somewhere else.

For ponder, to what lost paradise would we fain return? That we feel dissatisifed with or "alienated" in the daily world of our experience is evident enough. The world in which most of us live and breathe and have our being is, for the most part, neither created by God nor crafted by humans, only stamped out by computerized robots. It is a world of artefacts and constructs, deriving its purpose and value from satisfying our whims and so perpetuating the vicious illusion that we are the *Herrenvolk*, the source of all meaning and value, fully entitled to dominate and exploit the rest of the creation. It is not a place where anyone could feel *at home*. Only in rare moments of insight, though, do we glimpse the truth that the experiential depersonalization and alienation of our world are not accidental. They are what enables us to exploit and devastate nature without qualms. To get what we want we give up the home for which we long.

What, though, is the nature to which we would return? It is surely not nature constituted as an inert reservoir of raw materials, the counterpart of the production-driven world of producers/consumers. You cannot return to the green peace of the forest with a chainsaw, a skidder and the thought of another payment due. Had Henry David financed his year at Walden Pond by working for a lumber company, it would have been a bust.

Is it then the nature of subsistence farming and pasturing to which we would return? Through the millenia between the domestication of cattle and wild grasses and the invention of the steam engine, humans did share a basic perception of nature rather different from ours. All records suggest that Nature [*sic*] tended to present itself to them not as inert material but as a living partner, not always kindly but definitely personal. The relation of our herder/farmer ancestors to their domestic animals sums it up: they used them, often cruelly, just as they used each other, but they recognized that there is a respect and a consideration due in exchange for the service. Be it fodder, fertilizer or patience, Nature demands its own.[5] Is that the Nature to which we would return?

In moments of fond fancy, we do dream of it, and, as a source of personal restoration and inspiration, it is surely invaluable. I know no greater priviledge than spending seasons close the land, at Walden Pond, at the Outermost House, in a New Hampshire clearing. As a global ecological strategy, however, it becomes rather problematic. Even if there were sufficient arable land to resettle billions of humans in the self-sufficient communes of anarcho-syndicalist dreams, can you imagine the massive bureaucracy, army and police required to resettle them – and keep them resettled? And would we really want to do without novacaine and the New Hampshire State Police, with marauders abroad on the land? After all, for millenia humans did live as herders and farmers upon the land – and did everything in their power to escape that condition. The nature of the agricultural age is hardly a paradise to be regained.

Or is that nature something earlier still, the nature of the hunters and gatherers, as Paul Shephard forcefully argues?[6] Is it the rise of agriculture rather than of the cities that marks the definitive alienation of humans from their natural nurturing ground? Again, all evidence does suggest that our hunter/gatherer ancestors did experience nature differently than their herder/farmer descendants. The hunters and gatherers survived in total dependence on nature's unpredictable bounty, vast and mysterious. It was Schleiermacher who identified the *feeling of ultimate dependence* as the core of all religious experience – and to the hunting/gathering peoples nature does seem to appear as *deus sive natura*, to be worshipped, feared and placated rather than used.

Still, is Nature so experienced the lost paradise to which we would return? Hardly. Dependence on the unpredictable makes us acutely uncomfortable. Besides, hunting and gathering could appear nature-friendly in great part because the North American continent supported perhaps one percent of its present population. Three hundred and sixty million hunters/gatherers would pick it clean in a short order. Nor does the incessant tribal warfare typical of hunter/gatherer cultures, knowing neither mercy nor quarter, suggest an antelapsarian state.

It seems quite possible that some catastrophe, natural or human made, will drive the remnants of humanity to primitive agriculture or to hunting and gathering and that the rest of life upon this planet will be better off for it. I find it hard to believe, though, that remnant would thereby be freed of alienation from nature or that its life would be particularly idyllic. Perhaps Neil Evernden is right in describing the species *homo sapiens sapiens* as the *natural alien*,[7] a species exotic by its nature, having no natural environment.

Over the centuries, many romantic writers have reached that conclusion. Jean-Jacques Rousseau spoke for all of them when he declared the thinking human a depraved animal. In more recent idiom, we might say that it is reflective consciousness, misleadingly termed "reason," that individuates humans, cutting them off from the nourishing darkness of their collective unconscious. Perhaps *that* is the real alienation: that in our individuality – or perhaps in our rationality – we have lost touch with the deep prerational bond of all creation, the harmony of all being. Perhaps the nostalgia we feel so powerfully – and the futile hope to which it points – is a longing to break through the constraining armor of reflective consciousness and return to the spontaneity of prereflective emotion.

Deep ecologists like John Seed and Pat Fleming argue that powerfully and convincingly.[8] To me, it seems a rather terrifying form of romanticism because it effectively abolishes categories of good and evil, replacing them with a vague notion of being connected with a putative depth of being. Somewhere I have read of German soldiers who carried Hölderlin's poems in their cow-hide backpacks and so were, presumably, listening to the voice of Being, yet still were able to kill and guard death camps with a clean conscience. I have heard them speak of it, dismissing the distinction of good and evil as an Enlightenment prejudice. It was the elevating elation of combat, rising above the tedium of everydayness, that they considered authentic and important.

Nothing against spontaneity: no human act can be meritorious without it. And nothing against plumbing the depth of our common unconscious: life becomes shallow and brittle without it. A return to innocence, though, it is not. The prereflective spontaneity of human emotion is not free of ambiguity. Hate, envy, cruelty are also emotions, attested from the earliest beginnings of the human race. The experience of mortal combat does connect with some of the deepest movements of the human soul. War is a natural, perhaps the father of all, and death the great aphrodisiac. Peace is an achievement. Perhaps humans turned their back on the life of prereflective spontaneity precisely to escape its ambiguity, opting instead for a life in the responsibility of freedom.

Opting for a life in freedom: that is the positive metaphor, pointing to the same reality as Evernden's "natural alienation." Or perhaps not quite. If we remember we are speaking in metaphors, we can say that when humans first experienced the breakdown of their instinctual apparatus, they turned to custom and tradition as substitutes, to *ethos* and *mores* as the source of ethics and morality. The subsequent breakdown of tradition, whether in the conflict of cultures or under the impact of rapid change, leaves only three choices. One is to preserve the crumbling edifice of custom by force. That has been often tried and always futile, since force derives most of its effectiveness from custom. So much for "Asian values." The second is the counsel of despair: accept the breakdown with post-modern nonchallance and wait for The Invisible Hand (or *noch ein Gott*) to set things right. Invisible Hands, though, work on evolutionary time. Nature and culture alike might perish before they invent a remedy. So much for Invisible Hands. Finally, the third possibility is to accept, in conscious critical reflection, the responsibility for establishing the norms of living with each other and the natural world. That is the choice of life in the responsibility of freedom with which European history begins . . .

. . . and seems rather likely to end. For it was in their freedom that Europeans and their heirs set about exploiting the earth to the point at which they are likely to render the survival of their species, perhaps of all mammalia, problematic at best. Ironically, we have called it "progress." If "going forward" means hurtling to our doom, it is quite understandable that there are those who would rather go back, to auld acquaintance and the old ways, returning to paradise lost. Yet if our brief sketch is at all accurate, there *is* no return. And so the conclusion seems inescapable: *Nur noch ein Gott kann uns retten* . . . and, according to Zarathustra, *Gott ist tot*. The romanticism which sensitized us to the ecological crisis also makes it appear insoluable.

But is return the appropriate strategy? Not if there is no point to which to return. Then a different strategy may be appropriate, one of *accepting the responsibility* of our freedom. For better of for worse, we have managed to evade the safeguards which limit the impact of other species. With few instincts to guide us, we have become as free as a pilot taking the craft off autopilot – and as responsible for the impact of our decisions upon the Earth.

The fact is old, the recognition is new. For countless millenia, humans were so few, their powers so puny and their needs so modest that it did not seem to matter what they did. Nature would always repair the damage. There would always be more Nature, there would always be more room. Ecologically speaking, Genesis 9:1–2, "Be fruitful and multiply, and fill the earth. The fear of you and the dread of you shall be upon every beast of the earth, and upon every bird of the air, upon everything that creeps on the ground and all the fish in the sea; into your hands they are delivered," is an utter nightmare. It is also an expression of almost incredible innocence, of a humankind relishing its freedom while blissfully unaware of the weight of responsibility.

Since those words were written, three explosions have shaken our naivete. One was the nuclear explosion at Hiroshima. Few of us noted it at the time: there had been too many explosions in the war years just preceding.[9] The significance of that explosion was symbolic as much as it was military. It served notice that the relative impotence of the human species would no longer safeguard the Earth. We had devised a way to release a virtually unlimited amount of energy.

The second was the population explosion. There is little point in attempting to calculate how many humans the earth "can support." A species can be said to overpopulate when it starts exterminating other species rather than sharing a territory with them, when it begins to expand into territories where its survival requires excessive expenditure of energy and when it multiplies so fast that it can no longer pass on its survival skills to successive generations. Humankind today meets all three conditions, with horrendous results. We can no longer depend on our limited numbers to contain the damage we do to the Earth.

The third is the consumption explosion. For millenia, we could assume that we would run out of hunger long before the ocean ran out of fish. In our time, though, we have learned to expand not only our power and our numbers, but also our needs, creating ever new wants to absorb our excess production. In the overconsuming countries, we consume perhaps thirty-five times as much energy and generate thirty-five times as much waste per capita as humans had through the millenia. There may be only so much we can eat, but there is no limit to how much we can waste in the name of rising standard of consumption.

In the face of that drastic change in the relation of humans to the rest of the creation longing for an earlier, less burdened time is wholly understandable. It is, finally, the longing to shed the responsibility of our freedom, to have Nature, the collective unconscious or *nur noch ein Gott* – or, more dangerously, that God's self-appointed prophet – make the decisions for us. And therein lies the danger. For there are no returns. We can turn over our responsibility to one Invisible Hand or another while we go off to commune with Nature, but we cannot change the fact that there are too many of us, that we are too powerful and want far too much for the Earth to sustain. We have become a cancerous growth. Perhaps *Gaia*, the biosphere as a self-governing organism, can protect itself against us, whether by a massive immunological breakdown or by drastic climactic changes. It is, however, likely to do so on evolutionary time – and in rather unselective ways.

That is the point. Our species has proved far more destructive than any of our kin, but it has created values as well, values worth preserving. No, not the wonders of technology nor even Beethoven's *Ninth* or the collected works of

Shakespeare, much though I treasure them. Rather, there are such things as freedom, little known among other species, and personal safety. There is the ability to alleviate pain and to compensate for the contingencies of life. There is, most importantly, the ability to love and to grieve, the ability to care for the good of another and the common good. All that is worth preserving – and will not preserve itself. For these are not Nature's values from which we have deviated and to which we could return. They are a human achievement which we need to safeguard. It is up to us.

It is a daunting task. We need to accept responsibility for defusing the three bombs – for reducing human populations drastically yet humanely and for reducing the per capita consumption in the overconsuming countries no less drastically while raising the consumption in the undernourished countries to prevent a civilization breakdown, all the while seeking to preserve the values of respect and goodwill towards all. We need to recognize that we are but one species among many upon the Earth, with no special entitlement – yet with a special responsibility for our impact. Only then could we live at peace with each other and the Earth.

Can we do it? Perhaps not: if we simply let it happen, humankind might perish as other species before it, turning the earth over to insects. Or perhaps some shock will break the momentum of our greed and give us another chance. Or perhaps we will accept the responsibility for creating a new, sustainable life world. I do not know. This much, though, I do know: our future is not something to which we can return.

So, *should auld acquaintance be forgot*? Of course not, any more than Elisabethan English or the weaver's craft – or, for that matter, Leibniz's *Monadology*. It is the *auld lang syne* that gives life its rich texture and offers a mirror in which it can see its possibilities. It is only that nostalgia is not an adequate methodological model. Having opted for freedom, we need to recognize the radical openness of the future before us – and the radical responsibility we bear for it – that, like Solomon, we may discern between good and evil.

Philosophical Faculty, Charles University, Prague/CZ (Boston University, Boston, MA, USA, emeritus) (1995)

NOTES

[1] Presented at the conference of Northern New England Philosophy Association in Nashua, NH on October 28, 1995.

[2] Paul Ricoeur. *Le volontaire et l'involontaire* (Paris: Aubier, 1963), 420.

[3] Notably in *The Embers and the Stars* (Chicago: University of Chicago Press, 1982): 27ff, with its strong motif of a return to nature.

[4] Gary Snyder, *Staré cesty* (Prague: Volvox Globator, 1995) is Alexandra Hubáčková and Renata Vystrčilová's Czech translation of *The Old Ways*, a book I relished and cherished in English quarter of a century ago. It loses nothing in translation: it is universally true.

[5] So Keith Thomas, *Man* [*sic*] *and the Natural World* (New York, Random House, 1983), note esp Ch. 3.i–iii.

[6] Paul Shepard, *The Tender Carnivore and the Sacred Game* (New York, Scribners, 1973) is a tender and beautiful classic of the genre: ecotopy as return to humans the hunters/gatherers. Essential reading regardless of agreement. For a gorilla's view of the same, see Daniel Quin's delightful book, *Ishmael* (New York, Bantam, 1992), P. 250.

[7] Neil Evernden, *The Natural Alien* (Toronto, University of Toronto Press, 1985) presents alienation as a natural state of our species rather than as a deviation.

[8] John Seed *et al.*, *Thinking like a Mountain: Towards a Council of All Beings* (Philadelphia, New Society Publishers, 1988) is a handbook for workshops in identifying with the entire biotic community. See esp. pp. 45–52. This book, by the way, is not in any way related to Susan L. Flader's excellent older study, *Thinking like a Mountain: Aldo Leopold and the Evolution of an Ecological Attitude toward Deer, Wolves and Forests*, (Columbia, Mo., The University of Missiouri Press, 1974).

[9] An honorable exception was the German philosopher Karl Jaspers who, in *Die Atombombe und die Zukunft des Menschen* (Munich, Piper, 1958), pointed out the fundamental change it represented. His, though, was a solitary voice: to most of us, it was just one more weapon in "the arsenal of democracy."

ERAZIM KOHÁK

VARIETIES OF ECOLOGICAL EXPERIENCE[1]

This paper started out as a modest little survey of the varieties of ecological experience, with a deep formal bow to William James and a conspiratorial wink to Edmund Husserl. It ended up wrestling with that tar baby of ecological ethics, the place of (anthropocentric) moral categories within a (biocentric) evolutionary reality.

The story, in barest outline, is this. We humans, in our presence on earth – or in *constituting the life-world* – as hunters/gatherers, as herder/farmers and as producer/consumers generate three basic modes of experiencing nature – as an awesome presence to be placated and worshipped, as a working partner to be understood and respected, and as raw material to be used and exploited. That in turn leads to three modes of experiencing nature's agony under our impact – as simple fact, as personal failure and as inexorable fate.

The trauma of confronting the ecological crisis as fate opens a new possibility, that of constituting the life-world as moral agents, experiencing nature as a precious trust and the ecological crisis as a challenge to accept the responsibility for forging a mode of being compatible with the welfare of the Earth and all who dwell therein.

That recognition of moral responsibility for our impact on the Earth, however, leads to an apparent *ecological dilemma*. In traditional philosophical terms, the problem is this. In *nature (reconstructed) as "fact"*, the extinction of a species, be it whales, humans or elephants, is simply a fact of evolutionary life. Writers as diverse as Sir Charles Sherrington, M. Holmes Rolston III, and Annie Dillard have noted as much.[2] Ecocide can become morally grievous only in the context of *nature as experience*, that is, nature ordered in terms of value and meaning, not only of mathematical, spatiotemporal or causal relations. Yet is not the very idea of *nature as experience*, required by ecological ethics, irretrievably anthropocentric, making an *ethics* that would be *both biocentric and ecological* seem a contradiction in terms?

I

The term *nature as experience* is, admittedly, somewhat idiosyncratic. I should far prefer to speak of *nature as consciousness*, but that would be asking for trouble except among Husserl *afficionados*.[3] Edmund Husserl had the great

R.S. Cohen and A.I. Tauber (eds.), Philosophies of Nature: The Human Dimension, 257–271.

advantage that the German term for consciousness, *Bewußtsein*, lends itself to such lovely etymologising! Though Husserl himself never stooped to it, we can say that *Bewußtsein* is *Sein* that is *bewußt*, being that is known. The English term *consciousness* retains but a paltry trace of that. Still, with little shame and lot of daring, we might squeeze from it *con science, with knowledge* and so again *being that is known.*

For that is the point. Any putative Cartesian *res extensa*, lurking out there *an sich*, just waiting to imprint itself on our minds like Caesar's signet ring, would be qualitatively unintelligible, an aggregate of Lucretian atoms devoid of any but spatiotemporal relations. *What there is* is intelligible precisely as constituted – please note, not 'created', but constituted, given a constitution! – by the presence of purposive agency.[4] The meaningful whole, "nature," is likewise qualitatively intelligible not as the theoretical construct, "the objective world," understood as a set of entities from which consciousness has been abstracted, but rather as a complex of transactional relations of which experience, the functioning of purposive agency, is an integral aspect. Only thus can we speak of the meaningful whole *nature* rather than of an aggregate of *what is*. In that sense, *nature* is intelligible – and indeed, *is* at all – only as *Bewußt-sein*, being that is intrinsically known, that is, as consciousness – or, in our rather homemade terminology, as *nature as experience.*

Here let us stress sharply: this is *not* a matter of dealing with nature "as we interpret it" in contrast with nature "as it really is." Nature as experience *is* how nature really is. Experience is not an external afterthought, an ethereal observer or a passive waxen tablet. It is, as William James recognized (n.3, above), *ab initio* an intrinsic dimension of all living being. It is the putative "nature in itself" that is a theoretical abstraction from the sole experiential given, *nature as experience.*

That is why metaphysical speculation cannot be grounded in scientific theory but requires, as its propadeutic, a careful description of the meaning and value structures of *nature as experience.* Or, in less obscure terminology, before we can significantly reflect upon the ways humans should conceive of nature and of its present distress, we need to inquire into the ways they *perceive* or *experience* it. That is why, as the starting point for this inquiry, we take neither scientific theory nor the art of argumentation but rather a descriptive phenomenology of the ways humans experience nature – that is, prereflectively constitute *what is* as the meaningful whole, "*nature*" – and how they experience nature's distress of which we speak as the *ecological crisis.*

II

How do humans experience nature? What is it we take for granted about *what there is* before we begin to reflect upon it? Here the source of information is myth and story, anthropology and poetry as well as personal and social practice.[5] Having surveyed the evidence, we might then speak, with Ernst Cassirer, of mythico-religious and scientific modes, or with Morris Berman of the disenchantment and reenchantment of the world, or with Carolyn

Merchant of the organic and the mechanistic view of nature.[6] Each such symbolic system has its advantages and drawbacks. For myself, I find it more helpful to distinguish three rather than two modes of experiencing nature, *as sacred presence, as a community of persons* and *as a store of raw materials.*

To be sure, actual individual experiences can be as varied as factual individual experiencers. However, in this study we are not concerned with an empirical description but with a study of ideal possibilities – of the ways in which humans can in principle experience nature. That was the reason for our conspiratorial wink to Edmund Husserl. Using his terminology, we could speak of *nature as consciousness* and of studying its eidetic or *wesentliche* structures. To do that, however, we should have to take a mighty oath never to translate *eidos* or *Wesen* as "essence," as some reckless fellows have been known to do. Husserl never uses *Wesen* as a synonym of *Essenz.* The one synonym he uses is *ideale Möglichkeit*, ideal possibility, and only in that sense can we make use of his terminology.

The point is that ideal possibilities are not empirical generalizations. The *eidos*, 'something to eat', is not something we derive from the fact, 'stuffed avocado' by an operation of the active intellect or by a corresponding chemical procedure, much as we extract the essence of vanilla. Nor is it eternal, a-temporal "platonic" essence such as Rupert Sheldrake criticises.[7] Rather, the *eidos*, 'something to eat', is constituted – or, speaking with Anthony Steinbock, "generated"[8] – by our presence in the world as omnivores. It has, initially, no specific content – in Husserl's terms, it is an empty intention, a way in which we could *in principle* relate to the world. Only secondarily and contingently is it filled by a particular fact such as the stuffed avocado. In this study, however, we are not concerned with stuffed avocadoes or other factual givens. We wish to inquire into the basic modes in which humans in principle can experience nature and so it is ideal possibilities, not empirical generalizations, that interest us.

And one more *caveat*: such ideal possibilities are not a function of an arbitrary reflective judgement but of prereflective ways of being present in the world. Or, more precisely, our active, reflective ways of (re-)constituting our experienced world operate within the matrix of meaning structures which are a function simply of our (prereflective) presence. In a shopworn example, simply because we are phototropic, oxygen inhaling beings, we experience light, airy places as desirable and dark dank dungeons as distressing. On reflection, we might, for good and sufficient reason – for instance that there is an air raid in progress – choose to reverse that valuation. However, we make such conscious, reflective judgements always already within a prereflectively value-indexed life-world whose value indexing is a function of our presence as such, not of our (arbitrary) personal reflection.[9]

III

With those methodological *caveats* in place, let us return to the three possibilities we suggested earlier, the experience of nature as a sacred presence, as a community of living beings and as an aggregate of raw materials.

The first possibility articulates the experience of nature as the correlate of humans as hunters and gatherers. For millenia, bands of our ancestors survived precariously in fire-lit clearings amid the dark vastness of nature, wholly and very directly dependent on her bounty. Quite understandably, to such early hunters the nature around them would appear as vast and mysterious, the source of 'every good gift and every perfect gift' (James 1:17) yet also stern and awesome, sometime chastising, sometime blessing, always inscrutable, bountiful, self-renewing and always infinitely more powerful than humans. In terms of the direct lived experience which we can tenuously glimpse in token and imagination, nature must have presented all the aspects which humans traditionally ascribed to God – in Rudolf Otto's words, a *mysterium tremendum et fascinans*, eternal, infinite, omnipotent, mysterious, stern and merciful. When Spinoza spoke of *Deus sive Natura*, he was, perhaps unwittingly, calling up humankind's oldest experience of nature, as a sacred presence which it is appropriate to placate and worship.[10]

To romantic writers and readers, the experience of nature as a sacred presence evokes echoes of a golden age. Nor is such evocation simply escape literature. Those echoes help reorient humans distraught by the impersonal meaninglessness of the world of artefacts and constructs. The contrast between the experience of the seemingly antelapsarian hunter and the estranged denizen of the world of artifacts is an almost irresistible theme for reflection and fiction.[11] The transition to a pastoral/agricultural mode of sustenance might well seem like the original sin, a step from the Garden of Eden directly to Broadway and 42nd Street at 1:00 am on a hot Friday night.

Actually, the step is not direct. The experience of pastoral and agricultural peoples fits neither pole. Subsistance farmers do already manage the land and their extrahuman kin, the domesticated animals. Nature does not present itself to them as a mysterious presence, as to the hunters/gatherers, but neither is it simply the raw material which the city dweller encounters. Rather, it is intimately known kin. Keith Thomas documents the contrast sensitively in his *Man and the Natural World*.[12] On the one hand, subsistence farmers and herders treat their land and their domestic animals as property for their use. On the other hand, they name and speak with their animals, assume understanding on their part, recognize and respect their needs much as the needs of the land. They are no longer worshippers of a sacred nature: God, for them, has become nature's creator. However, neither are they simply nature's exploiters. Nature, though no longer divine, is very much animate, with a life of its own. In effect, they act out fully the ambivalence of empathy and use of which Jeremy Rifkin writes in *Beyond Beef*.[13] Thus when James E. Lovelock in *Gaia* speaks of the Earth as an organism, he is not only presenting a theory but reviving a mode of experience, much as Spinoza speaking of *Deus sive natura*.[14]

When humans experience nature prereflectively as a sacred presence, the posture of worship is equally prereflectively appropriate. When humans experience nature as animate, as a community of living beings, the posture of the good shepherd or good steward, familiar from the pastoral/agricultural metaphors of the Bible, is no less prereflectively appropriate. Much of the current environmentalist literature in fact seeks to revive or at least to recall in empathic memory the experience of nature as a sacred presence or as a community of living beings, the experience of gatherers and shepherds respectively. Ecological authors on the whole tend to agree with Arne Naess, the *doyen* of deep ecology, that we cannot hope to save nature until we learn to experience it once more as sacred and as living, though there is a whole range of opinion as to what that actually means.[15] Intuitively, the claim is highly persuasive. We need to love nature in order to care for it – and in caring for it, we learn to love it.

Still, though we may imaginatively recreate for ourselves the experience of the hunter and the shepherd, our own lived experience of nature is already prereflectively different. We do not experience nature as sacred or as animate in great part because we do not experience it directly at all. We have interposed a layer of technology between ourselves and living nature. As mediated by machine, *what there is* is constituted as a meaningful whole – 'enframed', if you wish, – only as raw material. We the machine users, have no place within the system except as consumers/exploiters. That is the third mode of experiencing nature: as solely raw material for the use of beings who are solely consumers/exploiters.

Defining that third mode of experiencing nature in terms of technology is actually somewhat misleading.[16] The basic shift, I would suggest, is not one from innocence to technology. After all, even a sharpened stick is a tool. Rather, it is a shift in the experienced relation between need and satisfaction. The hunter gratefully accepts satisfaction of need as Nature's/God's gift. The farmer and the shepherd manage it. The industrial age takes the next step: it manages the need. The change comes early. Lynn White notes that already a heavy plough, requiring four yoke of oxen, has that effect.[17] The landowner who invests in such heavy technology must make sure that there is sufficient demand to absorb his production. That landowner no longer produces to meet an existing need, as his peasant ancestor did. He must now produce to utilize the technology and must seek to generate demand to absorb product. The vicious circle typical for the technological age has set in: the need to generate an ever increasing demand or face a structural breakdown.

An age trapped by the logic of its own technology can no longer affort to experience nature as a living community. Empathy with cows or trees threatens jobs. The consumer age protects itself against it by reducing nature to the status of entities which do not evoke empathy or sympathy – a reservoir of raw materials. The hunter worshipped nature as sacred, the farmer managed it, working with it as with living kin. The consumer exploits it – and to do so, must learn to experience it as inanimate. That is the third variety of experiencing nature – as an aggregate of raw materials, devoid of value and meaning of its own, there for us to exploit.

IV

It is in the context of experiencing nature as raw material that the clash between nature's needs and human demands first appears as an *ecological crisis*. The romantics may have grieved the *Entgötterung* of nature, the conservationists its ravishing, symbolized by a seagull dying in an oil slick. It is the consumders, experiencing what there is as supply – rivers as power and sewer, woods as biomass, animals as biomechanisms – who experience an imbalance of supply and demand as a *crisis* in the literal sense of that dead medical metaphor. The survival of the system is in the balance.

The fact, first noted by the Club of Rome, is simple enough: there is a fundamental contradiction between infinite growth and finite Earth. The complex system within which affluent consumers live their lives is predicated on there being ever *more*! – or, in Heidegger's terms, on the world being infinitely ready at hand. When we confront the possibility that the world might prove unready, that its capacity for renewing resources and absorbing waste might be finite, the entire system suddenly appears at risk.

Within the context of the consumer/exploiter experience of nature, that contradiction is simply a fact: we want more, there is no more. Perhaps the most spontaneous response is equally factual: this is not so, it cannot be so, it may not be so. We simply will not let it be. There must be more. It is simple denial. That, as Kübler-Ross pointed out long ago, is how humans typically first respond to what they can neither avoid nor accept.[18] That is how Americans responded to *The Limits of Growth* or to Rachel Carson's *The Silent Spring*.[19] At times the response was patronizing, dismissing such reports as "unscientific." At other times it was angry, attacking ecological concerns as a Communist (or capitalist) plot or as a failure of nerve. Reading the rhetoric of a generation ago, it is difficult to avoid the impression that it is un-American to conserve. Americans make more. The pattern, though primitive, remains constant – a factual denial of a factual claim.[20]

Still within the context of nature as raw material, there is another variety of ecological experience which, again with Kübler-Ross, we could describe as bargaining. A patient confronting a terminal diagnosis tries to offer Death a bargain. If I switch to filter cigarettes, can I keep smoking? If I stop smoking, can I go on living? Or, in ecological terms, if we install catalytic converters, can we go on driving cars? If we recycle our plastic bags, can we go on consuming? If we don't speak of a crisis and concentrate on inventing ecological technologies, can we go on expanding our numbers and our demands?[21]

The variations of that theme are endless, but the mode of ecological experience is the same. The problem remains as external as nature itself. The ecological crisis is simply a fact. There is nothing wrong with the way we are living; it is nature that has failed to meet our demand. With a bit of ingenuity we can make her do so and go on overconsuming.

It is only when the fact will not go away that our mode of experiencing changes. Facts, above all, are supposed to be infinitely manipulable. When the ecological crisis proves stubbornly recalcitrant, it is in effect forcing us to admit that it is not simply a fact, ready at hand. It is as if nature had a will or a mind – or in any case, a life – of its own. The recalcitrance of the crisis leads to a rediscovery of nature as a community of autonomous beings.

Here we are dealing with a fundamentally different way of experiencing nature, originally typical of the pastoral/agricultural mode, and so with a fundamentally different variety of experiencing the ecological crisis. It now appears not simply as a failure of supply but as a breakdown in a relation among living beings.[22] It has become nature's distress – and our personal failure. The supply has not failed. A metaphor of interpersonal relations is to the point. Nature and we have become estranged, out of synchronization.

Much of our philosophical ecology in fact reflects the experience of the ecological crisis as a matter of estrangement and personal failure. The flannel-shirt conservationists among whom I feel most at home sense that as they encounter nature as living kin, beyond the powerline and the paved road. To them, nature is no longer a reservoir of raw materials but a community of kin in the midst of which humans appear as a disruptive presence or, in Neil Evernden's phrase, as the *natural alien*.[23] From that perspective, the key to the ecological crisis is the alienation of humans from the rhyme and rhythm of nature as a community of living beings. The solution then is a return to nature, whether in the literal sense of taking to the woods or in the symbolic form of replacing plastic with wood, synthetic foods with organic and so seeking to recover life's natural rhythm.

Deep ecology in its later stages stresses that intensely personal dimension in its recognition of alienation as not only one from the living nature around us but from the deep nature within us as well.[24] Here the basic claim is that not only the intervening layer of technology, but already the intervening layer of reflective reason has effectively alienated humans from their own nature, leaving them brittle and disoriented. The villain, on this reading, is not just technology and the turn to consumerism, but the Enlightenment and perhaps even an earlier turn to reason.[25] The ecological crisis now appears as a crisis of a Western civilization seen as founded upon a technical reason which seeks not harmony but control. The ecofeminists recognize in it the basic alienation of a culture based on masculine traits of reason, domination, conquest – and hopelessly out of touch with the vital sources of all being, with our collective identify. On that reading, the solution cannot be simply a return to the living nature around us. It needs to be a rediscovery of the deep nature within us, coming to terms with all that is spontaneous, sensitive, caring, nurturing within us – really, with life itself.[26]

If, though, reflective consciousness itself is the villain, tearing the seamless robe of nature, then the conception of nature as a community of living beings becomes itself suspect. It is far too individualistic. The exercises proposed in the manual of deep ecology, *Thinking like a Mountain*,[27] seek in fact to overcome that individuation and to recover the sense of the unity of all being. Such reconnecting with the unity of all being in turn leads back to the third, long forgotten mode of experiencing nature, as the presence of the sacred, and with it the third mode of experiencing the ecological crisis, this time as fate.

That experience has its structural counterpart in the experience of the futility of our efforts. Though we personally may seek to recover our roots in the living nature around us and the deep nature within, we as humankind, as the postindustrial socieity, do not care. Individually we may seek to economize and recycle, but the system drives relentlessly on toward economic expansion, celebrating each escalation of consumption as progress, inventing ever new products for those who have everything and seeking ever new markets among those who do not.[28] Persons may care, but the system does not.

And neither, it seems, does nature. All through the history of life upon this earth, one species has replaced another. Nature's way of maintaining a balance of life is to overproduce and overkill, mocking our attempts at conservation and preservation alike.[29] Time and time again, species have overexpanded and self-destroyed. Cancer, it seems, is not an anomaly but the metaphor of life. Life strives to secure itself upon this earth and, as it succeeds, it breaks out of its bounds and goes on to be fruitful and multiply – until it exceeds its possibilities and perishes of its own lack of self-limitation. Species have done this, and so have cultures. The great civilizations of America provide a classic example. We ourselves provide another. We cannot tell whether the Aztecs or the Mayas realized that their civilization is exceeding its food base, but we do know it about ours. For more than a generation we have been aware that if we do not succeed in defusing the population bomb, all our efforts at preserving the Earth will be in vain. We are no less aware that if we do not succeed in defusing the consumption bomb, we will destroy the Earth. We know we can survive only if we can stabilize human populations and human consumption at a sustainable level, yet we move inexorably toward destruction, seemly powerless to do anything about it.

Here it is hard to avoid a sense of the ecological crisis as fate. We might not want to adopt Martin Heidegger's fanciful metaphor of Beyng that plays hide-and-seek with humanity, yet the reality which it expresses presses in upon us. Humanity appears to have a built-in deep program which leads to destruction. The counterpart of the experience of nature as sacred presence is the experience of the ecological crisis as fate, so that we conclude with Heidegger, *Nur noch ein Gott kann uns retten*, now only a god could save us.

V

Therewith we have come a full circle. We have started out from the experience of human helplessness before the overwhelming presence of nature, and we have ended with the same experience, as if humankind's excursion into reflective reason and life in truth has been but a costly mistake. We want to protest, object, argue, yet history is full of examples of the futility of the best arguments against the power of lived experience. Even the most cunningly wrought argument is vain unless it can find its *Rechtsquelle*, its justification in the pure givens of lived experience[30] – and our experience of the ecological crisis as fact, failure and fate offers none. Unless . . .

. . . unless there were a new kind of experience, and that would be an audacious claim. The world began long ago, with a heigh, ho, the wind and the rain. There is nothing new under the sun, says the Preacher. Still, it just may be that the ecological crisis has opened up to us a new kind of experience, one of *responsibility both to and for nature.*[31]

That is indeed something new. Whether humans experienced nature as a sacred presence, as a community of beings or as a store of raw materials, they always assumed that nature can take care of itself while they need worry only about their responsibilities to each other. Only in our finest moments did humans recognize a responsibility to and for the common good as well. Responsibility for nature, though? Never. The most common trait of all varieties of ecological experience is irresponsibility. During the wars of my childhood, the radio regularly announced just how many thousands of brutto registered tons of British crude daring German submariners just sank into the Atlantic. We Czechs worried a great deal about Britain, upon whose survival our future depended. It never occurred to us to worry about the ocean and all who dwell therein. Nature, we assumed, is inexhaustible. It will take care of itself. That is not our responsibility.

For the most part, that is what we still assume. Occasionally, though, amid our panting after *more*!, we glimpse something different – that Nature is intensely vulnerable and that we are responsible both to it and for it. It is still not a common experience. The way we act suggests that, quite prereflectively, we remain convinced that nature is inexhaustible, that it will take care of itself. Still, occasionally we do seem to glimpse the new reality – that survival of the nature of which we are a part depends on our choices. It is not just another new theory. That is the new experience: of nature as our responsibility and of the ecological crisis as not a fact, failure or fate, but as a challenge.

That experience can become an occasion for a rather different conceptualization of our place in the cosmos. Reflective consciousness – the decision for life in truth, not in thoughtless habit or custom – need no longer appear as the original sin, an estrangement from nature from which we can be redeemed only by returning to the dark spontaneity of a collective unconscious. Life in truth can appear as humankind's natural role – or, in an older terminology, as its "destiny,"[32] the proper or "authentic" way for humans to be. The only condition is that humans recognize their life in freedom as also a *life in responsibility*, consciously accepting the task of preserving the earth, restricting their numbers and their consumption for the sake of preserving the harmony which makes life possible. If you must be human, then be Albert Schweitzer.[33]

VI

This, though, is where Bre'er Fox encounters the tar baby. Formally, it can be described as *the ecological dilemma.* Its first horn is the claim that, in a biocentrically conceived world, human values are simply the peculiarity of one species among many. A world so conceived is devoid of value and meaning. An ethic of ecological responsibility is possibly only in an anthropocentrically conceived world, where human moral norms prevail. In such a world, though –

and that is the other horn of the dilemma – humans are the source of all value and their interests override those of other species. In such a world, we might imaging a prudential ethic, but not an ecological one. Humans cannot be both the species that sets the rules for the world (anthropocentrism) – and at the same time just one species among many within that world (biocentrism). An ecological ethics is a contradiction in terms or, in Niklas Luhmann's words, static, noise in the channels of communication, getting in the way of The Invisible Hand.[34]

Afficionados of classical philosophy will readily recognize a problem with which all idealism has to cope. In traditional terms, the question is how consciousness can be both that which constitutes the world and a constituted reality within it.[35] How can the serpent swallow its tail? The answer is no less traditional. It is not individual psychological consciousness – Tom, Dick or Harry – that constitutes the world but rather the presence of consciousness as such. *What there is* is value laden and meaningfully ordered *in relation to* the presence of purposive agency in principle. Thus it is the *presence of consciousness as such* – Jamesian "mentality" – that enables us to think of components of the world as "good" or "bad." That is what Husserl called the *absolut fungierende Subjektivität* or, more generally, *transcendental subjectivity.*[36] It is the presence of consciousness as such ("transcendental subjectivity," not individual subjects) that gives to what there is the constitution of a (life)world. Particular thinking beings ("psychological subjects") are then a constituted reality within it.

However, since the *afficionado* argot is something of a briar patch to outsiders, it might be better if we deal with the problem in a more user-friendly terminology. How can humans be simply one species among others, so that their specific interests should yield to the interest of other species, and at the same time be the species whose (specific) moral rules are binding on all cosmos? Or, as we put it earlier, if in a biocentric world there are no values while in an anthropocentric world nothing overrides human interests, how can there be an ecological ethics?

Start with the first horn of the dilemma, the claim that a biocentrically constituted world – that is, one in which humans are but one species among many, with no pride of place or status – is devoid of meaning and value, a Cartesian world of *res extendes.* That, as I have argued elsewhere,[37] is plainly not so. Value is not a function of reflective consciousness but of purposive agency as such. A woodchuck may not reflect upon the gustatory or nutritional qualities of dew-damp grass at dawn, but its differentiated behavior clearly attest that it experiences that grass is desirable – that is, as good. Wherever there is life, wherever there is a purposive agency, there *what there is* is no longer value neutral, simply a fact, but value laden and meaningfully ordered. Value may be a relational reality, but it is not a merely anthropocentric one. Wherever there is life, there is value.

Nor – addressing the second horn of the dilemma – is that value simply an expression of my individual interest. Long before I, as an existing human, set out to reflect upon my individual value preferences, the life of which I am a contingent instance already establishes value preferences of its own. Or, in a wholly different idiom, it is the world, with us in it, that tells the stories.[38]

Whatever lives, down to the humblest of creatures, seeks to remain alive and strives to avoid destruction. There may be empirical instances in which this or that particular being, for good and sufficient reason, accepts or seeks death. There is, too, the phenomenon of individuals who sacrifice themselves for the well being of a community or of a species – Life with a capital L overriding individual life.[39] As an eidetic rule, though, it holds that whatever is alive wants to remain alive. Life – not simply my life, not simply this or that individual existing life, but life in principle – is a value for itself.

This has an interesting consequence which resolves the problem of that square circle, the intrinsic (or "inherent") value.[40] If value is a relational reality, how can there be a value which is inherent, not dependent on any relation? Quite simply, when that relation is internal. Because life is a value *for itself* – "for" entails it is relational, "itself" implies independently of any external relatum – it is also a value *in itself.* It is not humans as individual valuers who constitute the world as value-laden and meaningfully ordered. It is life as such and humans only as an instance of it. As individual valuing lives, humans are instances within the world they (co-)constitute as (an instance of) life as such. There is no ecological dilemma. It has perished of an undistributed middle, defining anthropocentrism transcendentally in premiss and psychologically in the conclusion – or less obscurely, treating anthropocentrism as the primacy of the moral law in one case, as the primacy of individual self-interest in the other. As responsible, we have a prima facie obligation not to destroy the Earth.[41] So much for the tar baby.

VII

Alas, as Bre'er Fox discovered to his grief, it is not so easy to dispense with tar babies. The ecological dilemma may not be logically compelling, but it does carry significant experiential clout. Polluting pays handsomely. However logically flawed, the ecological dilemma can be twisted by persons of ill will to entrap those of flawed insight. Dismissing ecological responsibility as human *hubris* while invoking an evolution 'red in tooth and claw' is highly effective rhetoric for all who have a vested interest in driving a gas-guzzler to a whale meat dinner eaten with ivory chopsticks, drilling for oil in the Arctic or mining the last grizzly range. Given that interest – and it is massive – we are likely to keep hearing of the ecological dilemma for some time to come, no matter how cogent the reasons we advance against it.

The ecological dilemma has a more legitimate experiential motivation as well. Paul Ricoeur spoke of it as *la tristesse du fini*, the sorrow of finitude.[42] The more we love, the more we become aware of the precariousness of all there is. We long to enfold the world with the angels' wings of our love and yet we know it will dissolve in our sheltering hands. Loving what is finite is so desperately agonizing! Over the millenia, humans have learned to be careful about loving because loving means losing.

Perhaps the most basic ecological experience is that of an audacious generosity, of daring to love all the suffering, perishing creation. Loving nature is something different from worshipping it, managing it or using it. It forces us to recognize that though we may save the elephants from extinction in our millenium, in due time they and we shall perish. It forces us to preserve life while aware of the inevitability of death.

The ecological dilemma, whatever its flaws, articulates the sorrow of finitude. Ultimately, there is no answer to it except the love which dares to cherish what it knows it will lose, as a mother loving a mortal child. But life is not about being forever. Its eternity is of a different order, a vertical dimension intersecting the horizontal line of passing and perishing. Life is about being true and good and beautiful, about a seagull soaring over the spray, not perishing in an oil slick. It is about mother elephant confidently caring for its cavorting young, not fearfully avoiding poachers. It is about humans cherishing this Earth, not despoiling it in a paroxism if greed.

Perhaps the ultimate answer to logic is poetry. Or perhaps not: I do not know, though I suspect any appeal to the ineffable of concealing a lazy reason, an escape from the responsibility I believe to be the human calling. Still, I am convinced that the ultimate ecological experience is the love that passeth if not all understanding, then at least all species boundaries. And just perhaps that is the revolution which environmental ethics needs.[43]

Philosophical Faculty, Charles University, Prague/CZ (Boston University, Boston, MA, USA, emeritus) (1995)

NOTES

[1] Presented as an invited paper at the 91st Annual Meeting of the American Philosophical Association, 28. 12. 1994 and reworked since in Czech and English. Another version appeared in *Environmental Ethics* 19.2 (Summer 1997): 151–73. I should like to thank my colleagues in the Northern New England Philosophical Association, especially Prof. Anthony J. Steinbock, then of the University of New Hampshire, to my friend Robert S. Cohen and to my colleagues and students at Boston University and at Charles University for their many valuable suggestions.

[2] The *locus classicus* may well be Charles Sherrington's description of the life cycle of the malaria bacteria in *Man on His Nature* (New York: Doubleday, 1953):266–75, though Annie Dillard's "Fecundity" in *A Pilgrim at Tinker Creek* (New York: Bantam, 1974) must run a close second. For a scholarly overview, see Holmes Rolston III, "Disvalues in Nature," *The Monist*, 75.2 (Apr 92):250–78.

[3] Said *afficionados* will readily recognize reference to Edmund Husserl's *Ideen I* (The Hague: Martinus Nijhoff, 1950) § 47, "Die natürliche Welt als Bewußtseinskorrelat" and chapters 2 and 3 of Section II generally. Jan Patočka provides useful elaboration in "The Natural World and Phenomenology" in E. Kohák, *Jan Patočka* (Chicago: University of Chicago Press, 1989): 239–74.

[4] I prefer to speak of *purposive agency* rather than of life (Schweitzer), subject (common usage), mind (James) or consciousness (Husserl) because I am concerned with a mode of functioning, not with some putative "mind-stuff." However, as William James has it, "the pursuance of future ends and the choice of means for their attainment are thus the mark and criterion of the presence of mentality" (*The Principles of Psychology*, vol. I (New York: Henry Holt, 1890): 8). Wherever purposive agency is present, we can speak of consciousness – though not necessarily of reflection.

[5] While most recent writers take the accessibility of prereflective experience to reflection blithely for granted, an inquiring reader might find Paul Ricoeur's classic analysis "Spéculation, mythe et symbole" and "Critériologie du symbole," *La symbolique du mal* (Paris: Aubier, 1960) much to the

point, together with Hans-Georg Gadamer's presentation of hermeneutics in *Wahrheit und Methode* (*Truth and Method*), New York: Crossroads, 1986): 235–73.

6 Ernst Cassirer, *An Essay on Man* (New Haven: Yale, 1944); Morris Berman, *The Reenchantment of the World* (Ithaca: Corness, 1980); Carolyn Merchant, *The Death of Nature* (New York: Harper and Rowe, 1983) and others. All, though, share a tendency to contrast an "organic" and a mechanistic conception of reality, usually blaming Descartes for the latter. Perhaps only Sherrington (Ch. 1 of *op. cit.*) notes the difference between nature as sacred and nature as created.

7 Rupert Sheldrake, *The Presence of the Past* (New York: Times Book, 1988).

8 Anthony J. Steinbock, "Generativity and Generative Phenomenology," *Husserl Studies* 12 (1995).

9 I take this to be the meaning of the much misunderstood and maligned *transcendental subjectivity* in Husserl's late work, in the *Krisis (Krisis der europäischen Wissenschaften und die transzendentale Phänomenologie*; Hua VI; The Hague: Martinus Nijhoff, 1960) and the posthumously published texts in *Zur Phänomenologie der Intersubjektivität III* (Hua XV; The Hague: Martinus Nijhoff, 1973). See also Erazim Kohák, "Knowing Good and Evil . . .," *Husserl Studies 10.2 (1993): 31–41.* – I take Jim Cheney to be making the same point in a different idiom when he says that "our world stories are, indeed, stories, but *stories told by the world – the world with us in it.*" "Intrinsic Value in Environmental Ethics," *The Monist* 75.2 (Apr 92): 232.

10 Rudolf Otto, *The Idea of the Holy* (Tr. John W. Harvey. London, Oxford University Press, 1950). – Ernst Cassirer's reflections on this topic in *op.cit.* are based largely on older German sources, esp. von Humboldt. For an annotated bibliography of sources in Native American thought, Annie L. Booth and Harvey M. Jacobs, *Environmental Consciousness* (Chicago: Council of Planning Librarians, 1988), though perhaps the finest evocation can be found in poetry and fiction, as in Elisabeth Marshall Thomas's *Raindeer Moon* (Boston: Houghton, Mifflin, 1987)

11 I failed to resist it myself in *The Embers and the Stars* (Chicago: University of Chicago Press, 1982), though the genre is as old as Rousseau and Virgil before him. For a superb analysis of the experience of nature as kin and of alienation from it, I know none finer than Paul Shepard's *Nature and Madness* (San Francisco: Sierra Club Books, 1982). Contrast esp. pp. 43 and 108.

12 Keith Thomas, *Man and the Natural World*, (New York: Random House/Pantheon, 1983), note esp. Ch. 3.i on interaction between humans and domestic animals.

13 Jeremy Rifkin, *Beyond Belief: The Rise and Fall of the Cattle Culture* (New York: Plume Books, 1992)

14 James Lovelock's insight ("In Quest for Gaia," *New Scientist* (Feb 75) might be distorted by his metaphor which does tend to invite colorful images of Mother Nature, gauze gown and all. Compare his *The Ages of Gaia* (New York: Norton, 1988). His thesis – that the biosphere is an internally related system capable of regulating its internal climate – is actually compatible with a whole range of metaphors, from kindly God to feed-back mechanism, though an organisms is the most obvious instance of such a system. Here older vitalists, as Hans Driesch, *The History and Theory of Vitalism* (London, 1914) are again interesting.

15 It may be useful to distinguish a wide and a narrow sense of the much abused term, *deep ecology.* As Arne Naess first proposed the term in "The Shallow and the Deep, Long Range Ecology Movement," *Inquiry* 16.1 (Spring 1973), it refers to any ecology which sees the root of our ecological problem in our conception of the place of humans in nature, not only in our technology, a sense for which the term *philosophical ecology* might well be more useful. In some later writings, such as the handbook, *Thinking like a Mountain: Towards a Council of All Beings* (Philadelphia: New Society Publishers, 1988) it has come to mean something akin to depth psychology, that segment of philosophical ecology which sees the root of the problem in the alienation of humans from the collective Ucs. of all life and so the solution in recapturing the experienced unity of all being. For ordinary use, Bill Devall's and George Sessions' definition (the eight principles listed in *Deep Ecology: Living as if Nature Mattered*, Salt Lake City: Peregrine Smith Books, 1985) appears to me most useful.

16 Here the *locus classicus* may well be Heidegger's "Die Frage nach der Technik" (*The Question Concerning Technology*, tr. William Lovitt; New York: Harper and Row, 1977), though Jan Patočka's Varna Lecture, contrasting Husserl's critique of the misuse of technology with Heidegger's view of technology as danger, may be a useful introduction (Jan Patočka, "The Dangers of Technicization in Science According to E. Husserl and the Essence of Technology as Danger According to M. Heidegger" in Erazim Kohák, *Jan Patočka* (Chicago: University of Chicago Press, 1989): 327–40).

17 Lynn White Jr., "The Historical Roots of Our Ecological Crisis," *Science* 155 (10.3.67): 1204.

18 Elisabeth Kübler-Ross, *On Death and Dying* (New York: Macmillan; 1969). A reader familiar with ecological literature of the past thirty years can readily recognize therein the five attitudes

Kübler-Ross identifies, denial, anger, bargaining, depression and acceptance. Only the posture of responsibility (below) is absent.

[19] Dixy Lee Ray and Louis Guzzo, *Trashing the Planet* (Washington, D.C.: Regnery Gateway, 1990) charge Rachel Carson with inciting "lyrical hysteria." For a less impassioned overview, Marco Hollingworth *et al.*, eds., *Silent Spring Revisited* (Washington, D.C.: American Chemical Society, 1987)

[20] There is an interesting parallel with the way America responded to the Surgeon General's report on the effects of smoking. The tobacco industry first commissioned a spate of studies questioning the scientific validity of the evidence. In the second round, the tobacco industry jumped on the band wagon, acknowledged the danger and used it to promote filter cigarettes, much as industry today promotes "ecofriendly" products. In the third round, domestic advertising stresses "freedom of choice" while the industry shifts its focus to third world markets. See following footnote for an ecological parallel.

[21] In a special advertising section, "The Environment: A Corporate Challenge," *Newsweek* 17 June 1991, a number of manufacturers promote high consumption products as jet aircrafts or luxury automobiles on the ground that they are 25% more energy efficient or contain recyclable parts. In the opening statement, Britain's Prince Philip is quoted as saying on the same page that "Humanity is making intolerable demands on the world's natural resources" and "Companies are entitled to maximize profits for the sake of their share holders . . ." – though, the ad. goes on, they should go about it more considerately. Such advertising, reminiscent of the push for filtered cigarettes, has become quite common.

[22] Daniel Quinn makes ingenious use of this: the guru in his *Ishmael* (New York: Bantam, 1992) is a gorilla who does not suggest a return to the life-style of hunter/gatherers but rather a shift from an attitude of Takers to that of Leavers, echoing Erich Fromm's distinction between *having* and *being*.

[23] Neil Evernden, *The Natural Alien* (Toronto: University of Toronto Press; 1985) points out that humans are exhibiting the behavioral traits of an exotic species, that is, one taken out of its natural habitat – overreproducing and overeating.

[24] While the variety of views here is staggering, there are some common emphases. See fine Habermassian treatment by John Dryzek, "Green Reason: Communicative Ethics for the Biosphere," *Environmental Ethics*, vol. 12 (Fall, 1990): 195–210.

[25] Morris Berman argues this most forcefully in *The Reenchantment of the World* (Ithaca: Cornell University Press, 1981). Note especially his comments on hermeticism and alchemy, pp. 69–113.

[26] Within the broad range of ecofeminist contributions, note Val Plumwood, "Nature, Self and Gender: Feminism, Environmental Philosophy and the Critique of Rationalism," *Hypatia* 6.1 (Spring 91): 3–7, 10–22 as well as that entire special issue of *Hypatia*; also Karen J. Warren, "Feminism and Ecology: Making Connections," *Environmental Ethics* 12 (1990): 121–46, and numerous other articles devoted to the topic in that journal.

[27] Aldo Leopold uses the phrase "thinking like the mountain" to mean taking into account the entire balance of nature rather than the interest of one species ("Land Ethic" in *)A Sand County Almanach, New York: Ballantine, 1949*). Susan L. Flande elaborated that theme in a fine volume, *Thinking like a Mountain: Aldo Leopold and the Evolution of an Ecological Attitude Toward Deer, Wolves and Forests* (Colombia: University of Missouri Press, 1974). Her excellent study is wholly unrelated to the manual cited in n. 15 above, *Thinking like a Mountain: Toward an Assembly of All Beings*, whose authors shift the meaning of the phrase to an emphatic identification with the prereflective collective awareness of all life, the "assembly of all beings" and to a retreat from domineering reason to the presumed innocence of sensitivity.

[28] See Note 20 above.

[29] See Note 2 above: Annie Dillard's chapter there cited stresses this point which some writers, as Niklas Luhmann, *Ecological Communication*, (tr. John Bednarz; Cambridge: Polity Press, 1989) take as an argument for turning ecology over to *The Invisible Hand.*

[30] Husserl *afficionados* will recognize the reference to Husserl's "principle of all principles" in *Ideen I* § 24. Here, however, we are giving it an experiential edge: theory is unconvincing unless it articulates a lived experience.

[31] The obvious reference here is Hans Jonas' classic, *The Principle of Responsibility* (Chicago: University of Chicago Press, 1984). However, the theme of responsibility for the survival of the world appears earlier, in the wake of the explosion of the first nuclear weapons when destroying the earth became a practical possibility. So noted Karl Jaspers, *Die Atombombe und die Zukunft des Menschen* (Munich: Piper, 1958).

[32] I take this to be the meaning of the much misunderstood "Wiener Vortrag" in Husserl's *Krisis der europäischen Wissenschaften*: that the decision to live not in tradition and habit, but in truth, that is, in freedom and responsibility, choosing our action in critical reflection, which Husserl

considered distinctively Western, is not the original sin for which the romantics took it but the full, *eigentlich* meaning of being human – though bearing in mind Husserl's critical comments about Enlightenment rationalism in § 6 of *ibid.*

[33] Or if you cannot be Albert Schweitzer, be Paul Taylor whose *Respect for Nature* (Princeton: Princeton University Press, 1986) faithfully develops Schweitzer's basic stance in the idiom of English kantian ethics. Failing that, be Duane Elgin, practising *Voluntary Simplicity* (New York: William Morris, 1981) or at least James Lovelock, *Healing Gaia* (New York: Harmony, 1991) by exercising moderation in using the three deadly C's – cars, cattle and chainsaws.

[34] See note 28 above.

[35] Husserl presents the problems in Krisis §§ 53–55. Note also excellent treatment by Earl Harris in his *Perceptual Assurance and the Reality of the World* (New York: Crown, 1974).

[36] *Krisis* § 72. *Afficionados* will recognize the passage in which Husserl sums up what he considers the great acheivement of phenomenology, "*die absolut fungierende Subjektivität zu entdecken, nicht die menschliche, sondern die als menschliche oder zuerst als menschliche sich objektivierende*" – "discovering the subjectivity which functions *ab solo*, not a human one, rather one which manifests itself (or perhaps which first manifests itself) as human" (my tr.).

[37] "A Dialogue on Value," *Journal of Speculative Philosophy* 5 (1991): 1–24.

[38] The idiom is Jim Cheney's, cf. Note 9 above. As long as we stress the two components of Prof. Cheney's assertion equally, that *the world tells the stories* and that it is *the world with us in it*, we are presentingm, in briar-free terminology, precisely what Husserl is presenting in his conception of transcendental subjectivity (transcendental = the world is telling; subjectivity = with us in it) in *Krisis III* and the *Nachlaß.*

[39] My thanks to Prof. Robert S. Cohen for calling my attention to Bert Holldobler and E.O.Wilson's *The Ants* (Cambridge, MA: Harvard University Press, 1990) which presents this phenomenon most forcefully.

[40] Tom Regan offers a survey of the problem in "Does Environmental Ethics Rest on a Mistake"? *Monist* 75.2 (Apr 92): 161–82. He concludes that it does, or, in our terminology, that an *intrinsic relational* reality is a contradiction in terms – which it would be, but for the possibility of value being a function of an internal relation.

[41]This I have argued at some length in "The Ecological Dilemma: Anthropocentric Values, Biocentric Reality" presented to Northern New England Philosophy Association, 21.11.1994; published in Czech, "Ekologické dilemma," *Československý časopis pro fyziku* 45.3 (October 1996): 242–47 also below, pp. 293–301.

[42] *Le volontaire et l'involontaire* (Paris: Aubier, 1963): 420.

[43] "Environmental ethics needs a revolution . . ." quoth Bryan G. Norton in "Epistemology and Environmental Values," *The Monist*, 75.2 (Apr 1992): 207–26. Recovering the experiential grounding of our petrified constructs, as Norton does in his reflection on Gifford Pinchott, may well be that revolution. Wouldn't it, though, be an irony worthy of Richard Rorty if the most revolutionary slogan proved to be "*Zu den Sachen selbst*"!

ERAZIM KOHÁK

NATURE AS PRESENCE AND EXPERIENCE

I

In traditional philosophical terms, the topic of my paper could be said to be the place of humans in the cosmos, considered from a metaphysical, epistemological and ethical perspective. Less traditionally, I could say that my concern is with postmodern ecological ethics. However, since neither of those summaries is particulary informative, let me begin by sketching the outline of my argument instead.

The purpose of my current work is to formulate a conception of nature and humanity which would recognize both nature and culture as legitimate objects of concern. That might seem a modest enough goal – except that most writers dealing with the topic have tended to assume that nature and culture exclude each other. To some, a concern for nature is a barnacle on the ship of progress and a threat to "property rights." To others, culture is a malignant tumor upon the body of nature and a threat to survival. Both can give good and convincing reasons for their respective positions. I would like to recognize the genuine insight of both and yet claim that a concern for nature and a concern for culture are both legitimate and can even be compatible.

I take as my starting point the way humans experience nature. Let me stress, I am not concerned with what nature in some unspecified sense *is*, or perhaps *really is*. Nor am I concerned with what humans *think* about nature, what elaborate theories we construct as philosophers or as physicists to account for its intrinsic wonder. Rather, I am concerned with the way humans *experience* nature, how nature presents itself to them or, if I may use the verb transitively, how humans *live* nature. I am concerned with all the half-conscious commonplaces of our daily life that seem so obvious to us that we seldom stop to examine them. I am convinced that the way humans deal with the rest of nature is far more powerfully affected by the way they *experience* or *live it* than by what they reflectively think it is or whatever it might be in the sight of an absolute observer. For that reason, my concern is *Sein als Bewußt-sein*, nature as experience – or, if you prefer, nature as *Lebens Welt*, the totality of all there is constituted as a meaningful whole be the presence of life.

My concern with it is not simply descriptive but rather unabashedly normative. I am daily confronted with evidence from unimpeachable sources that our present way of experiencing and dealing with nature is likely to destroy

R.S. Cohen and A.I. Tauber (eds.), Philosophies of Nature: The Human Dimension, 273–284.

the habitability of the earth for our species and all its mammal and indeed all its vertebrate kin in the not too distant future. My concern is to inquire how we ought to *experience* nature and what strategy we ought to adopt in order to enhance the prospect of a sustainable coexistence of the species *homo sapiens sapiens* – with its specific peculiarity, "culture" or "civilization" – with the rest of nature, with its necessity and spontaneity.

I would suggest that hitherto humankind tended to follow, broadly, two kinds of strategy, both wholly understandable yet catastrophic in their consequences. One I shall rather arbitrarily label the "Enlightenment" or "modernist" strategy, even though we encounter it already in antiquity, not only in Stoic though but also as acted out by the Roman Army's equivalent of the Corps of Engineers. Basically, it is a strategy based on the assumption that nature is an irrational and dangerous surd, devoid of meaning or value, and that humans, as the bearers of reason – albeit rather idiosyncratically understood – are called upon to conquer that surd and make it conform to their image. Once that meant building aquaducts, today the vision is more audacious: we shall retool our bodies with genetic engineering and turn over the management of the rest of nature to the Corps of Engineers and to General Motors. It is this strategy that has given us an environmental crisis of catastrophic proportions.

The second strategy I shall just as arbitrarily designate as "Romantic" or "traditionalist," for it is as old as the traditional romances of the Golden Age. It is the strategy which assumes that nature is the dark and fertile ground of our prereflective innocence. Standing out of that innocence in reason, in reflection and technology, now appears as humankind's original sin which humankind is called to expiate by rejecting the works of our mastering, domineering and individuating reason, returning instead to the receptive and nurturing womb of Mother Nature. There is no other hope. I rather suspect that as the "modernist" strategy led us into a catastrophic ecological crisis, the romantic strategy made that crisis seem insoluble, allowing us to sign with the aged sage, "*Nur noch ein Gott kann uns retten*." By a curious coincidence, such wizened resignation allows us to stop worrying, leaving the fate of nature up to that god's invisible hand as we go about our inalienable right of pursuit of property, exactly as we would according to the "modernist" strategy.

I believe both of these strategies stem from a genuine experiential grounding, from something seen, not just thought, but that both suffer from a fascination with their own insight that blinds them to a different reality and a different experience. I wish to point out that different, long overlooked reality and the newly discovered experience of our *responsibility for* nature. The *reality* of that responsibility is as old as the human turn to life in freedom rather than in instinct or custom. In that turn, we have, figuratively speaking, taken our culture from nature's autopilot and so became responsibile for its direction and consequences. However, the *experience* of that responsibility is something as novel as the mushroom cloud over Hiroshima and the dramatic rise of carbon dioxide in the atmosphere, both events within my lifetime.

I believe that in part at least we find formulating an adequate strategy for ecological survival so difficult because our attitudes are shaped by age old experiences of Nature's might and human weakness while the experience of Nature's vulnerability and of our responsibility is of a very recent vintage and far from universal. Certainly, it was always irresponsible of humans to let Nature, the Invisible Hand or *nur noch ein Gott* deal with the consequences of our wars and our greed. Yet for millenia nature seemed to vast and humans so few and puny that it did not seem to matter. Suddenly, we are waking up to the realization that we have become many and powerful – while nature as we know it is becoming an endangered species. With that realization, our experience of millenia has become obsolete and the strategies based upon it, whether exploitive or ecological, have become irrelevant.

Please do not get me wrong. I treasure the Romantic sense of harmony of all creation and consider empathy with the appresented other the foundation of all ethics. I reject, though, the Romantic longing for a return to some antelapsarian state, be it pretechnological, preagricultural, or preconscious, as deeply irresponsible. Similarly, I treasure the Enlightenment's sense of reason and its committment to liberty, tolerance and just plain decency. However, I reject its anthropocentrism and its illusion of human omnipotence as arrogant – and ultimately no less irresponsible than Romantic irrationalism.

If there is a solution for the ecological crisis, I believe it is not to be found either in returning to a simpler time or in progressing to yet more complex one. Rather, it lies in confronting the experience of Nature's vulnerability and *accepting the responsibility of freedom, consciously seeking to minimise the negative impact of culture on nature and disarming the intrahuman conflicts of injustice and greed that distract us from that task*. It is, if you wish, a strategy of culture in the service of both humanity and nature, predicated on a recognition of humankind as the most powerful and so the most responsible, but emphatically not a most privileged part of nature.

That, admittedly, is an outline for a book, not for a fifty minute paper. God willing, that book will come in due time. For now, let me use the time remaining to add a few words to at least some of the points in my all too brief outline.

II

How do humans experience nature? Existential philosophers of a generation ago were ever ready with grand and wholly unsupported pronouncements about "Man" [*sic*] and "his" [*sic*] mode of being, contrasting it with that of (other) animals, in total disregard of even animal *Völkerpsychologie*. More recent ecological and sociobiological research as well as older anthropological writings provide us with a rather more reliable body of empirical material which we an understand and use philosophically because we ourselves are human and so can interpret that third-person material phenomenologically, as lived experience. In light of what we can learn from von Humboldt, Cassirer or E.O. Wilson, how do humans experience nature?

The cultural diversity within the species *h. sapiens s.* is, of course, immense. Still, I believe that we can isolate some significant common traits. Perhaps the most important of these is that humans, as a species, experience the rest of nature both as a *Heimwelt* and as a *Fremdwelt*, both as an intensely familiar place of dwelling and as distinctly and irreducibly other. It is a *Heimwelt*, as close to us as our own life in flesh. That life is itself a natural process so that nature is never wholly other. As we experience it, it was meaningfully ordered and value laden, and not all the value in it stems from our wishes. A woodchuck, too, can cherish a dew-sparkled meadow at dawn. We can empathise with the life appresented in the woodchuck's body. Nature is we, it is kin and a comforting context. We are and can be so thoroughly at home in the world, as much as we are in our bodies.

Yet, at the same time, nature is other. It resists us, its order seems often incomprehensible and unheeding. Even our own bodies resist us and ultimately bring us down. In the metaphor of yesteryear, the soul may be willing but the flesh is weak. Or, as the apostle Paul would have it, it is not I but the law of my members that dwelleth in me. Apostle Paul also anticipates another recent pair of terms, *innerweltlich* and *in der Welt*. His way of putting it is that the human mode of being is one of being in the world, yet not of it.

That much, I think, we can say with some confidence of ourselves as a species – that our experience of nature is fundamentally ambiguous. We are can can be only as incarnate in it, yet our being is not reducible to it. Perhaps other primate species experience the same distinction – I do not know. There is, however, significant evidence that even the early hunters who don a totemic animal's mask and assume its name are not unaware of performing a conscious act of identification. *Sera y no sera* – though they will to become wolves and in that sense they do, it is a willed act and that is a fundamental difference. Wolves do not will to be wolves, they simply are that. At the other end of the spectrum, even the hypothetical cosmonauts daring boldly to go where no humans had gone before, into far reaches of the cosmos and of split infinitives, do so in *all'zu menschliche* bodies that constantly remind them that though not *of the world*, they are very much *in the world*, even aboard the spaceship *Enterprise*.

There is significant evidence of other universally human traits, ideal possibilities or *eide*, built structurally into being in the human mode. Thus beings aware of their own vulnerability will in principle value trustworthiness. Still, what is interesting for our purposes are the diverse modes of experiencing nature, clearly and basically affected by human culture, a given community's way of being in the world. Romantic ecowriters tend to use a binary distinction between pretechnological innocence (good, natural) and technological decadence (evil, unnatural). Modernistic ecowriters concur, only inverting the valuation. Both empirical and phenomenological research, however, suggest a more complex schema, distinguishing at least four basic culture-conditioned modes of being in the world and so of experiencing nature.

One is that of the earliest hunter-gatherers. To attribute to them an antelapsarian innocence or a presocratic unity with *Beyng* appears rather far-fetched. The shattered skulls of early hominids as well as the persistence and cruelty of intertribal warfare among hunter-gatherer peoples in historic times mitigate against that. Still, our hunter-gatherer ancestors did live in a total and direct dependence on a vast and mysterious nature. The experience of total dependence, as Schleiermacher pointed out, is the core of religious experience. It is not unreasonable to assume that to the early hunter-gatherers, precariously surviving at its mercy, nature appeared as *deus sive natura.* Not, to be sure, as the personal God of latter day Protestant piety: that is an embarassingly modern image even for a believing Christian. Rather, it seems that the hunter-gatherer experienced nature as the *mysterium tremendum et fascinans*, a mysterious power, both fascinating and terrifying, to be placated and worshipped rather than respected or controlled. One mode of experiencing nature is definitely as sacred, powerful and terrifying: *natura, sive deus.*

With the cultivation of wild grasses and the domestication of animals there are indications of change in the experience of nature as well. The dimension of dependence remains, though now in a context of mutual transaction. For the most part, nature no longer appears as god but rather as god's, the visible works of god, intermediate between the sacred and the human. The farming and pasturing peoples – my great-grandfather still one of them – seem to have experienced nature as a partner requiring honest dealing rather than worship. If the land is to yield its grain, it must be rewarded with manure. Again, all empirical evidence mitigates against the romantic view of bucolic harmony, Virgil's or Shakespeare's. These are not antelapsarian innocents: nature can be treated as cruelly and deceitfully as fellow humans. Still, it is experienced in a distinctive mode – as personal, a partner in the task of living.

A different mode of experiencing nature becomes evident in the culture of crafts and trades as it emerged in Europe with the rise of the cities in the thirteenth century. This is an experience we still marginally share. To the master cooper, nature is not a partner, as to his peasant ancestor. It has become a source of raw materials. Those raw materials, though, are still not soulless but endowed with value and calling for respect. The New England credo, "Use it up, wear it out, make it do, do without" is not just a reflection of poverty but also of a deep rooted experience of material as valuable, to be appreciated and respected. So is the box of "strings too short to save" – destined for making tinder. Here, an urban culture, driven by need and served by craft and trade, had its unwitting ecological ethic. Experiencing nature as a gift to be appreciated and used wisely is another culture-determined mode of experiencing nature.

The experience of nature changes again when production replaces need as the driving force of social interaction. Expensive technology – an early example is a plow requiring eight yoke of oxen – drives humans to produce to cover the cost of the technology, regardless of personal wishes, availability of material or need to be met. What must be met are payments – and that means generating artificial need regardless of consequences. The welfare of nature has become accidental to the one great need, meeting payments. In more usual language, it is reduced to a reservoir of raw materials – Locke's "bounty of nature," waiting to be plundered – and humans to its desperate despoilers. Here nature is experienced solely as a means, wholly devoid of any standing of its own, devoid of value, devoid of meaning – and so human relation to nature is drained of any ethical consideration. For the hunter, nature was aweful, sacred. For the farmer, it was personal, a partner. For the craftsperson, it was valuable, cherished and respected. For the producer, it is only there, to be used, abused or destroyed in the short term pursuit of property.

III

There is, expectably, a correlation between modes of experiencing nature and strategies towards it that appear appropriate to various cultures. Among the hunter-gatherers, living in awe before the vast and inscrutable power of nature, a strategy of placation and worship is both predictable and appropriate. A respectful partnership, even if not above covert duplicity, is equally appropriate among the farming and pasturing peoples who experience nature as a personal other. Crafts and trades people, smiths and spinsters and carvers, successfully follow a strategy of appreciative care. Only in the production-driven producer/consumer cultures do we encounter the strategy of rutheless exploitation, driven by greed and justified by "property rights," in equal disregard of the welfare of nature and of humankind as an integral part of it.

It is a frightening phenonemon, humans immune to empathy as they hurtle along asphalt ribbons in their toxic monsters past the mangled bodies of their fellow creatures. The vision of Route 128 at rush hour will ever remain with me as the ultimate metaphor of dehumanized horror. Interestingly, though, it is only in cultures that generate such monstrosities that we encounter an awareness of an ecological *crisis*. The hunters and gatherers may have feared incurring nature's wrath, the farmers and herders undernourishing their fields and flocks, the crafts and trades people wasting away their materials – but all remained confident that, as long as they do it rightly, what they do is itself meet and right. It is only the producer/consumer civilizations that become aware that their very cultural strategy is destructive and self-destructive, that precisely their success, not just their failures, spell death to nature and in turn to culture as well.

To be sure, the motive of that recognition is not at first empathy or generosity of spirit. The reports of the Club of Rome, *the Limits of Growth* and *Beyond the Limits*, were not motivated by grief and compassion for the sacred, personal or intrinsic values of nature, but simply by concern over the impact of the overuse of 'natural resources" on human – and specifically northern and western – material affluence. Still, the shock was no less real and the search for alternative stragegies no less earnest.

Interestingly, the response to the ecological crisis in the generation just past retraced the succession of human experiences of nature, only in reverse. Its earliest phase was appropriate to the thought of the producer/consumer civilization, reasserting a rather primitive version of the "modernist" strategy, committed to the "conquest of nature" for "the good of 'Man' [*sic*]." Concretely, that meant a quest for more powerful technologies to force nature to yield more and new resources and, philosophically, to a reassertion of the putative human "right" to own the earth. Humans are the *Herrenvolk*, nature is here "for 'Man' [*sic*]" and human wishes override all other considerations. A bumper sticker I saw recently sums it up: *Destroy the environment – your job depends on it*!

Given the pooling of accumulated data by researchers world wide at Bergen, Dobříš, Rio de Janeiro, Berlin, that initial attitude has today become untenable except in the neanderthal margins of the global society. A variant emerged, still in the modernist mode though closer kin to the crafts and trades mode of experiencing nature. Nature is still here for human use, but it needs to be used wisely. Recycling may be the most visible manifestation of the new mood, but its most substantive impact may come from the search for new, non-poluting technologies and sources of energy.

That strategy is clearly a significant step forward in its clear recognition that something is amiss, though it still assumes that what is amiss is not *what* we do, only *how* we do it. It encourages young people to love, not only exploit, this earth and all who dwell therein, but it does not challenge the basic, internal contradiction of the modernist strategy – that infinite expansion within a finite world must inevitably result in an explosion, no matter how may trees we plant or how many plastic bags we recycle.

That recognition represents the next step – the admission that the problem is not just that technology is inadequate, but that our whole approach to life and nature, predicted on the double assumption that greed and aggression are the meaning of life and that humans are a priveleged species whose whims override nature's needs, is hardly out of joint. Nature has its own needs and possibilities – and our mode of being in the earth is at odds with them. The problem is one of a broken relationship between humans and nature. That is a conception that harks back to the agricultural/pastoral experience of nature as a working partner. Humans have grown out of touch with nature, or, to use a more fashionable term, have become *alienated* from it. It is the mutuality that needs to be restored.

With that we are leaving the modernist frame of reference and entering a romantic one. Its basic vision is not one of a conquest of the natural surd but of a paradise lost and regained. As for the present, strategies of return to a Golden Age are appropriate – back to the land, back to nature, back to the collective unconscious, back to the world before the fall.

That schema comes in many variations. The Christian view of history, interestingly, is not one of them, except perhaps in popular or sectarian imagination. In Christian and Judaic theology, salvation is not a return to a paradise that once was. It is a new achievement. "Behold, I make all things new." The emphasis is unmistakeable: it is a *new* earth, a *new* Jerusalem. There is, however, a strong romantic strain in Engels' parallel between the primitive communism of the matriarchal society and the communism of the future – though a believing Marxist would probably label that vulgar Marxism. No one seems willing to claim the label, "romantic," yet a great many thinkers share the romantic devaluation of the present, a nostalgia for a paradise lost and a hope of its restoration. To beings struggling to retain their identity amid the flux of time, it is understandably appealing. These are, after all, fundamentally strategies of return.

In the environmental movement, that return is usually understood as a (further unspecified) *return to nature.* Nature – once "the visible works of God," more recently the works of natural necessity as distinct from the works of human freedom – is presumed to be whole, wholesome and healing, free of stain of human fall. Evil appears as the product of human alienation from nature. So, let us don our flannel shirts, strike out for Walden or New Hampshire and return to Nature!

As a source of personal renewal, I find that most appealing. The happiest years of my life – so far – were lived on a New Hampshire homestead, wood heated and kerosene lit. It is a great gift and a privilege I would wish for everyone. There is a *vis medicatrix naturae.* However, I cannot claim an antelapsarian innocence for the technology of Thoreau's *Walden* or for life close to the land. The problem is not only that low-density dwelling patterns destroy natural habitats as well as cultural amenities. It is, rather, one to which Jean-Jacques Rousseau, the father of modern romanticism, points with his claim that "The thinking person is a depraved animal." Even on my New Hampshire clearing, I remain a being of reason and of conscious decision, alienated from the spontaneity of my elemental being. Return to living nature may be fine and good, but it is not enough.

If return to nature in the concrete sense of flannel shirts, woodstoves and vegetable gardens is not enough, then perhaps we need to return deeper, not to the nature around us but to the nature within, to the deep, soft, dark warmth of the prereflective unconscious, the collective unconscious of all being. Again, as a personal strategy for a restoration of personal balance, such delving into the prereflective depth of one's unconscious might be quite healing. However, as an ecological strategy, I find it frightening. It calls for rather more faith in the benign purity of the prereflective unconscious than any evidence warrants. We need not accept Freud's assessment of aggression and sexuality as the sole archaic contents. We might, with Jung, posit social, sustaining dimensions in the unconscious as well. However, it would be difficult to claim we can trust our primordial emotions implicitly. Hate and fear are emotions as basic as love and trust. It was thinkers like Alfred Baeumler or Alfred Rosenberg and their party comrades who decried critical reason and extolled the "sound instinct of the German people."

As far as I can see, only one thing does follow necessarily from a return to the unconsious: the inability to choose, to act and to bear responsibility. Free action presupposes reflective reason: anticipation, deliberation, decision. If we rejected reason, we could only wait for things to happen, for passion to move us. Then, of course, it would hold that *nur noch ein Gott konnte uns retten.* Once we accept that view of ecology, the future of humankind and nature alike become a matter of infrahuman passion and suprahuman destiny, as it was for our earliest hunter–gatherer ancestors.

So we have come a full circle. We started out with our hunter–gatherer ancestor's experience of nature as divine and worked our way through the experience of partnership and that of a respect for nature to a posture of simple exploitation and so to the ecological crisis. In the quest for answers, we started out from masterful arrogance, retreated to considerate use, then looked for a return to living nature and ended up in a return to the collective unconscious, in mute wonder before the Beyng that has withdrawn from our technological world. Only a god, we are told, could still save us – and we are unbelievers.

IV

Counsels of despair, however, should always be treated with caution. Not that there are no desperate situations: clearly, there are, and our ecological situation may be one of them. However, given the human propensity for preferring excuses to efforts, it is wise to think twice before declaring the situation hopeless. In this case, a new factor enters in. Human experience, for millenia essentially local, is on its way to becoming global and so is capable of including three phenomena which exceed all tradition and all strictly local horizons – the nuclear explosion, the population explosion, and the consumer explosion. The nuclear explosion forced us to recognize that the earth and its nature are not something we can take for granted. Not only a meteorite from space, but humankind itself has acquired the ability to render the earth uninhabitable. The population explosion forces us to recognize that we are not only capable to destroying the earth but that we are actually doing it. Satelite photographs of burning Amazon forests make the point no less than the squallor of our cities. The consumptionm explosion among the overconsuming segments of the earth's population – thiry five times of what it was two centuries ago – both of itself and as a hope for others, makes the point that we are destroying nature not by accident but as an unintended but inevitable result of our civilizational strategy.

In the margin, let me comment briefly on those three explosions. The significance of the nuclear explosion is not just that some petty dictator might launch a nuclear exchange that will render much of the world uninhabitable, though that is frightening enough. Rather, it is the recognition that humans are capable of mobilizing practically limitless sources of power. The physical limits on human destructiveness may not have been removed altogether but they have been expanded most uncomfortably. It is the sheer power which the nuclear explosion represents, immense power in the hands of humans as frail and fallible as in the days of bow and arrow, that is frightening.

The consumption explosion similarly represents a removal of a natural safety device. As long as human economic activities were motivated by experienced need, they operated within an internal limit. Metaphorically stated, there is only so much a person can eat. When humans stop consuming when they are sated, their demands on nature remain finite. However, once human activities come to be driven by the needs to produce rather than consume, that limit is removed. Demand now reflects an artificial need consciously invented and fostered to absorb already excess production. An artificial need is infinite. The earth, though, remains finite.

Perhaps the population explosion requires most comment. The point is not whether the earth "can support" – whatever that would mean – six, twelve, or twenty-two milliards of people. I know no way of calculating that. However, we can establish criteria for overpopulation. A species can be said to overpopulate when it begins to displace other species from its habitat rather than sharing it. Secondly, a species can be said to overpopulate when it destroys the ability of its habitat to support it and begins to expand to habitats where its survival requires excessive expenditures of energy. Finally, a species can be said to overpopulate when its numbers increase so rapidly that it is no longer able to communicate its social survival skills to its young. Whatever the absolute number of humans on earth could support, we can speak of a population explosion because humankind meets all three of those conditions – with all their consequences.

Those three explosions, I would submit, constitute a qualitatively new mode of experiencing nature – as *vulnerable* – and of experiencing our position in it – as *responsible.* Both may always have been the case. It was primitive hunters who hunted the mammoth to extinction, and bronze age warfare that reduced the land of milk and honey to a desert. Nature was always vulnerable, we were always responsible. However, for millenia we could affort to ignore it. There were so few of us, our demands were so modest and our power so puny! The three explosions have changed that. Today we need a new strategy of dwelling upon the earth, one based on the experience of nature as vulnerable, not omnipotent, and of humans as responsible, not privileged.

Projecting such a strategy, this time based neither on nostalgia nor on arrogance, but on the recognition of nature's vulnerability and human responsibility, appears to me as the basic task of environmental philosophy. Fortunately, we do not need to create such a strategy *ex nihilo*. From the romantic strategy we can take over, gratefully, its sense of humans as dwellers, not masters, and its ability to empathise with non-human nature. We do not need to adopt the more fanciful imagery of the collective unconscious or of the Earth as Gaia, though it may be useful. The point is that we can accept the recognition that both joy and sorrow are indivisible, that humans are but one species among many and that our welfare is not separable from the welfare of nature as a whole.

However, I do not believe we can or should accept the romantic strategy of return to a golden age. There is no golden age in history: the ambivalence of being in the world yet not of it is a hallmark of our humanity from the start. There is no golden age in the depth of our unconscious: not only is that unconscious itself conflicted, but our freedom and the reflective awareness on which it rests are an intrinsic part of our being. Most of all, there are no returns. Hope must aim toward the future – or it dooms itself to failure.

That is the contribution of the Enlightenment tradition: the clear recognition that humans, without an adequate instinctual apparatus, bear the responsiblity for forming a coherent conception of the world and of our place in it. We *are* different: in our freedom and our power we have assumed the responsibility for our impact on our world. Humans are the species that must impose its own limits and cannot rely on instinct or inclination in so doing. The greatness of the Enlightenment tradition appears to me in its clear recognition of the responsibility of our freedom.

We cannot, though, accept from the Enlightenment the arrogance of its anthropocentrism. Though different, we cannot claim to be a privileged species – any more than any one of our groups can claim to be a privileged race or nation. Nor can we accept the conception of reason as a calculating ability. The reason that makes us free is a qualitative reason, the ability to form a coherent image of nature and of our place in it, in relations of value and meaning. Only such a reason can provide us with points of reference that we may discern between good and evil. For a species that must decide and choose its acts, there is no ability more important than that.

V

Can culture – and that means the human way of being, in freedom rather than in instinct or custom – be compatible with nature? Or put it differently: can the Earth afford humans – or is the human race an aberration, a cancerous growth which overflows its limits and destroys the host organism?

The triple explosion has made evident what, *in nuce*, was present already in the human turn to freedom: that humanity has effectively abolished the natural constraints on its cohabitation with the rest of nature. Numerous ecologists – as Lovelock and Margolis – have pointed out that nature, as a self-regulating system, can defend itself against that imbalance. The second report of the Club of Rome, *Beyond the Limits*, anticipates a population collapse within a century, on statistical grounds. Since nature works on geological time, it might be longer – and the damage humans do in the meantime that much greater.

Need it, though, be at all? I think not, if humankind comes to recognize its responsibility for nature's survival and orders its affairs accordingly, in effect placing its culture on the service of nature, humankind included.

It is, actually, no particular mystery what that would mean in practice. It would mean, first of all, disarming the population bomb by a systematic policy of reducing and stabilizing human populations – since tigers and childhood diseases no longer do it for us while a natural disaster would do it rather unselectively. Some ecologists concerned with habitats for all species estimate that reducing global population to 30% of its present state – approximately the population of the earth a century ago – would assure sustainability. Yes, Rome in the wake of barbarian invasions went from over a million to perhaps 17 000 inhabitants, but its culture went with it. We need to disarm the population bomb without sacrificing the distinctive achievements of humankind.

We need, secondly, to disarm the consumption bomb – to adopt a systematic policy of reducing individual consumption to perhaps 30% of the current level in the overconsuming sectors. That, incidentally, is hardly poverty level: it is about the level of material consumption in the United States in 1940 or Czechoslovakia in 1965. It is a level that covers needs – and offers some very essential luxuries, such as personal safety, medical care and education, as well.

Finally, we need to raise the levels of individual consumption throughout the globe to that same level. That is the only way we can disarm the third, nuclear bomb. As long as individual societies and the globe as a whole remain divided between an overprivileged, overconsuming minority and an impoverished and resentful mass, signing peace treaties will remain futile. There will be conflict – and as in all the wars past, from the clash of Assyria and Egypt to the wars in the Balkans, nature will be the prime victim. By some estimates, human losses in the Balkans will be made good in thirty years. For the environmental destruction, the optimistic estimate is in centuries.

To disarm the nuclear bomb, the population bomb and the consumer bomb: it means nothing short than a complete turn, from systematic maximization to a systematic minimization of human impact upon the earth. Can we do it? I do not know. Some ecological writers believe exhortation could be enough: that if humans only realize their situation and accept their responsibility, we can solve the environmental crisis. Others claim that it would take a middle size catastrophe, small enough to destroy humankind altogether but severe enough to shake it out of its lethargy. Others still think the task hopeless: humans love this earth as their *Heimwelt* but will accept its destruction because it is also a *Fremdwelt*.

I do not know. However, the absence of any guarantee of success does not absolve philosophy from the task with which Socrates had charged it, that of being the gadfly of the steeed of state. Besides, there is not only the Socratic charge. There is also the testimony of Jonah.

Philosophical Faculty, Charles University, Prague/CZ (Boston University, Boston, MA, USA, emeritus) (1994)

ERAZIM KOHÁK

THE TRUE AND THE GOOD: REFLECTIONS ON THE PRIMACY OF PRACTICAL REASON

The purpose of these pages is to sketch the outline of an extended argument which links the various studies that make up this book. Ultimately, I wish to claim that there is a fundamental asymmetry between life and death – of, if you wish, between being and not being – which endows life's world with a moral orientation not reducible to the preference and/or the consensus of individual subjects. Contrary to the claims of radical scepticism, ancient and modern, I believe this asymmetry and this moral orientation enable us to claim that at least some of our assertions can be non-trivially true, independently of our assent. Or, in another idiom, I wish to claim that what is said can be true or false – and that what is done can be right or wrong – because, prior to our reflection, something is good, something is bad, and something is evil. Scepticism, I believe, is ultimately false because this is a value indexed, not a value neutral cosmos.

Radical scepticism, to be sure, can be and in our time often is no more than a fashionable posture and an excuse for intellectual irresponsibility. When it is not, as in the case of great and morally earnest thinkers like Gorgias and Rorty, it is ineviably based on an ontological rather than merely an epistemological conviction. The truly great sceptics do not just happen to think that there are some difficulties with our cognition which might be overcome by a rigorous application of the scientific method or by deeper eidetic intuition. Rather, they doubt that there is anything to be known at all – or, more precisely, they doubt that what there is has any univocal meaning that we could mirror in our theories. Reality, as they see it, is of itself an undifferentiated and so unintelligible continum. Only in the act of knowing does it become articulated, reflecting the preferences of a knowing individual or a discoursing community. What we know then is reality as we have constituted it in our act of knowing. We cannot claim any ontological justification for our knowledge. For the great sceptics that does not mean that what we think or say does not matter. Quite the contrary, it matters a great deal, since there is no autonomously meaningful reality of which we could say, Galileo-like, that it goes on turning in spite of our denial. It means only what it says: that we must *create* our truth, *ex nihilo.* There is nothing to keep us from holding beliefs that will destroy us and, ultimately, no reason why we should not do so.

R.S. Cohen and A.I. Tauber (eds.), Philosophies of Nature: The Human Dimension, 285–291.

It is at this point that I wish to part company with scepticism. Even though, with Richard Rorty, I am convinced of the futility of seeking to construct a *theoretical* mirror image of reality, I do not believe reality which includes life can ever be *practically* – morally – neutral. I believe that, whether we admit it or not, there is a good and sufficient reason why we should not destroy ourselves or our world. I believe that *there is something good*: that reality is not infinitely plastic but has a meaning structure built inherently into it, a value orientation which we acknowledge – or not – but which we do not create. The fundamental asymmetry between life and death, upon which environmental initiatives build – hence the "Green" in my title – provides an archimedean reference point for beliefs of which we could say that they are not only noble but also in a significant sense *true* – or, in the words of the Book of Common Prayer, "worthy of all persons to be believed."

That, admittedly, is a rather large claim, calling for an extended argument. Its first step need be a clarification of the task of philosophy. I take philosophy, the proverbial "love of wisdom," to be neither a positive science not yet an instrument. Its task, I believe, is neither to describe the world nor to change it. Rather, it is to provide humans with a moral orientation, with an overall conception of reality and of the place of humans within it that would enable us to discern what to seek and to what to shun.

Humans, I believe, do not seek philosophical understanding out of idle theoretical curiosity. That is at most the motivation of home-grown speculations which their authors present as "metaphysical." Nor do humans seek philosophical understanding as a tool for resolving specific problems within a preconceived framework. The practical and mechanical arts can do that far more efficiently. We humans seek philosophical understanding because, of all the species, we have the most inadequate instinctual equipment. We cannot simply wait for an instinct to push us. We need to choose a course of action and decide in order to be able to act. For that, we need a reasonably coherent image of how things (in the broadest sense) fit together (again in the broadest sense, with a special emphasis on their moral coherence). Without at least a general idea of what it is all about we should become as katatonic as a Czech child asked to select from a menu written entirely in Chinese characters. Philosophy is not basically a road map; the sciences provide us with those. Philosophy is, rather, a proposal about goals, what to seek and what to shun. It is, most fundamentally, about values.

Therein, though, lies the rub. Value is intrinsically a relational reality: a good is ever a good *for* someone. Or, in order to avoid that most recent of philosophical bugbears, the "metaphysics of the subject," let us say that an X can be said to be good only with reference to a purposive activity. In a perfectly unchanging universe, devoid of all activity – like the frozen universe constituted by pure theoretical reason – nothing would be either good or bad. It would only be. Ours, however, is a value-laden universe, never simply an aggregate of *rēs extendes*, because it is constituted as a structured whole by the presence of purposive activity. And so the puzzle: how can a reality be relational, yet not relative? How can value be activity-related and yet not activity-relative? Or, more concisely still, how can something be good only in relation to someone – and yet not be independent on someone's preference?

Here a phenomenology of value experience is very much in order. How does value in fact present itself to us? We need to ask that not in the empirical sense of why we value this rather than that but in principle: why do we experience our world as value-indexed, as meaningfully ordered in terms of the good, of its absence and of its negation?

One dimension of value experience is rather obvious: we are present in the world as purposive agents and so encounter the world as facilitating or as hindering our activity – and so as good and bad in that sense. That, we should note, is a dimension of value experience far more basic than philosophical reflection or, for that matter, than any conscious reflection. Long before we think, we already breathe – and so encounter fresh air as good and the urban mix of coal dust and carbon monoxide as noxious. When we reflect upon it and come to use such categories as "good" and "noxious," we are only naming values which had been constituted prior to our reflection simply by our purposeful activity.

There is a second, formal discussion to our value experience as well. We encounter our world as good or bad not only in relation to our activity but in relation to our expectations. Because purposive activity essentially projects itself towards an anticipated future state, purposive agents are present in the world with expectations. The guardian dog expects his reward, the human hikes out from the end of the tram line in search of fresh air. The reality they encounter meets or fails to meet their expectations and so presents itself as better or worse. By its expectations as by its actions, life constitutes the world as value-indexed.

Here, though, we encounter another rather basic problem. If life were worthless in itself, how could it constitute other realities as valuable? Something useful in perpetuating a worthless activity would not become worthy thereby. Nor would a worthless reality become worthy by living up to a worthless standard. A value-less banknote does not become valuable by being the perfect specimen of a value-laden model, a perfect Confederate bill. Nor does that Confederate bill become valuable because it equals one Confederate war bond. Only an intrinsically valuable reality can constitute its world as valuable. Life can constitute what facilitates it and what comes up to its expectations as valuable *only if life itself is valuable.*

Here we have two options. One is to argue that life is valuable because it in turn serves some other, greater good. Corn serves pigs, pigs serve humans, humans in turn serve God – or perhaps Nature, as Spinoza would have it. Nothing is valuable in itself; whatever is, is valuable only instrumentally, in terms of its subordination in a hierarchy or perhaps in a circle of mutual dependence. However, that option is vulnerable to all the familiar arguments against infinite regress. It does not help explain why one link of a chain moves to be told that it is being pulled by another link, whether we extend that chain to infinity or close it in a circle. Endless regress explains nothing, only acknowledges the fact of change.

A different interpretation may be more adequate. It starts with the most elementary of recognitions, that whatever is alive seeks to remain alive, that everything would rather live than die. A sprouting bean reaches for the sun, a plant for water, a wounded animal licks its wounds, humans seek to preserve their life under the most gruesome conditions, perhaps even a boulder resisting a force testifies to the same basic, utterly fundamental rule of all life: whatever is alive wishes to remain alive, whatever is in being wishes to remain in being rather than be not. Certainly, animals which rely on understanding and judgment as canines, higher primates and most notably humans, may at time judge the prize of surviving not worth the price. At times humans do commit suicide, aging elk give up the struggle and face the wolf pack. Those, though, are voluntary decisions, for good and sufficient reason overriding the rule which remains the rule: all other things being equal, whatever lives strives to remain alive; whatever is seeks to continue being. Self-destruction is an exception, self-preservation the rule.

This very common experience and observation suggest first of all that *life is a value for itself.* Life is not simply a value-neutral reality: it is, *ab initio*, also a value for itself. The reason why the presence of life, or purposive activity, can constitute the world as valuable and value-indexed is that is is most basically a value for itself, something life strives to preserve. Life relates to itself as to something both desired and desirable. There is a value dynamic built into the very structure of life as it does so. Wherever life is present, the universe – or, more exactly, life's world, *Lebenswelt* – is intrinsically value-laden and value-indexed. To live is good and to perish is bad, to destroy evil, as negation of the primordial good. That is the most basic value indexing. Life is good, because it relates to itself as a value. Death is bad because it negates that good.

Here we encounter the fundamental asymmetry in the structure of being. Life is intrinsically good because it relates to itself as to a value. Death, though, is not its negative counterpart. It does not relate to itself, it cannot even *be* of itself. It can be only as a negation of life. What is alive wishes to remain alive and so constitutes value. What is dead does not wish anything, constitutes nothing – it just is, though strictly speaking even that is not true. What is dead *is* no more. The lifeless being – the boulder of our earlier example – still *is*, in Husserl's imaginative term, a "dormant monad." Death is not: its reality is parasitic on being as its corruption, as a negation of what is. To kill is not to transpose a being from one state to another. It is to negate a being in all its being, to end it, to deny it, utterly and absolutely. There are not two states, life and death, being and non-being. There is only one state, life, and its corruption, its negation. To speak of death as the counterpart of life, the yang of life's yin, is fundamentally misleading. Death has no autonomous reality: its sole reality is as a negation of life.

That is also the limit of all value relativism. We may legitimately debate the relative merits of broccoli and cauliflower or of Mozart and Bach, but we cannot debate relative merits of life and death. There is no debate because there are no two realities to choose from. There is only one reality – life which, as self-related, is intrinsically good and the source of all goodness – and its negation. We have reached the moral rock bottom on which all further value systems are built, the rock bottom of good and evil.

Here, though, we would do well to pause and shed light on a fundamental error. There is a sense in which death can legitimately be said to be a part of life or, in the time-honored Biblical metaphor, one of God's angels. Limitation is an intrinsic part of finitude. Life which overflows all limits – a cancer can serve as an example of such bourgeoning growth – becomes destructive. The basic error is the assumption that because death is necessary it is also good.

Here Aristotle's distinction between natural and violent movement might prove helpful. It is a distinction we have largely forgotten in almost all other instances, but cannot ignore in the case of death without baleful consequences, including, *inter alia*, a crisis of our health care system. There is such a thing as a "natural" death, the ending of life full of years, there are the realities of the food chain. Yet even such death is never good of itself. What is good is life; natural death can be said to be good not in itself, but only derivatively, as a part of life which alone is good in itself.

Violent death, still in the sense of Aristotle's distinction, is a different matter altogether. It is not a part of life, an integral part of life's rhythm and so derivatively good. It is a *negation* rather than a completion of life, intended to destroy life rather than to embrace it.

Violent death, we might note, is a distinctly human phenomenon, just as murder is distinctly human. In the order of nature – and even in the lives of humans, to the extent to which humans remain a part of that order – death remains a natural phenomenon, an inevitable but in a sense unintended by-product of life's striving for itself. An owl does not will to kill. It wills to eat, and kills in the process. The will to kill is human. So is the will to ignore killing – or in a somewhat fanciful metaphor, to decriminalize it. Loosing an unaimed mortar round into Sarajewo is an act of willed killing, profoundly evil because it is not an expression of a will to life but of the surplus of hate humans carry within them, erupting as the will to destroy. Releasing toxic waste into a river is equally evil: it is an instance of the "decriminalization" of killing whose motive is nothing loftier than a desire to increase profits by cutting waste disposal costs. Humans do violence to nature's order both by decriminalizing death and by willing it. The fundamental error then is one of using the legitimate justification of natural death to justify violent death.

Yet the two are not the same. The fact that every one of the six million would in time have died naturally does not excuse the holocaust. The fact that every one of the fish would in time die anyway does not excuse pollution. The value-free order of nature, as it appears to the disinterested observer – in which life and death are equally parts of the grand scheme of being – does not override the moral order of the participant, the order of practical reason. Humans – and all living beings – are always primordially participants, acting and so living in a world value-indexed by the dynamics of life. The order of pure theoretical reason is always an abstraction. The world of a living being is always primordially a life's world.

Here we are touching on a second crucial point. The first crucial point was the asymmetry of life and death, constituting the asubjective moral order of the life-world. Because life is a value for itself, life is good, prereflectively, independently of preference, simply good – and the destruction of life is bad, willed destruction of life evil. The basic moral command is the affirmation of all life in its variety and multiplicity, within the limits imposed on any individual life by the needs of other lives. Acting ethically then means, spoken with Aldo Leopold, to "think like a mountain," making our decisions for the good of the community of all beings, not of one privileged species only. Speaking with Albert Schweitzer, we might speak of it as reverence for all life – or we might paraphrase Immanuel Kant and say we are bade "so to act as to treat all beings as (also) ends, never as means merely." That is not just one ethical opinion among many. It is the moral order built into the stucture of being by the fundamental asymmetry of life and death, of being and not-being. To cause unprovoked injury would remain morally wrong even if all humans were to approve of it.

The second crucial point is the primacy of the life world and its order or, in traditional terms, a primacy of practical over theoretical reason. As theoretical reason reconstructs the world, abstracting purposive life from it, it may well appear that the world is primordially a value-free, lifeless system of forces and matter in motion, as some thinkers believe, it has been for millenia before life, a late phenomenon on the cosmic scene, introduced value indexing and before humans, even later phenomena, gave that value order a moral significance.

Such theoretical ordering, though, is hopelessly artificial, a reconstruction in retrospect. The reality amid which we live is always, inevitably a life's world, the world of our life, value-indexed by life's presence and morally oriented by our presence in it. We can, per impossibile, imagine a world devoid of life, a once and future reality as putatively once it was or once it shall be, the world as constituted by disinterested theoretical reason. When we do so, however, we are not "uncovering" some putative "true reality." We are building up an imaginative construct. What would be true in it – for instance, that it would be value free, undifferentiated, equally compatible with any possible interpretation – is not true of the world of our reality which is always, necessarily and primordially a world of practical reason, value-indexed by life's presence and morally oriented.

That, finally leads us to the third crucial point – that our philosophical claims are not arbitrary but must be adequate to lifeworld's pre-given ordering. Unlike in the world as reconstructed from the standpoint of theoretical reason, in the world of our lives, the world of practical reason, *there is a truth*, the basic truth of the goodness of life, of the badness of its negation and of the evil of the will to negate. Granted, that basic truth is always more than any finite set of propositions. Our philosophical assertions cannot claim to *be* the truth, only to point to it. Their relative adequacy is a function of their effectiveness in pointing to it, that is, of bringing about an awareness of life-world's order. Thus the proposition that the life-world is a community of all life whose basic moral order is one of respect for life is not *the* truth. Yet it points to the truth more adequately than the proposition that the world is a value-free aggregate of particles in mechanical ordering. For all of Rudolf Carnap's theoretical brilliance, here Karl Jaspers provides a more adequate conception of what truth is all about.

With that we can come to our conclusion. Relativism – and radical scepticism – apply to the world as reconstituted by theoretical reason because that world is value-free and devoid of any but mechanical order. Such a world construct, however, is not an adequate conceptual schema for orienting in our world, which is a world of purposive lives, a world of practical reason, value indexed and meaningfully ordered. The value-indexing and meaningful ordering of the world of practical reason is not arbitrary, contingent on individual preference or community consensus, because it is based on the asymmetry of life and death: life is a value for itself, so intrinsically good and worthy of respect in its multiplicity and harmony. Thus not all statements about it are equally valid. Their relative validity is a function of their ability to articulate and convey to humans an awareness of that basic reality, the asymmetry of life and death and the moral obligation to subordinate the interests of our species to the broader interest of all life upon this earth.

Philosophy has come to grief whenever it took its task to be either one of describing the world or one of changing it. The quest for wisdom is more fundamental: it is a quest for a moral orientation within it, and for that practical reason is primary. The claim of theoretical reason that there is neither good nor ill because such reason does not recognise these categories denies the possibility of such orientation – and so of reasoned action. The recognition of the green goodness of life and of the parasitic badness of death is the basic step toward generating such an orientation. The opposition between those two claims may be as asymmetrical as that between death – and life.

Philosophical Faculty, Charles University, Prague/CZ (Boston University, Boston, MA, USA, emeritus) (1993)

ERAZIM KOHÁK

THE ECOLOGICAL DILEMMA: ETHICAL CATEGORIES IN A BIOCENTRIC WORLD

Ecological ethics has grown of a double root, one factual, the other moral. The factual root is the recognition, amply attested by the successive reports of the Club of Rome and by ecological conferences from Bergen to Rio, that the commitment of global civilization to ever expanding levels of consumption on a finite planet cannot but destroy its livability for humans and as well as for other species which have inhabited it since the extinction of the great reptiles. Whether we consider the wholesale destruction of biodiversity, the destruction of forests with the attendant deterioration of the atmosphere or the destruction of the ozone layer, the conclusion is the same. Humans have become a cancer, a life-form which has lost its capacity for self-regulation and which goes on multiplying exponentially until it destroys the host organism, in our case, the Earth. Civilization as we know it is both self-destructive and destructive in ways which the planet cannot sustain. That is simply a fact, in itself morally neutral. Our species is destroying itself and other life with it. Civilizations have self-destroyed before. Civilizations do. So have species. Species do, too. So?

The other root of ecological ethics is moral – the recognition, equally clearly presented by authors as diverse as Albert Schweitzer, Aldo Leopold, Rachel Carson and a host of contemporary ecological writers, that this is grievously wrong, profoundly immoral. Over the past three decades, that recognition has grown from spontaneous grief over the agony of our planet into a philosophical position committed to what, with Hans Jonas, we could call *the imperative of responsibility*. It claims that, in light of the factual destruction all around us, we humans can no longer consider just our own short range interests, letting nature look after itself, as we have for millenia. There are too many of us, making too great a demand. As Peter Singer argued forcefully in his *The Expanding Circle*, we need to extend ethical considerations to other life forms and to all natural environment if we are to survive at all.

Nor is that enough. Once we recognize the applicability of ethical, not merely prudential categories to other life and other being, the time-honored distinction between humans (as ends, subject to ethical considerations, as for isntance in Kant) and other beings (as means for human gratification only, entitled to no respect in themselves) becomes untenable. Willy-nilly, we argue ourselves into a *biocentric* position which holds that humans are but one among the many species inhabiting this planet and that human interests and

R.S. Cohen and A.I. Tauber (eds.), Philosophies of Nature: The Human Dimension, 293–301.

categories can claim no privileged position among them. "For the good of humanity" simply is not a good enough justification any more. The world does not exist just "for the good of humanity." That is only a human prejudice, the natural egocentrism of our species. If our species is only one among many, then human desire for ivory – or for additional pastures – can claim no higher status than the elephant's desire for life, liberty and the pursuit of happiness. Human prejudices have no privileged validity.

Here, however, is the catch. Are not ethical considerations one of those human prejudices, not shared by other species? Many writers have argued that. The *locus classicus* may well be Sir Charles Sherrington's thoughtful study, *Man [sic] on His Nature*. Altruism, understood as the ability to take the wellbeing of others into consideration in making decisions, enters upon the scene only with humans. Ethical self-restraint is a thoroughly anthropocentric category. Annie Dillard, in her *Pilgrim at Tinker Creek*, shows it in all its merciless beauty. Her chapter on "Fecundity," vividly presenting the massive overproduction and overkill which nature uses to maintain a balance, is particularly striking in its fascinating horror. Nature knows no pity or, for that matter, no cruelty. It fosters a mad, senseless proliferation of life which it then destroys in particularly gruesome ways. Watch a chipmunk attacking a frog paralyzed by fear: yes, chipmunks are omnivores. The chipmunk knows no pity, any more than the owl who devours it in its turn. That is simply the way it is, neither good nor bad. Ethical categories are a human peculiarity which has no place in a biocentric universe.

Hence the dilemma. We can adopt a thoroughly biocentric position, knowing ourselves for but one of the species that dwell upon this earth, along with brother bear and sister coyotte. It is a noble position. However, if we assume it, then we can no longer say that it is morally wrong to destroy the planet. That is just our prejudice. In a biocentric universe, moral categories simply do not apply. There is only the hard reality that species come and species go, having their time in the sun and displacing each other in their turn. Perhaps we shall exterminate the whales and the elephants – it is rather likely – as we have already exterminated the dodo and the carrier pidgeon, and then shall perish ourselves in our turn, making room for the new species that will eventually evolve. Perhaps it will evolve out of insects, say, ants, one of the few species to have colonized the entire globe. So we, too, once evolved when the disappearance of the dinosaurs gave mammals their chance. Perhaps it is time for mammals to give insects a chance. Perhaps the all-devouring greed of the consumer society is simply Nature's way of giving insects their chance. Privately, we may well grieve it. I love porcupines and otters; I hate to see them perish. That, though, is just a private, subjective emotion, with no moral standing in a biocentric universe. Altruism or, for that matter, ethics, which Aldo Leopold defined as voluntary self-restraint for the good of another, is an anthropocentric phenomenon which has no place in a biocentric universe.

There is, of course, the other option. We can adopt an anthropocentric position and claim a privileged validity for the categories of our species. There is a hint of that even in Sherrington: we are what nature strives to become, the product of evolutionary effort.

For all other species, the passing and perishing which is the rule of evolution may be simply a fact of life. For humans, in a human perspective, the ecological devastation of the Earth is morally heinous. We have a moral obligation to restrict our use of fossil fuels, our ozone destroying emissions and, most of all, to limit both our numbers and our individual and collective demands so that life as we know it may survive upon this planet. Save the whales! Save the woods, save the Earth, save our souls!

Such a position is, surely, far more emotionally satisfying. However, it does make the moral imperative to preserve the planet a function of *human* preference. Humans are the source of all value. It is only in a human world that saving the whales is imperative. If that is what humans find satisfying, it also becomes obligatory. Taking that position has some distinct advantages. For one, since we are concerned solely with human preference, within an anthropocentric perspective, we can ignore the awkward question to which ecological efforts tend to lead in a biocentric context, namely, "What are you trying to do: freeze evolution at the year 2000, like King Canute tried to stem the tide?" The whole matter of our impact on the biosphere is irrelevant. We are concerned only with human preference, and, for the most part, we would hate to see the whales become extinct.

That, though, only sharpens the dilemma. *If humans are the source of all value*, the measure of all things – which we must assume if we wish to make human categories applicable to all being – *how can we justify subordinating their interests to the interests of other species*? If humans are the source of all value, how can we justify restricting Japanese fondness for whale meat, Chinese desire for rhinoceros horns or American infatuation with motor cars, fossil fuels and carbon monoxide just that whales may live?

Hence the dilemma. If anthropocentric categories are privileged, then human interests override those of other species, and we might just as well drive a gas guzzler to a whale steak dinner. However, if biocentric categories are privileged, then the ecological destruction of the Earth and the self-destruction of our species is only a fact, not a tragedy – and we might as well drive a gas guzzler to a whale steak dinner. One way or another, we might as well conclude with Heidegger that "*Nur noch ein Gott kann uns retten*" – "Only a god could save us" – and leave all ecological considerations to Her, as the greediest consumers and the worst poluters have insisted all along. Leave it to God, leave it to The Invisible Hand! After all, is not trying to save the Earth sheer anthropocentric *hubris* which must yield to biocentric humility?

The ecological dilemma is not just an academic exercise in logical reasoning. It is also the rationalization used by all who would defend their comfort or their profit against any restraint that the survival of the Earth may demand of them, even in matters as trivial as driving an energy efficient automobile rather than the customary insolent chariot. What are you trying to do, freeze evolution? Change human nature?

On the face of it, the ecological dilemma seems so invincible that it is tempting to turn away, like the Epicureans and the Stoics confronting the progressive disintegration of the glory that was Rome or like Comenius amid the horrors of the Thirty Years' war, from *Labyrinth of the World* to *The Paradise of the Heart* and to seek a *Centrum securitatis* within. Within the ecological movement today, there is a strong tendency to leave behind the insoluable problems of the dynamics of expanding economy and to seek solace, Thoreau-like, in the great green peace of the wilderness.

As a strategy of personal renewal, a return to nature has a deal to offer. So does the modern equivalent of the Epicurean strategy, a return to the primaeval depth of our soul, reconnecting with the deep sources of our collective unconscious, the dark, nurturing source of being. Unable to change the world, humans have ever turned inward, seeking to survive by changing themselves. As the full extent of the ecological predicament, its roots in our entire cultural orientation, become ever more evident, yesteryear's activists, despairing of changing the world, are turning inward, seeking to escape to the grand harmony of nature without and the naturing richness of nature within. As for the survival of the Earth, *Nur noch ein Gott kann uns retten . . .* Certainly recycling our plastic bags is not going to do that when the whole dynamic of the economy and of demographics alike drive relentlessly to further expansion.

In some ways, the debate within the ecological movement today between the practical ecologists and proponents of deep ecology is reminiscent of the debate between the social democrats and the Communists at the turn of the century. Back then, the social democrats advocated political involvement to bring about concrete improvements in the condition of the workers, old age pensions, medical insurance, affordable housing, job security, all that Marx summed up as "trade union consciousness." The Communists dismissed all such piece-meal reformism with boundless scorn, insisting that the flaw was not just in this or that practice but was built into the very nature of the system. All reform, they insisted, is vain until a fundamental revolution transforms the capitalist mind-set of our society. *Nur noch eine Revolution kann uns retten*! In the meantime, the worse, the better: with every deterioration the revolution is one step closer.

It was a persuasive argument then, and it is so again today. The flaw really is built into the very nature of the system. Exploitation of the earth, just as exploitation of the laboring class, feeds upon itself, turning ecological morality into eco-business. However, given our experience with the last revolution, we might well hesitate to resign ourselves to waiting for another one or, for that matter, to waiting for Godot. Is the ecological dilemma really so inescapable?

The ecological dilemma on the one hand claims, with Hobbes, that in a civil society no writ can run against the king – or, less metaphorically, that in an anthropocentric context all values derive from human desires and so no limits can be put upon the latter in the name of any other being.

On the other hand, it claims, again like Hobbes, that there are no moral rules in the state of nature – or, again less metaphorically, that in a biocentric context only considerations of survival obtain. Species come and species go; perhaps it is time we went. One way or another, the conclusion is that moral categories are irrelevant to ecological considerations.

Are they? Come, let us reason together. The ecological dilemma depends on a tacit of Hobbesian assumption which a Locke would not share and which we would be well advised to examine critically. It assumes that moral categories are intrinsically anthropocentric – that is, *that good and evil – and moral obligation to do the one and to avoid the other – are simply human ways of constituting what-is as a meaningful whole*, applicable only to human dealings with other humans, not to humanity as a whole, as a biological species, in relations with other species. Less obscurely, it is the assumption that there really is neither good nor bad but thinking makes it so. It is *thought which constitutes moral categories. Therefore moral categories are applicable only within human thought, not to it*. Or such is the claim. How strong is it?

Quite marginally, we might note that the tacit assumption that non-human beings do not think is factually highly questionable. William James effectively denied it already a century ago in his definition of "mentality" in *Principles of Psychology*. Since then, empirical research has demonstrated the presence of thought at least in higher mammals such as wolves, bears and dolphins. That, however, might require no more than that we extend Kant's category of rational beings from humans, Martians and angels to other *animalia* without affecting the core of the argument, the claim that in an anthropocentric world non-rational (or "means") beings have no claim to moral consideration while in a bio- or ontocentric one there are no moral categories to require such consideration.

Whatever may be the case with animal thought, I believe it is a different aspect of the ecological dilemma we need to consider, namely, the assumption that good and evil are merely individually subjective categories, contingent on the presence of thought, whoever all may be doing the thinking.

Let us admit, first of all that good and evil (and moral predicates generally) do indeed indicate relational realities, as John Lachs insists. Something can be said to be good or bad or otherwise value-indexed only in relation to some activity. In a universe perfectly at rest, in the absence of any movement, any change whatsoever, there is no conceivable way we could make sense of value terms. And, while we are at it, we might as well also admit that using words in order to label particular relations as good or bad is indeed a function of thought, since that is no more than a tautology. The question is not whether humans communicate their value-indexing by using words, since clearly they do. It is, rather, whether the relations so indexed can be distinguished from each other independently of such labeling or whether it is only the act of linguistic labeling which makes them distinct from each other. Is reality (in relation to an activity) intrinsically good or bad, or is it intrinsically neutral, becoming good or bad only upon being labeled as such?

Some value relations are clearly a function of labeling and of arbitrary individual preference (relations of "I like") or of socially generated conventions (relations of "people like.") Granted. However, even in the case of reflecting humans, convention and preference are not where the process of value-indexing begins. Even in the absence of reflective thought, the context of human activity, what-is, constituted as our *Umwelt*, is not value neutral – or else babies would not cry before the age of reason. Less frivolously, because humans enter the world as purposive agents, they orient their surrounding world as good and bad simply in virtue of their purposive presence, their agency, quite independently of any reflective awareness. That agency may be as minimal as breathing, at the moment a child draws its first breath, yet even that is enough. Though the child may not be reflectively aware of it, it is already experiencing some aspects of the world as desirable, others as noxious, some as a help, others as a hindrance. By the time human beings come to reflect upon it, they have already thoroughly value-indexed their experienced world. *Value, though a relational reality, is not relative in the "relativistic" sense because it is not a function of arbitrary preference. Rather, it is defined by the structure of our situation, independently of reflection and of reflective preference.* Like it or not, some things sustain, other things hinder the activity of living. There is good and bad in our lives whatever we think of it.

Nor is that true of humans alone. A woodchuck (*marmota marmota*) at its grazing may be wholly innocent of the *Nichomachean Ethics* and not even know its Latin name, yet its grazing still constitutes the grass as good and the spilled gasoline as bad as well as the whole meadow as a meaningfully ordered, value-indexed whole. Though it lacks words so to label it, the woodchuck's varied response testifies that the relations structuring its world are indexed in terms of desirability. Though relational, value is not arbitrary because value relations are not that. They are initially constituted prior to and independently of the entrance of reflective awareness (and so of freedom) upon the scene. In Paul Ricoeur's long-ago terminology, their constitution is involuntary. They are a function of the agent's purposive presence, not of the agent's arbitrary will.

Afficionados of Husserlian phenomenology will find it interesting to note that that description quite precisely fits the reality of which Husserl speaks as the *absolut fungierende Subjektivität* in *Krisis IIIB* § 72. It is *subjectivity* – not a "subject," but subjectivity, a focused network of meaning relations – *but it functions* – that is, constitutes what-is a meaningful whole within which differentiated response is possible – *really* ab solo, *of itself, not as a function of an arbitrary act of a putative reflecting subject.* Husserl's *Lebenswelt* – or better, *Lebens Welt, die Welt des Lebens*, "life's world" – is a meaningful whole constituted not by a reflective act but by the purposive presence of life, functioning simply as agency – the absolutely functioning subjectivity which, to make it a bit more obscure still, Husserl chose to call "transcendental."

That is relevant in our context in part because to a true *afficionado* Husserl is always relevant, but in part also because it helps explain how value can be a relational reality, yet at the same time both non-arbitrary and *interspecifically applicable.* The goodness of love – or generically, relation of care and respect for a significant other, by whatever name we call it – is not a function of a given person's preference. Actually, many people like being nasty. In spite of that, such relation is good for any being whose mode of being involves transaction with others, whether that being is a human, a coyotte or a Canada goose.

We are left, though, with a different problem: does not the ability of a relation to constitute value depend on the intrinsic worth of one of the relata? Being *good for* something does not yet make an X *good* unless that something is itself – and since an infinite regress solves nothing, that means good in itself. And, if that is so, are we not right were we started, forced to admit that a relational value is wholly arbitrary unless one of the relata is valuable intrinsically? And is not the very idea of an intrinsic value – that is, of a non-relational relation – a contradiction in terms?

Yes and no. Yes, in the sense that all relational value does need be anchored in a value which is autonomous – or, for *afficionados* and lovers of obfuscation, "transcendental" – and independent of any existing entity. No, since such value, though intrinsic, need not be that square circle, a non-relational relation. While a value *for* . . ., it can also be a value *for itself*.

Actually, being a value *for itself* is not something nearly as dramatic as Jean-Paul Sartre would have us believe. It is not even a distinctive trait of the human mode of being. Actually, it is the most generic trait of all life. Every living being, in its strenuous effort to remain alive – grass reaching for water along hard-trodden path, the mosquito seeking to avoid the descending hand, the prisoner seeking food – testifies that its own life is a value for it. Yes, there may come a moment when a particular individual individually decides that the struggle is no longer worth it, that the time has come to give up. That, though, is an individual judgement and decision within the structural matrix of life's quest for self-preservation. The prereflectively given rule of all life is that life is a value for itself – and as such, a value in itself, internally, quite independently of the existence or of the acts of any other being whatever. *Life is good in itself because it is good for itself. Wherever there is life, there is value.*

That is the point: a biocentric cosmos is not a value-neutral one. Though, as far as we currently know, apart from humans and their peculiar modes of communication no one goes around calling things "good" or "bad," the reality of good and evil is present whenever life is present. As good for itself, life is good in itself, and whatever sustains life in its conflicted diversity, in the spirit of Aldo Leopold's classic definition, is good. Humans, the species which differs from other species in its conscious recognition of good and evil and in its moral obligation to choose one and abstain from the other, cannot free itself of its responsibility by adopting a biocentric perspective. Though the obligation to choose the good and avoid the bad is distinctively human, the categories of good and bad are as universal as life itself.

That, I believe, disarms one horn of the dilemma – and, consequently, the other horn as well. If we choose the anthropocentric option, claiming that human moral categories are privileged, we, being human, are still obliged to act according to them, choosing responsibility, respecting all value. The only restriction which follows from anthropocentrism is that a porcupine is not obliged to act morally, not that humans are not so obliged in relation to the porcupine. Morality demands love and respect to all being, because, as good for itself, it is good in itself. Simply in terms of the logic of the argument, the ecological dilemma proves to be something of a paper tiger.

The strength of an argument, though, is seldom in the theoretical cogency of its theses but in the way it gives voice to lived experience. Certainly, there are people who use the ecological dilemma quite callously in the service of greed, much as tobacco manufacturers once used the distinction between causal and statistical evidence. However, they can do so not because the ecological dilemma is so logically rigorous but because it expresses something of which, paradoxically, precisely the effort to save an endangered nature made us aware – the finitude of all being. The more we strive to assure that whales, elephants or the pensive forests of northern Bohemia do not perish in our current paroxism of greed, the more we become aware that, ultimately, we shall fail. Species come and species go. Regardless of our efforts, the time will come when there are no more whales or elephants, when there are no more forests in Krkonoše. All who have striven to resist the ravages of time know that experience. Paul Ricoeur once wrote of it as *la tristesse du fini*, the melancholy or sorrow of all that is finite. The ecological movement is not suffering from a logical dilemma. It is suffering from the sorrow of finitude. Precisely because, like the physician, it has sought to defy Death, precisely because it has been successful in so many ways, it has had to face the fact of its ultimate failure. Species come and go, and the time of the whales, the elephants and the humans will come, too. It is that recognition – and not the putative cogency of an argument, which leads to resignation and to the current turn to personal solace rather than public solutions.

And yet, writes Robert Frost "when to the heart of man [*sic*] was it ever less than a treason . . . to bow and accept the end of a life or a season"? Is the fact that humans are mortal reason for committing suicide or for withholding medicine from another? Surely life is not about being forever, it is about being lived. The question is not whether the world will ultimately perish, as surely it will, but whether our presence upon it will have been a help or a hindrance to the living of life in all its conflicted harmony. The fly I refrain from killing will surely die . . . but is that any reason why I should be its murderer?

The ecological dilemma, I believe, is finally based on a willful misunderstanding. Biocentrism does not mean adopting the theoretical perspective of the natural sciences which abstract all value-indexing from the world a priori and then claim to have discovered a value-free reality. There is no value-free reality, wherever there is life. In turning to a biocentric perspective, we remain within a meaningfully ordered, value-indexed world, life's world. We are not denying the reality of value, only claiming it is constituted by life as such, not by human reflection only. Nor does anthropocentrism entail an absolute precedence of human whims, but quite the contrary, the applicability of human moral categories to *all* human transactions, including those with the non-human world.

Yes, whenever humans contend with Death, whenever they dare love, cherish what is finite, they necessarily encounter the sorrow of finitude, the recognition that love's labor is in the last instance always lost. But death does not mean that life is not worth living, and the ultimate demise of the universe does not mean that nature is not worth preserving. Nor, most emphatically, does it excuse destructive behavior on the part of humans. Because we are human, we are bound by moral considerations in all our actions, even though non-humans may not be similarly bound. The true meaning of anthropocentrism is that I, as human, am not free to kill a tiger even though that very same tiger may feel no compunctions about killing me. In turn the true meaning of biocentrism is that human values are only a special instance of the value-structure of all life's world. Our special obligation does not entail special privileges.

There is no ecological dilemma. There is only the human predicament – that the more intensely human love, the more intensely they feel the pain of finitude. The greatness of ecology is its willingness to extend that love to all creation. Its ethics needs to be anthropocentric in the rigorous moral demand it makes on humans, and biocentric in the respect it extends to all life. That is not a dilemma, not even a contradiction. It is, however, a challenge.

Philosophical Faculty, Charles University, Prague/CZ (Boston University, Boston, MA, USA, emeritus) (1993)

ERAZIM KOHÁK

CREATION'S ORPHANS: TOWARD A METAPHYSICS OF ARTIFACTS

The purpose of these pages it to focus philosophical reflection on the humblest denizens of our life world, the lowly artifacts. Nor do I wish to speak of the more picturesque of their kind, crafted by hand and endowed with a derivative humanity by long caring use, the birchen bowl, the long-cherished enamel coffee pot, the teddy bear or the earthenware pitcher in which, as Heidegger would have it, the potter's hand and the fruit of the vine focus the richness of the earth, the vastness of the sky, the glimse of the holy and the bittersweet awareness of mortality.[1] I am concerned with, so to speak, the "second generation" artifacts, the anonymous objects stamped out in their thousands by automated machines, distributed *en masse* and discarded in the same manner – plastic cups, rental typewriters, last Christmas' video games and cheap tin ashtrays – that invade our life world like anonymous aliens.

Artifacts in that sense – "objects" in Heidegger's terminology, as contrasted with things – have always posed something of a problem for Western thought. Angels, puppy dogs, trees and boulders, even the products of arts and crafts, are much easier to deal with. In some basic, deeply felt sense, they *belong*, they have their place in the order of things. To a religious believer, they are our fellow creatures, lovingly crafted by our God to declare God's glory and worthy of respect as such. Even to a person for whom God is not a living presence, they remain an intrinsic part of nature – or perhaps Nature – endowed with a function, serving their role, and so belonging. There is something personal about them. But artifacts? Garrish machine-made baubles of chrome and plastic, they lack even the aesthetic justification of a work of craft or art. What of them?

Western philosophic thought, if it deigned to notice them at all, always treated artifacts with despect. Heidegger speaks of *Zeuge*, usually rendered "gear," whose being is exhausted by their being ready or unready to hand and noted only in use, devoid of intrinsic reality of their own.[2] Gew-gaws, junk, baubles, widgets were not created, they did not grow, they were not even crafted. For the most part, they happened as accidental by-products of human intention, designed and produced not as an end but as a means only. At the same time, though, such artifacts have increasingly come to play an important, at times even a dominant role in human lives. No sooner do humans satisfy

R.S. Cohen and A.I. Tauber (eds.), Philosophies of Nature: The Human Dimension, 303–314.

their most basic needs for food, shelter, and security than they turn, as often as not, to the accumulation of widgets to give a continuing purpose to their lives. In the quaint language of a century ago, it is a "fetishism of commodities." Though we may flatter ourselves that the glory of America is its freedom and justice, empirical evidence suggests that it is the plethora of widgets that the world most envies us: an American ball point pen, not what is written with it, is the status symbol beyond these shores. Nor, closer to home, do the purchasing patterns of American consumers contradict that evidence. Philosophers may be too busy to notice it, but merchants know it well: it is the lowly widget, soon to end in a yard sale or a dumpster, that is the unmoved mover of our society.

Nor is the increasingly prominent presence of artifacts confined to our collections of fetishes. Even more prominent are the artifacts which increasingly displace physical objects in our effective environment. In Milan Machovec's metaphor, we walk on asphalt, not on the good earth, look up at neon, not the starry heaven.[3] Our life world is being invaded by beings which remain intrinsically alien to us. Here again a metaphor might be helpful. An alien, let us say, is a being whom I cannot address as "Thou," recognize as a fellow being endowed with an intrinsic worth, intrinsic dignity of his – or her or its – own. But the artifacts with which we are increasingly furnishing our world are precisely that. They come into being solely to serve as tools, as means rather than as ends. Increasingly, we live in an effective life-world which is devoid of intrinsic meaning, intrinsic dignity, a world of concrete cubicles arbitrarily lit by artificial light, a world in which only an occasional potted plant or a caged animal serve as reminders of the living world whose part we once were. What is more, there is every indication that, if humankind survives at all, it will be increasingly in man-made environments. To the extent to which philosophy, in spite of the temptation to become a *techne*, remains the Socratic attempt at understanding, it might face no task more urgent than to formulate a strategy for being human in an increasingly dehumanized context. How do we respond to a world devoid of meaning, anonymous and artificial?

Cautiously at best, and often with a distrust tinged with fear. Unless we are prepared to resign our own moral humanity and accept the role of producers and consumers within an automated system, we cannot but feel ill at ease in such a world and seek refuge, in fact or in fancy, in a world that is still God's, nature's and the craftsman's world. "There still is night," I wrote, nearly a decade ago,

> "down where the long-abandoned wagon road disappears amid the new growth beneath the tumbled dam, deep virgin darkness as humans had known it through the millenia, between the glowing embers and the stars . . . Here a human can dwell at peace with his world, his God and himself . . . In the global city of our civilization, girded by the high tension of our powerlines, we have abolished the night. There the glare of electric light extends the unforgiving day far into a night restless with the eerie glow of neon . . . The world of artifacts knows neither a law nor a rhythm . . . We ourselves have constructed that world for our dwelling place, replacing rude nature with the artifices of *techne*, yet increasingly we confess ourselves bewildered strangers within it."[4]

Against that background, it is easy and entirely understandable to link the defense of our personhood with the defense of the natural. Thousands of readers are rediscovering Tolstoy and Thoreau, Annie Willard's *Pilgrim at Tinker Creek* has gone into its tenth printing, craft shops do booming business and the dividing line between personalism of the human world and personalism of the world of nature grows increasingly tenuous. Not without reason: as Hannah Arendt understood clearly, depersonalization of our life world goes hand in hand with the depersonalization of human relations. The strategy of linkage – reaffirming our personhood by respecting the "personhood," now in the sense of the intrinsic worth and dignity of all creatures – represents a cogent alternative to the strategy which the West has followed ever since the Enlightenment, that of the "conquest of nature" and of replacing rude nature with the works of artifice.

Nor is that strategy alien to the spirit of personalism. The core of personalism has always been the recognition, in Borden Parker Bowne's phrase, of "Person as the ultimate metaphysical category." More generically, personalism is an affirmation of the ontological and epistemological primacy of "personal" categories not simply in describing the community of humans, but in describing the ultimate nature of reality.

American personalists have not typically resorted to that idiom, though other writers, such as Max Scheler, often have. Ironically, the clearest delineation of the position may well be in a book whose intent, I believe, was not personalistic, Edmund Husserl's *Ideen II, Phënomenologische Untersuchungen zur Konstitution.*[5] Husserl, like Scheler, notes that the contents of our experience are capable of being subsumed under several distinct categorical schemata. One is that suggested by the realm of the inorganic, centered on the metaphor of matter in motion and relying on cause and effect as its basic explanatory categories. Another is that suggested by the realm of living beings, centered on the metaphor of need and satisfaction and relying on teleological explanations. The third is that suggested by the realm of the works and interactions of humans, centered on the metaphor of the Person, the moral subject, and relying on categories of understanding rather than explanation – and an understanding in moral terms, within the context of a moral law. This is the realm that Husserl, Scheler and their spiritual kin designate, interchangeably, as "*geistig*" or "*personalistisch.*"

Personalism, however, goes a step beyond the recognition of the plurality of categorial schemata. By itself, such a plurality could be simply acknowledged as reflecting the irreducible plurality of symbolic forms, as Cassirer tends to do.[6] Personalism recognizes that while the rule of pure reason may be "both-and," the demand of practical reason is "either/or." The realms interact, raising the question of the relative ordering of categorical schemata. What categories, for instance, shall prevail in the interaction of humans with the realm of the organic? Should we argue that the vital categories of need and satisfaction here override moral categories, so that efficiency of production – conversion of grains into animal proteins for human consumption – overrides all categories of compassion and respect for our fellow creatures, so that the most unspeakable cruelty is permissible in dealing with "biomechanisms"? Or shall we argue that moral categories are primary and govern human interactions not only with each other but with the non-human world as well?

Personalism represents nothing less than the affirmation of the ontological epistemological and ethical primacy of "personalistic" categories. While Western thought since the Enlightenment, if it acknowledged the legitimacy of moral categories at all, treated them as a special case theory appropriate to a special realm only, the noumenal for Kant, the emotional for Schleiermacher, personalist thinkers like Scheler affirm them as fundamental. Describing the world in "personalistic" categories of value and meaning, governing its interrelationships in terms of respect and compassion, represents, for personalist thinkers, not a special perspective on an ultimately meaningless reality, but an articulation of the ultimate structure of reality itself. It is the categories of the natural sciences that present a partial, special perspective. Reduced to an aphorism, personalism could be said to be the view that ethics is the general, physics merely a special science. Or again, to speak of the cosmos as a community of persons is a factual description, to describe it as an aggregate of matter in motion a special perspective from a particular viewpoint.

So defined – and it is, admittedly, the most radical definition I could muster, short of doing outright violence to the tradition – personalism does indeed yield a perspective continuous with a recent rediscovery of the moral sense of nature. In a barest outline we could say that the cosmos is a society of Persons governed by a moral law: the command of compassion and respect between beings of intrinsic moral worth. We become aware of it, thanks to the double gift of reason and empathy, in our relationship with our fellow humans, but it governs our relationship to all created beings. It is in recognizing the intrinsic personhood of all creation that we affirm and fulfill our own personhood.

A reading of recent intellectual history and our present predicament follows from that basic perspective. The decision of the Enlightenment to substitute a strategy of conquest for respect and compassion, reducing the non-human world – in the age of colonialism, even non-European humankind – to a reservoir of raw materials, humans transformed the community of humans from an intrinsic part of a meaningful world to an untenable anomaly in a meaningless one. As step by step they – we, really – have substituted artifacts, devoid of intrinsic worth and dignity for natural objects, they created an illusory image of themselves as the makers of all things and the source of all meaning. Technology fulfilled a part of the serpent's promise, "Ye shall be like Gods" – for, in a world of artifacts, that is the role humans assume. But the burden has become too great, the situation too anomalous. There is no room for a moral subject in a world of artifacts and so, increasingly, humans have sought to resign their humanity and to conform to the ways of their anonymous, impersonal products, the artifacts.[7] The strategy for reversing the trend is thus appropriately one of "back to Nature," rejecting the anonymous world of artifacts and rediscovering the moral law and our own personhood in the context of natural beings, endowed with intrinsic dignity.

There is, I believe, enough truth in such a reading that it must be taken most seriously. Yes, I believe, this is ultimately a cosmos, a community of beings endowed with an intrinsic worth and dignity, and governed by a moral law. Yet, I believe, it is in recognizing our fellow beings, human or not, as Persons that we become aware of the moral law, thanks to the double gift of reason and empathy. Yes, in a world of artifacts, devoid of a dignity of their own, we become blinded to the moral law and acquire a wholly distorted vision of ourselves – and it is the world of natural beings that can recall them to us. All that appears to me true, for good and sufficient reason, and, in the language of the Prayer Book, "worthy of all men to be believed."

And yet I find the romantic conclusion that condemns the world of artifacts as a work of the devil and commends a return to "nature" as the sole strategy for preserving our personhood deeply problematic, and that for several reasons. One is historical: American personalism, especially in the writings of Borden Parker Bowne, did not include any clearly defined ecological ethic. Bowne fully shared the "technological optimism" of his age: though a farmer's son, he was a man of the city, not a ruralist, and welcomed the precipitous technological change of his time – which he witnessed at its most horrendously rapacious – as "progress." The non-human world appeared to him as less than fully real, and he was fully prepared to integrate his personalism within a 19th century strategy of a "conquest of nature." Bowne clearly saw something that we do not see.

A second, related problem is ethical. For all the problems it has brought on, technology has clearly improved the lot of humans. There is something deeply disingenuous about the romantic image of the mediaeval peasant or the eighteenth century craftsman. Their lives, for the most part, were not poems of harmony with Nature. They were nasty, short and brutish: human life had little value and that of animals none at all. Both the peasant and the craftsman were ruthless – albeit relatively inefficient – exploiters of the natural environment, driven by exploitation they suffered. Technology has humanized – as well as dehumanized – our lot.

A third problem is metaphysical. What, finally, would a "return to Nature" mean for humans? A return to the technology of crafts and subsistence agriculture? The problem is not that the human has, throughout his history on this earth, been a maker of artifacts, a *homo faber*, but that, as free, the human is intrinsically an "un-natural" being. Knowledge and decision stand between him and nature. Even nature's moral law he does not obey "naturally," instinctively, automatically: he must recognize it through reason and empathy, and *choose* to obey it. Whatever return is called for – and however nature can aid it – it cannot be simply a return to nature.

And, finally, there is the practical problem. Though the Enlightenment turn may have been unfortunate, it has become irreversible. Short of a global catastrophy which would wipe out all but a handful of the human inhabitants of this earth, humankind, in its present numbers, can survive only as a race of city-dwellers, calling upon the resources of ever more sophisticated technology and, consequently, living in an ever increasingly "artificial" world. Were the only solution to the dehumanization of technological civilization a return to nature, the problem would be insoluble.

All these are serious problems, and especially so because I believe that the problems which the nature-oriented component of personalism seeks to address are real and critical – and that its basic analysis is sound.

Let me start with the Enlightenment which today appears, to writers as diverse as Emanuel Rádl and Peter Singer,[8] as almost the original sin of modernity but which did not appear problematic to Borden Parker Bowne. On Rádl's and Singer's interpretation alike, that turn appears as a bolt out of the blue, an explosion of *hybris* which drastically altered the course of civilization from one of symbiosis to one of conquest. I, too, have so interpreted it, yet such a reading is problematic. The Enlightenment, finally, was in a real sense a rebirth, a rediscovery of themes present earlier in the history of Western civilization. In this case, it was a rediscovery of a Stoic motif. The late Roman Stoics, Marcus Aurelius as an example, anticipated the motif of a deep seated conflict between humans and nature. They conceived of humans as bearers of the divine *logos*, of reason, and of nature as dark, threatening and fundamentally irrational. The metaphor of struggle or battle appears not infrequently in Stoic writings, and is given a moral significance: the task of humans, the bearers of reason, appears as one of subduing nature's irrationality and imposing reason's rule upon it. In making straight the road – and the Stoics built with a vengeance, roads, bridges, hospitals – humans conceived of themselves as imposing moral rules on nature's chaos. Given the technology available to them, their impact was minimal and was soon wiped out by the invading Goths and Vandals – was that a return to Nature? – but their theme was one of conquest – and its motivation eminently moral.

When Enlightenment thinkers took up the Stoic themes, their motivation was still the same, fundamentally moral, though their technology was different. That technology, though, did not spring full-grown from the forge of Vulcan. Only in nineteenth century fantasies do "means and modes of production" determine social development: in reality, the order is reverse. What the Enlightenment had that the Stoics lacked was centuries of Christian heritage, with its insistence on the intrinsic worth of concrete, individual, embodied humans. After all, the Christians believed that this created world, though marked by passing and perishing, was so precious that God gave God's only begotten Son for its sake, and Godself became incarnate. To the Stoics, Reason may have been precious, but individual human lives were not and so the idea of labor saving would have held no particular appeal to them. Enlightenment thinkers, steeped in Christianity, saw it differently. To them, sparing needless human drudgery appeared as a moral command. Undoubtedly, motifs of human greed contributed powerfully to the momentum of the technological development, but those are perennial motifs. The distinctive Enlightenment combination was that of rationalization and labor saving, giving the distinctive Enlightenment direction to the perennial human striving for the goods of this world.

Seen against this background, the Enlightenment turn to "conquest of nature" appears in a different light, no longer simply as an expression of human rapacity but as the acting out of a moral impulse as well. It was this latter aspect of the industrial revolution that enabled even thinkers like Bowne, keenly aware of the primacy of the personal, to perceive the explosion of technology as "progress" – in spite of the wholesale devastation it also entailed. There may be, as the success of the Amish testifies, a yet more excellent way, but the late nineteenth and early twentieth century saw around itself a humankind altering what for millenia had appeared unalterable, famine, pestilence, poverty, drudgery and all that condemned the majority of humankind to the short, nasty and brutish life of precarious drudges. The luxury of concernn for the environment and the personalistic quality of life was bought for us precisely by the technology which successfully challenged the dark, dehumanizing aspects of nature of which the Stoics had been so keenly aware. The plight of the third world should serve as a reminder of what our own lot had once been. Though it might be possible to argue that the strategy of restricting needs and rationalizing agriculture, acted out hy the Amish, would have been preferable had humankind followed it consistently since the early seventeenth century, the fact is that it did not. Today, the reversal of that strategy – wholesale abandonment of technology – would produce nothing short of a disaster for the vast majority of humankind.

Nor – and that is the metaphysical problem – would it be easy to argue that the generation of artifacts represents a betrayal of authentic humanity. Rousseau's claim that "a thinking man is a depraved animal" may have been useful for its shock value, but it would be hard to defend. What we see around us suggests far more that it is the unthinking humans, their life no more than a series of conditioned responses to random stimuli, that are the depraved animals of their species. The mindless mob, the consumer wholly absorbed in the lock-step of need and satisfaction, extended by easy steps to the level of greed and gratification, that is what depersonalization is all about. Human freedom, which is precisely what transforms all human acts into artifice, product of free choice and responsible decision, is the definitive aspect of human nature, of human being as Persons. Rousseau's romantic conception of "return of nature" as a return to a putative pre-personal, pre-moral, instinctive level of being human would represent not a return to but a dramatic denial of such nature as we can attribute to humans. The ways we choose to use our power of artifice may be deeply problematic, but the power itself is not: technology, however alien it may have become, is also an authentic expression of being human.

And so also the last point: though the direction of our technological development may be susceptible to rechanneling, though we can set different priorities for our technological striving than the mindless quest for ever greater surplus affluence for a particular, rather limited community of the privileged, it is no longer within our power to reverse it. The original Enlightenment decision was not the original sin of technological civilization, solely an outburst of arrogance and greed held it check in previous centuries by the Church. It was a moral, morally oriented decision addressing intensely human problems which the mediaeval Church had ignored. The development which flowed from it, for all its flaws, had also an ethical value, holding out a promise of freeing humankind from the bondage of drudgery and disease. It was not "unnatural" but an expression of something definitive about the "nature" of humans: their ability to respond to ideal stimuli and to envision and produce alternatives to the ageless state of humans and societies. Nor, finally, is it a decision that we can reverse, even if we should want to. It is something we must live with – and, given its ethical aspect, can live with . . .

Unfortunately, even though all this may be true, it does not change the reality of the problem. If anything, such considerations make the problem more acute because they point out that humankind cannot escape the consequences of its technological turn by a romantic return to nature with which it had comforted itself ever since the problems of technology made themselves felt in the wake of the Napoleonic wars. Still the problem is real: humans living in a world of artifacts, of dead objects devoid of worth and dignity, devoid of a life and a rhythm of their own, easily lose the sense of the dignity of being and of the moral law that governs it. It is reason reflecting on the empathic identification with the Other that opens us to an awareness of that moral law. In a world in which the Other is an "it" and reason becomes the servant of technology, that awareness becomes exceedingly difficult. It would be easy, as Scheler – or, if you wish, St. Francis before him – say, to be a personalist in a world of nature: the only argument needed would be, *circumspice*! In a human-made world, a technologically produced world that is not God's but "man's," that argument would support depersonalization.

Should we then give up the attempt to remain Persons, moral subjects in a morally ordered cosmos, as a left-over of an earlier, pre-technological era? Or, as the existentialists of a generation ago, defiantly assert our personhood against the impersonality of our world? Looking over the literature of the past fifty years, those appear to be the options for dealing with technology – either reject technology and return to nature, (the Thoreau option,) or defy technology in a Promethean grandeur (the Sartre option,) or finally surrender to technology, becoming anonymous machines ourselves (the Watson option.) Personalism represents a significant contribution not in aligning itself with one of those options, though both "the Thoreau" and "the Sartre" lend themselves to it, but by presenting a distinctive option of its own, that of the personalization of the world of artifacts.

Here it behooves us to tread carefully, since "personalization of artifacts" all too easily conjures up images of painting faces and flowers on cruise missiles – and that is not what it is all about. Rather, much as the ecological movement in philosophy needs to generate a metaphysics of non-human living beings and an ethics of dealing with them, so a personalism of urban life must generate a metaphysics of artifacts and an ethics of living with them.

For far too long, we have treated artifacts as creation's orphans, and not without reason. If we set aside for the moment the special class of the products of crafts, artifacts are indeed beings which have no prima faciae root or place in the scheme of the creation – or, if you prefer, of nature. They were not created, neither did they grow or were crafted. They happened as the side effect of a specific intent. A carver may have intended this carving *in itself*, as an end, not as a means merely, and I may have appreciated it as such. No one, though, intended a computer or a cheap tin ashtray "an sich." The intention may have been data processing in the first case, litter control in the other. In either case, the artifact is a means, with no being, no intrinsic dignity of its own.

Or so it would seem – though it is not altogether so. Though for more than a century and a half, humans have now fancily applied the term "creation" to their production, the usage is inaccurate. Humans do not create: *creatio ex nihilo* is a privilege reserved to the Almighty God. The most humans can manage is to reshape, to mold a creation they receive as a gift. Their products are not creations, materialization of pure human intent. They embody and reshape other beings, even if it be no more than copper ore and fossils burned for energy. Artifacts, we could say, are uses to which humans put the created world – or, if you prefer a secular metaphor, "the bounty of nature."

That metaphysical recognition has an ethical consequence. Artifacts are not only products but also gifts, be it of God or of Nature. Their being has been bought at a price, be it of animals slaughtered, trees felled, ore mined. A gift, though, requires gratitude as a response. It is surely one of the most elementary prima facie obligations to treat a gift with respect. Though I might ordinarily ignore dandelions except as raw material for a salad, when a child hands me a wilted dandelion with an eager smile, "Look, beautiful flower, I brought it for you," I experience an immediate obligation to thread it in my buttonhole or place it in a glass with water. Though artifacts were neither created nor grown nor crafted, humans yet owe them the respect due to a gift.

It might, to be sure, sound farfetched to speak of my moral obligation to an aluminium beer can discarded by the roadside, yet that obligation is real. That aluminium, embodying both a prodigious amount of labor and a part of God's creation – or, in a secular metaphor, a non-renewable natural resource – is a gift. Though it might be my privilege to use that resource, that gift, it is immoral for me to waste it. Though in individidual cases a calculus of utility might be relevant, in principle I do not have a moral obligation to the discarded beer can to pick it up and deposit it in the recycling bin. I have an analogous obligation to the neat yellow-lined pad of well-made paper, to use it well, not to waste it, or to the coat in my closet, to wear it and care for it.

That recognition should lead to a related recognition – that I have an obligation not to neglect the artifacts that I have brought into my life-world. The reason has to do with the first metaphysical aspect of artifacts we have mentioned: artifacts are the uncreated beings, neither created nor crafted nor grown. They are wholly dependent on humans, their makers, for their meaning, and in producing and acquiring them humans undertake a responsibility. I can neglect a tree growing on my land: that tree has a life of its own, its own agenda. Though abandoned by me, it is not bereft. An artifact, the long-unworn coat in my closet, has no life except that which I give it through use and care. To leave it hanging there, unused until moths reduce it to rags, is, to be sure, immoral, not only because somewhere there is a person who is cold without it, but more fundamentally because I have taken of God's gifts or Nature's bounty not for use, but for waste. There is, though, a more far reaching consideration still: having no life of its own, the artifact not endowed with life by use and care becomes "dead" matter, impersonal, alien. It need not be that: it could become a friend of years of use – I have such coats, tools, pots. Unused, uncared for, however, it becomes alien and impersonal.

Here, I believe, we are at the core of the matter. Contrary to Rousseau's romantic intuition, we do not find ourselves strangers in a world invaded by aliens simply because we have utilized our God-given – or "natural" – ability to produce artifacts but because we have failed to live up to the responsibility which this ability entails. We have produced and abandoned our products, we have acquired and abandoned our acquisitions. It is in being abandoned that they have become alien, impersonal, in turn threatening our own being as persons, rendering it anomalous. A well used, cared for artifact does not become impersonal, and in turn does not depersonalize its user. It can, like my own well worn coat, enhance the quality of a person's personhood. It is the unused, uncared for, abandoned artifact that becomes anonymous and depersonalizing.

Perhaps a metaphor might underline the principle. It is not the Christmas tree, lovingly decorated, around which a family gathers, that is an alien in the world of humans, even if it is of the spring-loaded, aluminium variety. It is the Christmas tree discarded in an alley, blown by the wind, that becomes that. It could have been cut up for kindling, shredded for mulch, or used amid the underbrush to provide shelter for small animals. Only as abandoned did it become alien litter. It is not the use we make of the world but the disuse, the overflowing dumpsters and crowded closets, that are symptomatic of the decay of our civilization. We suffer not from production but from neglect.

Might that not, though, be a distinction without a difference? Might not our incredible waste, our voluminous production of useless surplus of artifacts, be an intrinsic feature of a technological civilization, built into its very nature? After all, a generation ago, in the 1950s, it was a commonplace to argue that waste is the pillar of affluence, that it is built-in waste, the disposable item, that fuels the economic cycle – and that to repair instead of discarding is to stand in the way of "progress."

If that were so, it would be a damning indictment of our civilization. I do not, however, believe that it is so, at least not necessarily. To be sure, living amid American affluence, it is difficult to imagine how we could possibly do justice to the flood of artifacts with which we surround ourselves to "keep the economy functioning." We have so much more than we can possibly hope to use and care for. We are not, though, alone: we can get help. The few hundred million who "have everything" are an island in the midst of billions who have practically nothing. We can get help: there are more than enough naked humans in the world to empty our closets and provide every one of our disused coats with someone to use and care for it. Our surplus of widgets might pose more of a problem, but where is it written that widgets are what we must produce? Could we not choose to use our resources to produce, say, fuel efficient cooking stoves for countries where shortage of fuel threatens to become a cause of starvation?

Of course we could, but here the argument becomes circular, since the word, "choose," introduces a new dimension. We could choose, but then our choices could not be the unthinking result of the interplay of unintended economic equations but of a personal, ethically guided decision. At this point, the ethics of living in a world of artifacts becomes directly continuous with the ethics of personal being as such, and the argument returns to the basic personalism.

Still, the circle was not in vain. Having travelled from the personalism of the community of persons through a personalism of nature to a personalism of artifacts and back we can no longer avoid the recognition that personalism cannot restrict itself to issues of personal growth and interpersonal relations. That may well be the starting point: the recognition of the central theme of the growth of humans to the full stature of Persons, or moral subjects, as a central theme *sui generis*, not reducible to psychological or somatic considerations alone. That theme, central to the early years of American personalism and still its intrinsic part, leads directly to the broader issue of the growth of human communities into societies of persons, governed by a moral law of which we become aware in reason and empathy – in the recognition of our fellow human as a Thou, as a Person endowed with an intrinsic worth and dignity, and in the rational and critical articulation of this recognition and its normative consequences for human interaction.

The recognition, familiar already to Kant, that we cannot preserve our own moral personhood if we are willing to allow it to our human Other – if, in familiar words, we are not prepared to treat humanity, whether in our own person or in that of another, as an end and not as a means merely – leads to the theme which began to emerge in personalism in the 1970s as much as the considerations of social ethics emerged in it in the 1930s – that we cannot hope to preserve the moral personhood of our individual and social being if we deny it to the created world around us. To recognize living nature as a society of persons and our relations to it as governed by the law of respect no less than our relation to our fellow humans is not only a condition of our physical survival, but of our moral survival as well. Cruelty to animals is no less dehumanizing than cruelty to our fellow humans. Yes, there is a food chain – though it is worth nothing that the vision of the lion and the lamb lying down together is as old as Isaiah – but the food chain does not justify disrespect. For there are, ultimately, no asymmetrical relations. The callousness required to treat the other, be he human or animal, in no sense as an end but as a means merely deforms the perpetrator as well as the victim. Resorting to Martin Buber's terminology, there are no "I–it" relations: in any such relation the I also becomes an it.

That is what leads to the next step: if we are to survive as Persons, I's, in a world of artifacts, it is urgent that we are not content ourselves with treating the world as a realm of its, of entities which are means merely, devoid of all end value, of all intrinsic dignity and not worthy of respect. It becomes literally a matter of Person-al survival to approach the world of artifacts with appreciation, care and respect, to endow it with quasi-personal value much as the child does with a teddy bear. It means recognizing the value of the artifacts as embodying a part of God's creation or of "Nature." It means recongizing the obligation – and the need – of responsibility for humanizing it. To have is not only a privilege: it is also an obligation.

That recognition suggests a different strategy than the one of the mindless multiplication of artifacts we have hitherto followed. Affluence appears not only as an abundance of possession but also as an abundance of responsibility, and needs be weighed carefully. Can we live up to the additional responsibility of adding yet another widget to our crowded world? Every time we displace yet another part of nature with a mechanism of our own designing, we are taking on yet another responsibility. Think of the responsibility we took on when we replaced the rhythm of going forth and of resting marked by the day and the night with the arbitrariness of flicking a light switch! Are we ready for yet another responsibility? The answer is not inevitably negative, but the point is that the question must be asked. If it is not, if we continue to treat our products as a means and not also as fellow beings, we imperil our own humanity as well. Personalism cannot be selective: either personal categories are in truth fundamental and universally applicable – or they become untenable even among persons. If you doubt that claim, *circumspice*! Any philosophy that establishes a category of beings exempt from moral considerations can always include subcategories of humans within that exempt category, and invariably does.

Technology, enabling us to produce a secondary reality of entities that neither grew nor were crafted, entities that were not created, fulfilled a part of the Serpent's promise: we have, within that world, become like gods. We can survive only by being like God in the crucial respect – that God is love.

Philosophical Faculty, Charles University, Prague/CZ (Boston University, Boston, MA, USA, emeritus) (1994)

NOTES

[1] I am referring to Martin Heidegger's evocative description of the immediate encounter with the thing – in Buber's terms, as a "Thou," in Heidegger's *in der Nähe* – in "Das Ding," *Vortrage und Aufsatze*, (Pfullingen: Gunther Neske, 1954, 1978), especially pp. 158–63, written in the solitude of the Schwarzwald.

[2] This description reflects Heidegger's earlier, urban perception, in *Sein und Zeit* (Tubingen: Neomarius Verlag, 1957), pp. 66–76.

[3] Regarding Milan Machovec, see my *The Embers and the Stars* (Chicago: University of Chicago Press, 1984), pp. x, 225.

[4] *Ibid.*, pp. ix–x.

[5] Husserl's excursus into personalities appears to have been occasioned by his reading of Max Scheler's *Der Formalismus in der Ethik und die Materiale Wertethik*. It appears only in the second volume of *Ideen, Phänomenologische Studien zur Konstitution* (Haag: Martinus Nijhoff, 1952), in the second and third part of that work. Husserl's intent appears to have been to describe three "regional ontologies," as correlates of natural, psychological, and human sciences respectively, but in adopting the terminology of "personalistic" for the third he, perhaps unintentionally, gives it a special status. I have dealt with it in a paper delivered to the 1984 meeting of the Husserl Circle, "Husserl's Personalism in *Ideen II*."

[6] Cassirer, in his Classic, *An Essay on Man* (New Haven: Yale, 1944, thirtieth printing, 1979) does, I realize, present science as the symbolic form par excellence (pp. 207 ff), but the point is that this is accidental to his basic claim of plurality, stressed throughout his work.

[7] Cf. detailed documentation in *The Embers . . .*, pp. 3–26.

[8] Rádl's slender volume, *Útěcha z filosofie* (Praha: Čin, 1946) remains unpublished in English; Peter Singer's work, *The Expanding Circle* (New York: Farrar, Strauss and Giroux, 1981) is readily available.

ERAZIM KOHÁK: BIBLIOGRAPHY

BOOKS

Demokracie pro začátečníky a pokročilé *(Democracy for Beginners and Advanced)* Praha: Sociologické nakladatelství SLON; 1997
An informal introduction to the spirit and institutions of democracy.

Hesla Erazima Koháka *(Erazim Kohák's Aphorisms)* ***Praha, Nakladatelství Pokorny, 1995***
A set of 52 short answers to questions about religion and popularly conceived "philosophy" originally published in *Mlady svět* (Young World) magazine as weekly column

Člověk, dobro a zlo *(Of humans, good and evil)* Praha: Nakladatelství Ježek, 1994
Lectures in the history of moral philosophy in the spring semester of 1992 at the Philosophical Faculty of Charles University (Prague), examining Western conceptions of good and evil from antiquity to present times.

Život v pravdě a moderní skepse *(Life in Truth and Modern Scepsis)* Praha: 3. LF UK, 1992; 2d ed. Praha: Nakladatelství Ježek, 1994
Five lectures about epistemology presented at the Philosophical Faculty of Charles University (Prague) in 1990. Basic thesis is a socratic one: the univocity of truth itself but the solely relative validity of all propositions pointing to the truth.

P.S.: Psové *(Post scriptum: Dogs)* Praha: Nadace EVA, 1993, 2d ed., Praha, Panglos, 1996; 3d ed., 1997
Fifteen popular radio lectures broadcast by Czechoslovak Radio in 1992, devoted to topics in ethics and ecology.

Jan Patočka: filosofický životopis *(Jan Patočka: A Philosophical Biography).* Translated by Josef Moural. Praha: Nakladatelství H+H, 1993
A sketch of the development and basic traits of the philosophical thought of Jan Patočka. First part of English book, *Jan Patočka* (below) in the translation of dr. Josef Moural.

Dopisy přes oceán *(Letters Across the Ocean)* Praha: Státní Pedagogické Nakladatelství, 1992
Monthly radio programs devoted to the philosophy of democracy and of nature, first broadcast by Radio Free Europe in 1980-82; first edition entitled *Čertování s Míšou* (Romping with Mikey) and privately circulated in Czechoslovakia, disguised by imprint 'Sharon: St Michael Press, 1984'

Jan Patočka: His Thought and Writings. Chicago: University of Chicago Press; 1989
A systematic presentation of Patočka's philosophical thought supplemented by an English translation of twelve basic texts and a detailed bibliography of Patočka's works.

Krize rozumu a přirozený svět *(Crisis of Reason and the Natural World)* Praha: *samizdat,* 1986
An early attempt at interpreting Patočka's thought, based largely on Patočka's writings prior to 1968 and wholly superseded by the two books on the same topic cited above.

Oheň a hvězdy *(The Fire and the Stars)* Translated by Milan Šimečka. Praha: *samizdat,* 1985
An English translation of the first three chapters of *The Embers and the Stars,* below.

R.S. Cohen and A.I. Tauber (eds.), Philosophies of Nature: The Human Dimension, 315–326.

The Embers and the Stars. Chicago: University of Chicago Press, 1984, 2d ed. 1987
A philosophical inquiry into the moral sense of nature based on a phenomenology of the experience of life in a forest clearing.

Idea and Experience. Chicago: University of Chicago Press, 1978, 2d ed.1982
Edmund Husserl's project of phenomenology in *Ideen I.*

Národ v nás *(The Nation We Bear Within)* Toronto: 68 Publishers, 1978
An attempt at an articulation of Czechoslovak national identity and of Masaryk's ideal of Czechoslovakia at a time of rigid 'normalization.'

Na vlastní kůži *(From Personal Experience)* Toronto: 68 Publishers, 1973
A dialogue with Heda Kovály, widow of a prominent Communist intended as a quest for common ground between the democratic and the (formerly) Communist opposition in the wake of the Soviet occupation of Czechoslovakia in 1968.

The Victors and the Vanquished. New York: Horizon Press, 1973
Same topic wholly differently presented for English-speaking audiences.

ANTHOLOGIES AND BOOK TRANSLATIONS

Jan Patočka: Heretical Essays in the Philosophy of History *(Kacířské eseje o filosofii dějin).* Chicago, Open Court Press,1997

Jan Patočka: Introduction to Husserl's Phenomenology (*Úvod do Husserlovy fenomenologie*). Chicago, Open Court Press; 1997)

Jan Patočka: Body, Community, Language, World (*Tělo, společenství, jazyk, svět*). Chicago, Open Court Press; in press - scheduled for autumn, 1997
Translations of three book-length works of the Czech philosopher Jan Patočka, of which the first is most notorious, the second most useful and the third most philosophically excellent of all his many writings.

Závod s časem: texty z morální ekologie *(A Race Against Time: Anthology of Writings in Moral Ecology)* Praha, Torst; 1996
An anthology of texts from western philosophical ecology in Czech translation, prepared by Erazim Kohák, Rudolf Kolářský and Igor Míchal.

Masaryk on Marx. Lewisburg: Bucknell University Press, 1972
An English translation of extensive selections from T.G. Masaryk's *Die Grundlagen des Marxismus.*

Paul Ricoeur: Freedom and Nature. Evanston: NWU Press, 1966.
An English translation of French philosopher Paul Ricoeur's book, *Le volontaire et l'involontaire.*

CHAPTERS IN BOOKS

"Of Dwelling and Wandering," in Leroy S. Rouner, *The Longing for Home;* Notre Dame, IL/USA, Notre Dame University Press, 1977:30-46

„Milý Václave" (The Significance of Václav Havel) in Anna Freimannová, ed., *Milý Václave*; Nakladatelství Lidové Noviny, Praha, 1977:148-57

„Ženy jako občanky" (Women as Citizens) in Eva Hauserová, ed., *Růže mezi trním;* Praha, Zelený Kruh, 1996:16-19

"A Requiem for Utopia" in Nicolaus Mills, *Legacy of Dissent;* New York: Simon and Schuster, 1996:389-402

"Faces of Democracy" in Sondra Myers, ed., *Democracy is a Discussion: The Handbook;* New London, CT, The Toors Cummings Centre, 1996; 2d ed, Washington, D.C., USIA, 1997:2-5, 34-35, 46-48;

"Speaking to Trees", tr. into Finnish by the editors, in Yrjö Uurtimo + Vesa Jaaksi, eds. *Ympäristöfilosofia: johdatus keskusteluun* (Introduction to Environmental Philosophy); Tampere, FIN, Tampere University Press; 1997

„Filosofická ekologie po dvaceti letech," ('Philosophical Ecology after 20 Years'), in Erazim Kohák *et al*, *Závod s časem;* Praha, Torst, 1996:7-19

„Člověk, dobro a zlo," ('Of Humans, Good and Evil'), in *ibid.*, pp. 154-88

„Velvet a ekológia v Riu" ("Velvet and Ecology at the Rio de Janeiro Conference")

„Filozofické poznámky na námet ekológie" (‚Philosophical Notes on the Topic of Ecology') both trans. into Slovak by Mikuláš Huba, in Mikuláš Huba, ed., *Ponovembrové Slovensko;* Bratislava, STUŽ, 1995:12-13 and 64-65

"Knowing Good and Evil..." in Gavroglou et al., *Science, Mind and Art*, vol. III, The Hague, Kluwer; 1995:243-54

"The Real and the Rational" in C.C.Gould and Robert Cohen, eds., *Artifacts, Representations and Social Practice;* Dortrecht: Kluwer; 1994:397-417

"Red War, Green Peace" in Frederick Ferré and Peter Hartel, eds., *Ethics and Environmental Policy;* Athens, Ga.: University of Georgia Press, 1994:163-77

„K hegelovským předpokladům amerického pragmatismu" ('Hegelian Assumptions of American Pragmatism') in Milan Znoj, ed., *Hegel v Čechách, na Moravě a v Americe: A Festschrift for Milan Sobotka*, Praha: FF UK, 1993:5-17

„Tři téze o Masarykovi" ('Three Theses about Masaryk')

„Masarykův Jan Hus" ('Masaryk's Jan Hus') both in V. Winkler, ed., *Filosofické rozpravy*, Praha: Impuls, 1993:45-71

"Selves, People, Persons: An Essay in American Personalism" in Leroy Rouner, ed., *Selves, People, Persons;* Notre Dame, IN: Notre Dame University Press,1992:17-36

„O život v pravdě" ('To Live in Truth') in Milan Machovec, ed., *T.G. Masaryk a naše současnost*, Praha: Academia, 1992:379-405

"Czech Philosophy in samizdat" in Markéta Goetz-Stankiewicz, ed., *Good-bye, Samizdat!* (Evanston, IL: NWU Press, 1992:217-21

„Křesťanství a reformace" ('Christianity and the Reformation') in Zdenek Kratochvíl, ed., *Laudatio Z.H.*, Praha: FF UK, 1992:72-75

"The Gift of the Night" in C.J. Hřibal, *The Boundaries of Twighlight*, Minneapolis: New Rivers Press, 1991:265-67

„Stolování s Míšou" ('At the Table with Mikey') in Karel Hvížïala, ed., *Benefice*, Toronto: 68 Publishers, 1991:140-44

"Making and Writing History" in Stone and Strouhal, eds, *Czechoslovakia: Crossroads and Crises*, London: Macmillan, 1989:188-206

"Il significato filosofico della 'Primavera di Praga'" ('The Philosophical Significance of the 'Prague Spring") in Francesco Leoncini, *Che cosa fu la Primavera di Praga*? Roma: Pietro Lacaita, 1989:91-126

"The Munich *Diktat*: A Moral Point of View" in Igor Lukeš, ed., *The Munich Dictate 50 Years After*, Boston: ISCIP Publications, 1989:1-8

„Hledání přirozeného světa" ('The Quest for the Natural World') in Václav Havel, ed., *Hostina: filosofický sborník*, Toronto: 68 Publishers; 1989:74-85

"The Crisis of Reason and the 'Natural' World" in Kaelin and Schrag, eds., *American Phenomenology*, Dortrecht: Kluwer, 1988:243-47

„Masaryk und die Monarchie: Versuch einer Demythisierung“ ('Masaryk and the Monarchy: An Attempt at Demythologization') in Josef Novák, ed., *On Masaryk,* Amsterdam: Rodopi, 1988:97-110

„Dějiny a smysl“ ('Of History and Meaning') in ed. anonymous, *Pojetí českých dějin;* Praha: *samizdat*, 1986:3-15

"Salvation as a Philosophical Category" ('Spása jako filosofický pojem') in Robert S. Cohen *et al.*, eds., *Studies in the Philosophy of J.N.Findlay;* Albany: SUNY Press, 1985:395-407

"To Live in Truth" ('Žít v pravdě') in Milič Čapek a Karel Hrubý, eds., *T.G. Masaryk in Perspective,* Washington: SVU Press, 1981:37-63

„Úvod“ ('Introduction') in Jan Patočka, *Dvě studie o Masarykovi,* Toronto: 68 Publishers, 1980:9-33

„Speaking of the Devil: A Modest Methodological Proposal“ in Alan M. Olson, ed., *Disguises of the Demonic,* New York: Association Press, 1975:48-57

„Povolání: český spisovatel“ ('Occupation: a Czech Writer') in F. Šturm, *Egon Hostovský,* Toronto: 68 Publishers, 1974:97-106

"Marxism and Terror" ('Marxismus a teror') in Miroslav Rechcígl, ed., *Studies in Czechoslovak Culture and Society,* New York: SVU Press; 1974:443-60

"Being Young in a Postindustrial Society" in Irving Howe and Michael Harrington, eds., *The Seventies: Problems and Proposals,* New York: Harper & Row, 1972:151-59

"Social Change and Social Work" in Sugata Dasgupta, *Social Work and Social Change,* Boston: Beacon Press, 1968:xi-xvi

"Masaryk and Plato in the 20th Century" in Miroslav Rechcígl, ed., *Czechoslovak Contribution to World Culture,* The Hague: Reidel, 1964:521-37

PHILOSOPHICAL AND SCHOLARLY ARTICLES

„Demokracie do nepohody“ (Democracy for Stormy Weather) *Pohledy* (Praha) 5.1-2 (March 97):1-3

„Masarykova Evropa“ (Masaryk's Europe), *Mosty* (Bratislava) 6.10 (3.3.97):3

„Možnosti a meze metafory“ (Possibilities and Limits of Metaphors) *Křesťanská revue* 63.10 (Dec 96):267-70

"The Phenomenology of Nostalgia," *Terra Nova,* 2.1(Dec96):6-14

„O smysl filosofické ekologie“ (The Significance of Philosophical Ecology,) *Listy* 26.4 (Jun 96):61-65

„Ekologické myšlení: pokus o sebereflexi po třiceti letech“ (Ecological Thought: Towards Self-Reflexion after 30 Years) *Veronica* (Brno) 10.1 (Jun 96):12-16

"The Search for Europe," *Dissent* (New York) 43.2 (Spring 1996):

„Je ekologie věda?“ (Is Ecology a Science?) *Veronica* (Brno) 10.1 (Apr 96):35-36

„Ekologické dilemma: etické kategorie v biocentrickém světě“ (Ecological Dilemma), *Československý časopis pro fyziku*, 45.3 (Oct 1996):242-47

"The Faces of Democracy–Looking at the 21st Century," *The Kettering Review,* Fall 1995:50-63 [vol. and no. not given]

„Předpoklady demokracie“ ('Presuppositions of Democracy"), *Křesťanská revue* (Praha) 62.7 (Sept 95):173-77

„Druhy ekologické zkušenosti“ ('Varienties of Ecological Experience'), *Filosofický časopis* (Praha) 43.6 (Dec 1996):899-920

"Hledání dobra: osudy utilitarismu" ('In Quest of the Good: The Fortunes of Utilitarianism'), *Mosty* (Bratislava) 4.5(31. 1. 1995):3

"Is Deep Ecology Deep Enough?" *The Trumpeter* (Canada) 11.3 (Summer 94):121-22

"Being Human, More or Less," *Human Studies* (Boston) 17.2 (Summer 94):1-19

„Konec ontologie?" ('The End of Ontology?'), *Filosofický časopis* (Praha) 42.2 (April 94):4-5

„Občan a příroda" ('Nature and the Citizen'), *Literární noviny* (Praha) 5.5 (3 Feb 94):1-3; 5.6 (10. Feb 94):4; 5.7. (17 Feb 94):4-5

„Ekologie, radikalismus, demokracie" ('Ecology, Radicalism, Democracy'), *Veronika* (Brno) 8.1 (Jan 94):1-6

"Democracy in the Post-Communist World," *Civic Arts Review* 6.1 (Winter 93):4-11

„Filosofická ekologie po 20 letech" ('Twenty Years of Philosophical Ecology') *Filosofický časopis* 41.6 (Dec 93):931-46; reprinted in *Universum* (Praha) 5.16 (Winter 94/95):79-94

„Lidská práva, správnost přírody" ('Human Rights and Nature's Rightness'), *Česká mysl* 42 (Sept 93):96-107

"Speaking to Trees," *Critical Review* 6.2-3 (Summer 93):371-88

"Knowing Good and Evil," *Husserl Studies* 10.1 (Summer 93):31-42

"Ashes, Ashes... Central Europe after Forty Years," *Daedalus* (American Academy of Arts and Sciences), 121.2 (Spring 92):197-215

„Maly průvodce po demokracii" ('A Brief Introduction to Democracy'), Supplement to *Listy* 23.3 (May 92):1-20

"Central Europe's Post-Captive Minds," *Harper's* 284.1705 (June 92):15-20

„Za agathocentrickou ekologii" ('The Case for an Agathocentric Ecology'), *Vesmír* 71/2 (Apr 92):92-93

"Speaking of Persons: Mirror or Metaphor?" *The Personalist Forum*, 7.2 (Fall 91):41-58

"Why Is There Something Good, Not Merely Something?" *Journal of Speculative Philosophy*, 5.1 (Spring 91):10-20

„Po stopách amerického personalismu" ('Tracking American Personalism'), *Studia humanistica* 1.3 (Apr 91):70-89

„Hovory se stromem" ('Speaking to Trees'), *Filosofický časopis* 39.6 (Dec 91):903-13

"Can There Be a Central Europe?" *Dissent* 37:2 (Spring 90):194-97

"Human Rights and Nature's Rightness" *Lyceum* 2.2 (Fall 90):22-36

„Deklarace přirozenych povinností" ('A Declaration of Natural Duties'), *Studie* (Roma) 20.3 (Fall 90):165-78

„28.X., filosofický vzato" ('Philosophical Significance of Czechoslovak Liberation'), *Paralelní akta filosofie* (*samizdat*) 4.10 (Spring 89):21-40

"Personalism: The Next Hundred Years," *The Personalist Forum* 4.2 (Fall 88):43-52

„Filosofie a filosofie dějin" ('Philosophy and the Philosophy of History'), *Paralelní akta filosofie* (*samizdat*) 3.6 (Spring 88):26-43

"Jan Patočka: A Bibliography," *Proměny* (New York) 24.1 (Spring 87):158-84

„Jak číst Patočku" ('Notes on Reading Patočka'), *ibid*, 142-58

„Pojem 'národní filosofie'" ('The Conception of a National Philosophy'), *Proměny* (New York) 23.4 (Dec 86):76-89

"Crisis of Rationality and the Natural World," *The Review of Metaphysics* 40.1 (Fall 86):79-106

"A Reply to Galen Johnson" [re. review of *The Embers and the Stars*] *Man and World* 19:2 (May 86):119-21

"A Reply to John Compton" [re. review of *The Embers and the Stars*] *The Personalist Forum* 2.1 (Spring 86):56-60

„O víře v Boha" ('About Faith in God'), *Studie* (Roma) 15.4-5 (Fall 85):402-14; reprinted in *Hlas národa* (Chicago) 11.26 (June 86):1,12-13

"Jan Patočka's Search for the Natural World," *Husserl Studies* 2.2 (Summer 85):129-39

"Creation's Orphans: Toward a Philosophy of Artefacts," *The Personalist Forum* 1.1 (Spring 85):22-42

„Masarykův *Jan Hus*: naše demokracie a otázka náboženská" ('Masaryk's *Jan Hus*: Religion and Our Democracy'), *Proměny* (New York) 21.1 (Mar 84):402-14

"Forest Lights: Notes on the Conceptualization of the Unconscious," *Journal of Religion and Health* 22.1 (Spring 83):49-57

„Pojem 'Evropa'" ('The Idea of Europe'), *Zpravodaj* (Schweiz) 15.9 (Sept 82):20-24

"Anti-Gorgias: Being and Nothing as Experience," *Human Studies* 4.3 (Jul 81):209-22

"*Idea and Experience*: A Summary," *The Monist* 63.3 (July 81):oo

"Erazim Kohák Replies" [a reply to John Brough's review of *Idea and Experience*], *Man and World* 14.3 (fall 81):345-49

"The Gory Pen: Reflections on Creativity and Violence," *TriQuarterly* 52.3 (Fall 81):104-20

"Sigmund Freud by Moonlight," *Journal of Religion and Health* 19.4 (winter 80):260-67

„O život v pravdě" ('To Live in Truth'), *Proměny* (New York) 17.4 (Oct 80):73-86

„Co měsíc dal" ('The Gift of the Moon'), *Studie* (Roma) 15.3 (Fall 80):260-64

„Die Erheblichkeit des Humanismus in dürftiger Zeit" ('Viability of Humanism in a Dry Season',) *Auslegung* 7.2 (Spring 80):184-205

„Patočkovy masarykovské studie" ('Patočka's Masaryk Studies"), *Proměny* (New York) 16.3 (fall 79):88-96

„Hermeneutika podezření, hermeneutika naděje" ('Hermeneutics of Suspicion, Hermeneutics of Hope'), *Svědectví* (Paris) 15.58 (summer 79):277-85

„Dvakrát o smyslu života" ('Twice About the Meaning of Life'), *Studie* (Roma) 15.1 (Jan 79):50-60

„Věřímt' i já" ('Still I believe'), *Proměny* (New York) 15.4 (Oct 78):1-13

"The Rational and the Irrational as Experience," *Idealistic Studies* 8.3 (Sept 78):220-32

"Religion and Socialism: The Relevance of Religion in a Socialist Society," *Dissent* 25.2 (Apr 78):174-85

"The Person in a Personal World: An Inquiry into the Metaphysical Significance of the Tragic Sense of Life," *Die unabhängige Zeitschrift für Philosophie* 1.1 (Summer 77):51-64

„Otázka české státnosti" ('The Problem of Czech Statehood'), *Proměny* (New York) 14.3 (Jul 77):3-12

„O Rusku a Evropě" ('Concerning Russia and Europe'), *Proměny* (New York) 14.1 (Jan 77):12-17

"Physics, Meta-Physics and Metaphysics," *Metaphilosophy* 5.1 (Jan 74):18-35

"Computers and Commissars," *Journal of Social Philosophy* 4.1 (Jan 73):20-36

"Possessing, Owning, Belonging," *Dissent* 19.3 (Summer 72):453-63

„K české otázce" ('Concerning the Czech Question'), *Svědectví* (Paris) 11.43 (Mar 72):456-59

"I, Thou and It," *Philosophical Forum* 1.1(NS), (fall 68):36-72

"Existence and the Phenomenological Epoché," *Journal of Existentialism* 8.29 (fall 67):19-47

"Edmund Husserl's Letter to Arnold Metzger," *The Philosophical Forum* 21(1964):48-68

"T.G. Masaryk's Revision of Marxism," *Journal of the History of Ideas* 25.4 (Oct 64):oo

"Religious Knowledge and Religious Reality," *The Philosophical Forum* 19(1962):40-50

"The Road to Wisdom: Lessons on Education from Plato's *Laches*," *Classical Journal* 56.3 (Dec 60):123-33

ESSAYISTIC WRITINGS

„Ze života (ostatních) zvířat" (About the Other Animals), *Literární noviny* (Praha) 8.16 (23.4.97):8

"Consolidating Freedom in Central Europe," *Dissent* 44 .2 (Spring 97):21-26

Commentaries in the daily *Svobodné slovo*, later *Slovo:* „Filosof na Hradě" 16.12.1996, „Trh je podmínkou..." 1.1.1997, „Lidská je štědrost..." 10.1.1997, „Ochranáři středního doletu", 4.97, „Ekologicky šetrné ministerstvo..?" 2.5.2997, „Vítězství demokratické levice...", 7.5.1997; ongoing)

Commentaries in *Magazín Mladé fronty dnes*, „Tři kameny na Javorině," Feb 97, „Můžeme si dovolit rušit tratě?" Apr 97, „Myšlenka české státnosti", Jun 97, ongoing

"Who was Masaryk?" *New Presence* (Prague) 2.1 (Jan 1997):28-29

„O víře a nevíře" (Faith and doubt), *Mosty* (Bratislava) 5.51-52 (17.12.1996):4

„Stromeček" (The Christmas Tree), reprint from *P.S. Psové* in *Obnova vesnice* 1996.3:62

„Predpoklady demokracie medzi vol'bami" (Democracy between Elections), *Ekológia a život* (Slovensko) 5.1 (Jan 1996):26-27

„Jak okrádali...?" (How did they rob us?), *Listy* (Praha) 26.3 (May 96):43-44

„Ticho na Kubě" ('Silence in Cuba'), *Mosty* 5.17 (23.4.1996):1

„České dráhy - privatizace nebo transformace?" (Czech Railways: Privatise or Transform?), *Mladá fronta dnes*, (28. března/March 1996):11

„Kdybych byl i posledním..." ('Even Were I the Last...') Rozhlas (Praha) 1996/24 (1.4.1996):27

„Stojí vám to za to?" ('Is it Worth It?'), *Mosty* (Bratislava) 5.12 (19.3.1996):1

„O duši Evropy: Francie, Czechoslovakia, Germany" ('What is Europe about: France, Czechoslovakia, Germany'), *Sondy* (Praha) 6.6 (19. Feb 96):1; reprinted *Mosty* (Bratislava) 6.1 (1.1.1997):4-5

„Obyvatelé a dobyvatelé" ('Dwellers and Conquerors'), *Literární noviny* (Praha) VII.11 (13.3.96):4

„Za císaře pána" ('For the Emperor' - a polemic with Otto von Habsburg), *Nová přítomnost* (Praha) 1996.1:26-27

„Bylo to jinak" ('Everything Was Otherwise'), in Jan Šimsa, ed., *Evangelický kalendář 1996* (Praha: Kalich; 1995):132-137

„Proč s kým se máme rozloučit?" (Should Auld Acquaintance Be Forgot?) *Mosty* 4.52 (27.12.1995):3

„Neznámí vojíni" ('The Unknown Soldiers'), *Listy* (Praha) 25.7 (listopad/Nov 95):60-64

„Osmadvacátý" ('Czechoslovak national holiday'), *Listy* (Praha) 25.6 (říjen/Oct 95):38-43, přetištěno/reprinted in *Mosty* (Bratislava), 4:43 (24.X.1995):4-5

„Desatero života na zemi" ('An Ecological Decatelogue"), *Listy* (Praha) 25.5 (září/Sept 95):3-7

„Jde o jinou orientaci této civilizace" (Reorienting this Civilization) *Rádce spotřebitele* (Praha) I.3 (Jun 95):4-5

„Etické a ekologické meze ekonomiky" ('Ethical and Ecological Limits of Economics'), written for *inFORUM* (Praha) 1995.6:8-9,16; reprinted in *Mosty* (Bratislava) 4.26 (27.6.1995):9

„Aby dlouhé byly dnové tvoji na zemi" (In memoriam: Josef Vavroušek, *Mosty* (Bratislava) 4.23 (6.6.1995):14

„Pepa a Pét'a - co znamenají?" ('Necrologue for Josef Vavroušek') *inFORUM* 4/95:6

„Město snů a vzpomínání" ('City of Dreams and Memories'), *Mosty* (Bratislava) 4.16 (18.4.95):4

„Filosofická ekologie po dvaceti letech" ('Philosophical Ecology Twenty Years Later' - reprint), *Revue Universum* (Křest'anská akademie Praha), 4.16 (Winter 94/95):79-94

„Dvojznačnost soukromého vlastnictví" (The Ambiguity of Private Property) *Mosty* (Bratislava) 4.12 (21.3.95):3

„Náš člověk v Kostarice" ('Our Man in Costa Rica'), *Mosty* (Bratislava) 4.11 (14.3.95):4,

„Demokracie se bez oborů neobejde" ('Democracy Can't Do Without Unions'), *Mladá fronta dnes* (7.3.95):8

„Betonový odkaz Vladimíra Iljiče" ('Lenińs Concrete Heritage'), *Mosty* (Bratislava) 4.4 (24.1.95):1

„Hesla Erazima Koháka," *Mlady svět* XXXVI (1995) - A weekly column devoted to popular presentation of basic philosophic and theological ideas. Also as book, illustrated by V. Renčín. Q.v.

„Sám proti sobe" ('Mine own enemy'), *Mosty* (Slovakia) 3.44 (Nov 94):3

„Až na dno blahobytu" ('To the Drags of Affluence'), *Listy* (Praha) 24.4 (Jul 94):47-50

„Perché ci separiamo?" ('Why are we parting?'), *L'Europa ritrovata* (Roma) 4.1-2 (Jun 94):1-9

„Dopisy přes oceán" ('Notes Across the Ocean'), *Mlady svět* (Praha) 36.19 (6. May 94):57-59 [reprint from book of same name]

„Přidaná hodnota" ('Value Added'), *Literární noviny* (Praha) 5.17 (28 Apr 94):3

„Separační myty" ('Myths about the Separation'), *Mosty* (Slovakia) 3.16 (19 Apr 94):3

„Obět' a zasvěcení" ('Sacrifice and Dedication), *Listy* (Praha) 24.2 (Mar 94):25-26

„Sekularizace: Víra vstupující do světa" ('Secularization: Faith Entering the World'), *Protestant* (Praha) 3.3 (Mar 94):4-9

„Sluníčko se usmívá" ('The Smile Button'), *ibid., loc. cit.*

"Varieties of Ecological Experience" ('Druhy ekologické zkušenosti'), *Forefront* (Colorado) 1.1(NS) (Spring 94):19-21

„Skalické dozvuky" ('Echoes of Skalice'), *Mosty* (Slovakia) 3.12 (22 Mar 94):3

„Křest'anství a ekologie" ('Christianity and Ecology'), *Anno Domini* (Praha) 5.3 (Mar 94):4

„Pějme píseň dokola" ('Let us Raise a Song'), *Mosty* (Slovakia) 3.10 (8 Mar 94):1

"Political Correctness in Prague," *Dissent* 41.1 (Winter 94):35-38

P.S. Psové, excerpted and serialized in *Mosty,* vol. 2 (1993) - „Život" 9 (2.3.):3, „Dikobrazi" 31(3.8.):11, „Beránci" 6(10.8.):11, „Vlci" 33(17.8.):11, „28. říjen" 43(26.10), „Svátky" 46 (16.11.):11, „Stromeček" 52(28.12):11

"Oh, the Devil of It. . ." *Desert Call* (Colorado) 28.3 (summer 93):11

„Počkejme sedm let" ('Let us Wait Seven Years'), *Listy* 23.3 (Jun 93):31-33

„Tři monology o exilu" ('Three Monologues about Exile'), *Listy* (Praha) 23.2 (Jan 93):44-48

„Na téma 'levice'" ('Speaking of 'the Left'), *ibid.*, 10-11

„Jiskry v popelu: Clinton a Amerika" ('Sparks Amid the Ashes: Clinton and America'), *Listy* 23.2 (Mar 93):5-16

„Čtyři drobnosti" ('Four ecological marginalia'), *Sociologické aktuality* (Praha) 3.3 (Mar 93):11

„Proč se rozcházíme" ('Why We Are Parting'), *Listy* 22.6 (Nov 92):21-25

"Tattered Velvet: A Country Falls Apart" *Dissent* 39.4 (Fall 92):441-45

„Maly průvodce po demokracii" (A Pocket Guide to Democracy), serialized in *Mosty* (Bratislava) vol. 1, nos. 1-10 (28 April through

„Král Karel by rozuměl" ('King Charles Would Understand'), *Mosty* 1.4 (19 Mar 92):9

„Krívanie na pravú nohu" ('Limping on the Right Leg'), *Kulturny život* (Slovakia) 26.9 (27 Feb 92):1-2

[„America, America" - ten columns about America, weekly in *Právo lidu* from February to May, details not available]

"Laudatio Carolinae" ('In praise of Charles University'), *Zprávy SVU* 33.6 (Dec 91):13-15

„Nemusím, mohu" ('I don't have to, I can'), *Listy* 21.6 (Dec 91):5-8 [excerpt from *Dopisy přes oceán*]

"La sociedad de los quebrantados" ('Community of the Shaken'), *La Vanguardia* (Madrid) 31 Jan 89

„Jan Hus: dlaczego to ważna sprawa?" ('Jan Hus: Why Does It Matter?') *Tygodnik powszechny* (Krakow) 15 Jan 89, p. 1; English tr., T.R. Davies, *RCDA* 28.2 (Spring 1989):333-37

"Twenty-first Century Blues" *Dissent* 35.3 (Summer 88):56-70

"The Relevance of Tolstoy" *Dissent* 34.1 (Winter 87): 5-9; partial reprint, **"Nuclear Power's Faustian Bargain,"** *Harper's* 274.1644 (May 87); French tr., **"La pertinence de Tolstoi,"** *Lettre internationale* 5.15 (winter 87):78

"Dealing with the Nuclear Threat" *Dissent* 33.3 (Summer 86):357-60

"A Letter from Europe" *Dissent* 32.4 (Fall 85):407-11

"The Power and the Glory: Roman Catholic Church in Central Europe" *New Oxford Review* 50.9 (Nov 83):6-13

"Politics and Ideology" *Partisan Review* 50.4 (Fall 83):487-502

„Jak proti komunismu?" *České slovo* 33.10 (Oct 83):5-6

„České vánoce" ('A Czech Christmas'), *České slovo* 33.1 (Jan 83):1-2

„Umění nesouhlasit" ('The Art of Dissent'), *Svědectví* (Paris) 16.59 (June 80):204-08

„Právo na práci" ('The Right to a Job'), *Právo lidu* 83.2 (Apr 80):5

"Fire in the Ashes" *Dissent* 26.4 (fall 79):484-90

"On the Failure of Marxism" *New Oxford Review* 46.8 (Oct 79):17-20

„Možnosti bezmocných" ('The Possibilities of the Helpless'), *Právo lidu* 82.1 (Jan 79):4

"Americans Redefine the Social Contract" *SPAN* (New Delhi), (Spring 78 [further data unknown]

"Big Brother, Brown Brother" *Dissent* 25.1 (Jan 78):108-12

„Rozhovor o humanitním ideálu" ('A Discourse about the Ideal of Humanity'), *České slovo* 40.1 (Jan 78):3 and 40.2 (Feb 78):4

"The Road Less Travelled" *Harper's* 255.1531 (Dec 77):21-28

"Philosophie en samizdat: *Les Essais hérétiques* de Jan Patočka" ('Philosophy in *samizdat*: Patočka's *Heretical Essays*'), *Les Cahiers de l'Est* 3.9-10 (Spring 77):153-58

"The Taste of Québec" *Commonweal* 104.3 (Feb 77):76-79

„Ein Bericht aus Rom" ('A Report from Rome'), *Schweizer Monathefte* (Zürich) 56.12 (Mar 77):1067-75

„Těm, co přijeli do Vídne“ ('To the Escapees in Vienna'), *Svědectví* (Paris) 14.54 (fall 77):214-19

„Amerika Jimmy Cartera“ ('Jimmy Carter's America'), *Listy* (Roma) 9.1 (Jan 77):41-45

"Italia sinistra" *Dissent* 23.4 (Oct 76):328-32

"Roman Spring" *Harper's* 253.1517 (Oct 76):356-61

"European Freedom, European Defense" *Dissent* 23.3 (Jun 76):oo

"Ancient Enemies" *Harper's* 253.1515 (Aug 76):14-25

„Dialog pokračuje“ *Listy* (Roma) 6.5 (Aug 75):30-34

„Tři téze o Masarykovi“ ('Three Theses about Masaryk'), *Svědectví* (Paris) 13.50 (fall 1975):225-41

"Democracy and American Power" *Dissent* 22.4 (Oct 75):356-61

"The Devil and Leszek Kołakowski" *Dissent* 22.2 (Spring 75):196-99

"Progress and the Good Lord" *Commonweal* 101.3 (Oct 74):57-61

"Czech Socialists Find Their Way" *Dissent* 21.3 (Summer 74):19-30

"Welfare State and Socialism" *Dissent* 21.1 (Spring 74):440-44

„K čemu básník?“ ('The Uses of a Poet'), *Svědectví* (Paris) 12.47 (spring 74):539-43

„Dialog přes barikádu“ ('A Dialogue Across the Barricade'), *Proměny* (New York) 11.1 (Jan 74):70-79

„Zornym úhlem poražených“ ('Through the Eyes of the Vanquished'), *Listy* (Roma) 3.1 (Feb 73):46-49

„Co dělat?“ ('What to do?'), *Proměny* (New York) 10.2 (Apr 73):46-49

„Tulák a grófinka aneb společenské základy demokracie“ ('The Tramp and the Countes, or, the Social Foundations of Democracy'), *Svědectví* 12.45 (spring 73):39-48

„Rosja: nastepnym razem požar?“ (Polish tr. "Russia: Fire Next Time?") *Aneks* 3.2 (summer 73):59-73

„Pauzy a trendy“ ('Lacunae and Momenta'), *Svědectví* 11.43 (Dec 72):546-48

"Maoism in East Europe“ *Commonweal* 95.12 (Dec 71):oo

„Augusti a pieroti“ ('Red Clowns, White Clowns'), *Svědectví* (Paris) 11.42 (Nov 71):320-28

"Russia: The Fire Next Time?" *Dissent* 18.5 (Oct 71):oo

„Co nového na levici?“ ('What is New on the Left?'), *Svědectví* (Paris) 11.41 (Jun 71):165-74

„Co očekáváme?“ ('What do we Expect?'), *České slovo* (München) 17.10 (Jun 71):1-3

"Small Truth, Big Lie" *Dissent* 18.2 (Mar 71):178-82

"Being Young in Post-Industrial Society" *Dissent* 18.1 (Jan 71):30-40

"The Humanization of Power" *The Humanist* 31.1 (Jan 71):20-27

"Revolution and Sanity" *Dissent* 17.5 (Sept 70):453-56

"A Requiem for Utopia" *Dissent* 16.1 (Jan 69):41-46

"Revolution, Reformation, Reform" *Commonweal* 91.13 (Dec 9):378-82

"Progress and the Proletariat" *Dissent* 16.6 (Nov 69):505-19

"Notes to a Practical Revolutionary" *Perspective* 1.2 (Fall 69):8-19

"Turning on for Freedom: Curious Love Affair of Sex and Socialism" *Dissent* 16.5 (Sept 69):437-44

"Czech Church in Limbo" *Commonweal* 90.22 (Sept 69):591-94

"The Wolf Has Grown Eager for the Lamb" *The Village Voice,* 11 and 18 Sept 69 [Pseudonym Michal Doubrava, misprinted as Dobry]

"Theology Behind the Iron Curtain" *Encounter* 26.3 (fall 65):362-72

„O církvi a státu" ('Concerning the Church and the State'), *Husův lid* (Chicago) 25.3 (Mar 64):36-45 [pseudonym Vlastimil Zajíček]

„Exil a čas" ('Exile and Time'), *Sklizeň* (Hamburg) o.o (fall 59):oo

„Nevrátíš-li se. . ." ('Should you not return...'), *Husův lid* (Chicago) 20.6 (Jun 59):oo

„Moci světa tohoto" ('The Powers of this World'), *Husův lid* (Chicago) 20.4 (Apr 59):oo

„Vyvolený národ" ('A Chosen People'), *Husův lid* (Chicago) 19.11 (Nov 58):oo

„Svět lidský a Boží" ('Human World, God's World'), *Husův lid* (Chicago) 19.9 (Sept 58):oo

„Začarovaný kruh" ('The Vicious Circle'), *Husův lid* (Chicago) 18.11 (Nov 57):oo

„Úvaha velikonoční" ('Easter Meditation'), *Husův lid* (Chicago) 18.4 (Apr 57):oo

„Do nového roku" ('Into a New Year'), *Husův lid* (Chicago) 18.2 (Feb 57):oo

„Americký materialismus" ('American Materialism'), *Husův lid* (Chicago) 17.10 (Oct 56):oo

„Čemu důvěřujeme?" ('Wherein do we Trust?'), *Husův lid* (Chicago) 17.7 (Jul 56):oo

„Evangelíci a svobodné Československo" ('Protestants and Free Czechoslovakia'), *Husův lid* (Chicago) 17.5 (May 57):oo

„Otázka návratu" ('The Question of Return'), *České slovo* (München) 1.3 (fall 55):1-5 [Pseudonym Václav Zhor]

BOOK REVIEWS

„Křest'an, Evropan, Čech" (review of Karel Skalicky, *Blochova filosofie naděje* (Ježek, Praha, 1995) and *Za naději a smysl* (Zvon, Praha 1995), *Mosty* 5.40 (1.10.1996):13

„Soukromý život psů" (Review of Czech ed. of E.M.Thomas, *The Hidden Life of Dogs*) *Nová přítomnost* 2.7 (Jul 96):25

„Himalajská *Babička*" (Review of Helen Horberg-Hodge, *Dávné budoucnosti*), *Poslední generace* 5.7 (Jul 96):25

„Čteme s vámi," (Review of E.M.Thomas, *Soukromý život psů*), 4.5 (May 96):11

„Vandrovní knížky mistrů cechu bohosloveckého" ("The Song of the Wandering Theologians,") *Nová Přítomnost* I.9 (15.9.95):3-5

„Demokracie pro všední den" ('Everyday Democracy' - review of J. Plamínek, *Řešení konfliktů*) *Mosty* (Bratislava) 4.33 (15.8.95):3

"Služebník krále Václava" ('King Václav's Servant') [Podiven, *Češi v dějinách Evropy*] *Listy* 22.2 (Mar 92):42-46

"Opožděné přednášky prof. Masaryka" ('Prof. Masaryk's Delayed Lectures') [Masaryk, *Přednášky v Chicagu*] *Právo lidu* 82.2 (May 79):2

"R.I. Sugerman, *Rancor against Time*," *The Review of Metaphysics* 38.4 (Jun 85):908-10

"Jan Patočka, *Éssais hérétiques sur la philosophie de l'histoire*," *The Review of Metaphysics* 26.2 (Dec 82):oo

"Aaron Gurwitch, *Human Encounters in the Social World*," The Review of Metaphysics 34.4 (Jun 81):609-11

"Sidney Hook, *Philosophy and Public Policy*," *The New Oxford Review* 48.1 (Jan 81):26-28

"Borden Parker Bowne, *Representative Essays*," *The Review of Metaphysics* 35.2 (Dec 81):412-14

"A. James Gregor, *The Fascist Persuasion*," *Sociology* 1.9 (Aug 74):205

"C. L. Sulzberger, *An Age of Mediocrity*," *World Affairs* 136.4 (April 74):385

"Feliks Gross, *The Revolutionary Party*," *Sociology* 2.2 (Apr 74):41

"P. Golan, *The Czechoslovak Reform Movement*," *Contemporary Sociology* (May 73):179-81

"George Lichtheim, *Collected Essays*," *Dissent* 20.4 (Fall 73):485-86

"Schneider, *The Idea of Culture*," *Sociology* 1.1 (Oct 73):20-21

"Ideology and Terror: Milovan Djilas, *The Unperfect Society*," *Dissent* 16.6 (Nov 69):358-61

"Susan Sontag, *Existentialist Prolegomena*," *Commonweal* 90.o (May 69):oo

"Martin Heidegger, *Essays in Metaphysics*," *The Philosophical Forum* 19 (1961):oo-oo

"Micheline Sauvage, *Socrates and the Conscience of Man*," *B. U. Graduate Journal* o.o (Apr 61):oo

"Milič Čapek, The *Philosophical Impact of Contemporary Physics*," *Zprávy SVU* 3.8, (Dec 61):oo

TRANSLATIONS

Jan Patočka, "Spiritual Crisis of European Humanity in Husserl and Masaryk," in J. Novák, ed., *On Masaryk* (Amsterdam: Rodopi, 1988):97-110

Václav Havel, "An Anatomy of a Reticence," *Crosscurrents* (Ann Arbor, MI) 5 (86):1-24

Milan Otáhal, "The Manuscript Controversy in the Czech National Revival," *ibid.*, 247-78

Jan Patočka, "Edmund Husserl's Philosophy of the Crisis of the Sciences and His Conception of a Phenomenology of the Life-world," *Husserl Studies* 2.2 (summer 85):139-55

Emanuel Rádl, *Consolation from Philosophy.* Scripta for classroom use at Boston University, 1982

Edmund Husserl, "Letter to Arnold Metzger," in McCormick and Elliston, eds., *Husserl: The Shorter Works* (Notre Dame: University of Notre Dame Press, 1981):357-64

Paul Ricoeur, "Methods and Tasks of the Philosophy of the Will" in M. Ballard, ed., *Essays in Husserl's Phenomenology* (Evanston: NWU Press, 1965):oo

INDEX OF NAMES

Boldface indicates a chapter by that author;
'n' indicates a note number.

R.S. Cohen and A.I. Tauber (eds.), Philosophies of Nature: The Human Dimension, 327–331.

Boston Studies in the Philosophy of Science

150. I.B. Cohen (ed.): *The Natural Sciences and the Social Sciences.* Some Critical and Historical Perspectives. 1994 ISBN 0-7923-2223-1
151. K. Gavroglu, Y. Christianidis and E. Nicolaidis (eds.): *Trends in the Historiography of Science.* 1994 ISBN 0-7923-2255-X
152. S. Poggi and M. Bossi (eds.): *Romanticism in Science.* Science in Europe, 1790–1840. 1994 ISBN 0-7923-2336-X
153. J. Faye and H.J. Folse (eds.): *Niels Bohr and Contemporary Philosophy.* 1994 ISBN 0-7923-2378-5
154. C.C. Gould and R.S. Cohen (eds.): *Artifacts, Representations, and Social Practice.* Essays for Marx W. Wartofsky. 1994 ISBN 0-7923-2481-1
155. R.E. Butts: *Historical Pragmatics.* Philosophical Essays. 1993 ISBN 0-7923-2498-6
156. R. Rashed: *The Development of Arabic Mathematics: Between Arithmetic and Algebra.* Translated from French by A.F.W. Armstrong. 1994 ISBN 0-7923-2565-6
157. I. Szumilewicz-Lachman (ed.): *Zygmunt Zawirski: His Life and Work.* With Selected Writings on Time, Logic and the Methodology of Science. Translations by Feliks Lachman. Ed. by R.S. Cohen, with the assistance of B. Bergo. 1994 ISBN 0-7923-2566-4
158. S.N. Haq: *Names, Natures and Things.* The Alchemist Jābir ibn Ḥayyān and His *Kitāb al-Aḥjār* (Book of Stones). 1994 ISBN 0-7923-2587-7
159. P. Plaass: *Kant's Theory of Natural Science.* Translation, Analytic Introduction and Commentary by Alfred E. and Maria G. Miller. 1994 ISBN 0-7923-2750-0
160. J. Misiek (ed.): *The Problem of Rationality in Science and its Philosophy.* On Popper vs. Polanyi. The Polish Conferences 1988–89. 1995 ISBN 0-7923-2925-2
161. I.C. Jarvie and N. Laor (eds.): *Critical Rationalism, Metaphysics and Science.* Essays for Joseph Agassi, Volume I. 1995 ISBN 0-7923-2960-0
162. I.C. Jarvie and N. Laor (eds.): *Critical Rationalism, the Social Sciences and the Humanities.* Essays for Joseph Agassi, Volume II. 1995 ISBN 0-7923-2961-9
    Set (161–162) ISBN 0-7923-2962-7
163. K. Gavroglu, J. Stachel and M.W. Wartofsky (eds.): *Physics, Philosophy, and the Scientific Community.* Essays in the Philosophy and History of the Natural Sciences and Mathematics. In Honor of Robert S. Cohen. 1995 ISBN 0-7923-2988-0
164. K. Gavroglu, J. Stachel and M.W. Wartofsky (eds.): *Science, Politics and Social Practice.* Essays on Marxism and Science, Philosophy of Culture and the Social Sciences. In Honor of Robert S. Cohen. 1995 ISBN 0-7923-2989-9
165. K. Gavroglu, J. Stachel and M.W. Wartofsky (eds.): *Science, Mind and Art.* Essays on Science and the Humanistic Understanding in Art, Epistemology, Religion and Ethics. Essays in Honor of Robert S. Cohen. 1995 ISBN 0-7923-2990-2
    Set (163–165) ISBN 0-7923-2991-0
166. K.H. Wolff: *Transformation in the Writing.* A Case of Surrender-and-Catch. 1995 ISBN 0-7923-3178-8
167. A.J. Kox and D.M. Siegel (eds.): *No Truth Except in the Details.* Essays in Honor of Martin J. Klein. 1995 ISBN 0-7923-3195-8
168. J. Blackmore: *Ludwig Boltzmann, His Later Life and Philosophy, 1900–1906.* Book One: A Documentary History. 1995 ISBN 0-7923-3231-8
169. R.S. Cohen, R. Hilpinen and R. Qiu (eds.): *Realism and Anti-Realism in the Philosophy of Science.* Beijing International Conference, 1992. 1996 ISBN 0-7923-3233-4
170. I. Kuçuradi and R.S. Cohen (eds.): *The Concept of Knowledge.* The Ankara Seminar. 1995 ISBN 0-7923-3241-5

Boston Studies in the Philosophy of Science

171. M.A. Grodin (ed.): *Meta Medical Ethics*: The Philosophical Foundations of Bioethics. 1995 ISBN 0-7923-3344-6
172. S. Ramirez and R.S. Cohen (eds.): *Mexican Studies in the History and Philosophy of Science*. 1995 ISBN 0-7923-3462-0
173. C. Dilworth: *The Metaphysics of Science*. An Account of Modern Science in Terms of Principles, Laws and Theories. 1995 ISBN 0-7923-3693-3
174. J. Blackmore: *Ludwig Boltzmann, His Later Life and Philosophy, 1900–1906* Book Two: The Philosopher. 1995 ISBN 0-7923-3464-7
175. P. Damerow: *Abstraction and Representation*. Essays on the Cultural Evolution of Thinking. 1996 ISBN 0-7923-3816-2
176. M.S. Macrakis: *Scarcity's Ways: The Origins of Capital*. A Critical Essay on Thermodynamics, Statistical Mechanics and Economics. 1997 ISBN 0-7923-4760-9
177. M. Marion and R.S. Cohen (eds.): *Québec Studies in the Philosophy of Science*. Part I: Logic, Mathematics, Physics and History of Science. Essays in Honor of Hugues Leblanc. 1995 ISBN 0-7923-3559-7
178. M. Marion and R.S. Cohen (eds.): *Québec Studies in the Philosophy of Science*. Part II: Biology, Psychology, Cognitive Science and Economics. Essays in Honor of Hugues Leblanc. 1996 ISBN 0-7923-3560-0
     Set (177–178) ISBN 0-7923-3561-9
179. Fan Dainian and R.S. Cohen (eds.): *Chinese Studies in the History and Philosophy of Science and Technology*. 1996 ISBN 0-7923-3463-9
180. P. Forman and J.M. Sánchez-Ron (eds.): *National Military Establishments and the Advancement of Science and Technology*. Studies in 20th Century History. 1996 ISBN 0-7923-3541-4
181. E.J. Post: *Quantum Reprogramming*. Ensembles and Single Systems: A Two-Tier Approach to Quantum Mechanics. 1995 ISBN 0-7923-3565-1
182. A.I. Tauber (ed.): *The Elusive Synthesis: Aesthetics and Science*. 1996 ISBN 0-7923-3904-5
183. S. Sarkar (ed.): *The Philosophy and History of Molecular Biology: New Perspectives*. 1996 ISBN 0-7923-3947-9
184. J.T. Cushing, A. Fine and S. Goldstein (eds.): *Bohmian Mechanics and Quantum Theory: An Appraisal*. 1996 ISBN 0-7923-4028-0
185. K. Michalski: *Logic and Time*. An Essay on Husserl's Theory of Meaning. 1996 ISBN 0-7923-4082-5
186. G. Munévar (ed.): *Spanish Studies in the Philosophy of Science*. 1996 ISBN 0-7923-4147-3
187. G. Schubring (ed.): *Hermann Günther Graßmann (1809–1877): Visionary Mathematician, Scientist and Neohumanist Scholar*. Papers from a Sesquicentennial Conference. 1996 ISBN 0-7923-4261-5
188. M. Bitbol: *Schrödinger's Philosophy of Quantum Mechanics*. 1996 ISBN 0-7923-4266-6
189. J. Faye, U. Scheffler and M. Urchs (eds.): *Perspectives on Time*. 1997 ISBN 0-7923-4330-1
190. K. Lehrer and J.C. Marek (eds.): *Austrian Philosophy Past and Present*. Essays in Honor of Rudolf Haller. 1996 ISBN 0-7923-4347-6
191. J.L. Lagrange: *Analytical Mechanics*. Translated and edited by Auguste Boissonade and Victor N. Vagliente. Translated from the *Mécanique Analytique, novelle édition* of 1811. 1997 ISBN 0-7923-4349-2
192. D. Ginev and R.S. Cohen (eds.): *Issues and Images in the Philosophy of Science*. Scientific and Philosophical Essays in Honour of Azarya Polikarov. 1997 ISBN 0-7923-4444-8

Boston Studies in the Philosophy of Science

193. R.S. Cohen, M. Horne and J. Stachel (eds.): *Experimental Metaphysics*. Quantum Mechanical Studies for Abner Shimony, Volume One. 1997 ISBN 0-7923-4452-9
194. R.S. Cohen, M. Horne and J. Stachel (eds.): *Potentiality, Entanglement and Passion-at-a-Distance*. Quantum Mechanical Studies for Abner Shimony, Volume Two. 1997 ISBN 0-7923-4453-7; Set 0-7923-4454-5
195. R.S. Cohen and A.I. Tauber (eds.): *Philosophies of Nature: The Human Dimension*. 1997 ISBN 0-7923-4579-7
196. M. Otte and M. Panza (eds.): *Analysis and Synthesis in Mathematics*. History and Philosophy. 1997 ISBN 0-7923-4570-3
197. A. Denkel: *The Natural Background of Meaning*. 1998 ISBN 0-7923-5331-5
198. D. Baird, R.I.G. Hughes and A. Nordmann (eds.): *Heinrich Hertz: Classical Physicist, Modern Philosopher.* 1997 ISBN 0-7923-4653-X
199. A. Franklin: *Can That be Right?* Essays on Experiment, Evidence, and Science. 1998 ISBN 0-7923-5464-8

Also of interest:

R.S. Cohen and M.W. Wartofsky (eds.): *A Portrait of Twenty-Five Years Boston Colloquia for the Philosophy of Science, 1960-1985*. 1985 ISBN Pb 90-277-1971-3

Previous volumes are still available.

KLUWER ACADEMIC PUBLISHERS – DORDRECHT / BOSTON / LONDON

SPRINGER NATURE

GPSR Compliance

The European Union's (EU) General Product Safety Regulation (GPSR) is a set of rules that requires consumer products to be safe and our obligations to ensure this.

If you have any concerns about our products, you can contact us on ProductSafety@springernature.com

In case Publisher is established outside the EU, the EU authorized representative is:

Springer Nature Customer Service Center GmbH
Europaplatz 3
69115 Heidelberg, Germany

FSC
www.fsc.org
MIX
Papier | Fördert
gute Waldnutzung
FSC® C083411

Kolibri 360 GmbH
Ferdinand-Jühlke-Straße 7
99095 Erfurt, Deutschland
produktsicherheit@kolibri360.de